Marine Microbial Diversity as Source of Bioactive Compounds

Marine Microbial Diversity as Source of Bioactive Compounds

Editor

Khaled A. Shaaban

MDPI • Basel • Beijing • Wuhan • Barcelona • Belgrade • Manchester • Tokyo • Cluj • Tianjin

Editor
Khaled A. Shaaban
University of Kentucky
USA

Editorial Office
MDPI
St. Alban-Anlage 66
4052 Basel, Switzerland

This is a reprint of articles from the Special Issue published online in the open access journal *Marine Drugs* (ISSN 1660-3397) (available at: https://www.mdpi.com/journal/marinedrugs/special_issues/Microbial_Diversity_Bioactive_Compounds).

For citation purposes, cite each article independently as indicated on the article page online and as indicated below:

LastName, A.A.; LastName, B.B.; LastName, C.C. Article Title. *Journal Name* **Year**, *Volume Number*, Page Range.

ISBN 978-3-0365-4581-3 (Hbk)
ISBN 978-3-0365-4582-0 (PDF)

Contents

About the Editor

Khaled A. Shaaban

Dr. Shaaban obtained his B.S. degree in Chemistry (2000) from Mansoura University, Egypt. He received his M.S. (2005) and PhD. degrees in Organic Chemistry (2009) from the University of Göttingen (Germany) under supervision of Professor Hartmut Laatsch. He joined Prof. Jürgen Rohr's group to continue natural product research as a postdoctoral fellow (2009-2011) at the University of Kentucky prior to joining Prof. Ben Shen's group (2011–2012) at The Scripps Research Institute (TSRI), Florida (USA). Currently Dr. Shaaban hold a Research Assistant Professor position at the University of Kentucky, College of Pharmacy and also serve as the Director of the Center for Pharmaceutical Research and Innovation (CPRI) Natural Product Repository initiative. His expertise is in the area of natural product discovery, natural product biosynthesis and natural product-based lead discovery. Dr. Shaaban has over 20 years of experience in NPs-based drug discovery, including isolation of bacterial strains from soils, screening, scale-up fermentation, complex NP structure elucidation, biosynthesis and semi-synthesis.

Editorial

Marine Microbial Diversity as Source of Bioactive Compounds

Khaled A. Shaaban

Center for Pharmaceutical Research and Innovation, Department of Pharmaceutical Sciences, College of Pharmacy, University of Kentucky, Lexington, KY 40536, USA; khaled_shaaban@uky.edu

Citation: Shaaban, K.A. Marine Microbial Diversity as Source of Bioactive Compounds. *Mar. Drugs* **2022**, *20*, 304. https://doi.org/10.3390/md20050304

Received: 21 April 2022
Accepted: 26 April 2022
Published: 29 April 2022

Publisher's Note: MDPI stays neutral with regard to jurisdictional claims in published maps and institutional affiliations.

Natural products continue to be a major inspiration and untapped resource for bioactive drug leads/probes. Natural products (NPs) or biologically inspired NPs account for approximately 24% of currently approved drugs [1]. Concerning the discovery and development of anticancer and antibiotics, microbially produced candidates now account for a very high proportion of drugs that are commonly prescribed. Compared to other natural sources, marine microbial diversity has now become a potential source for drug lead discovery. Oceans and seas cover more than 70% of the earth's surface and are massively complex, containing diverse assemblages of life forms. Marine bacteria, fungi and other microorganisms develop unique metabolic and physiological capabilities that enable them to survive in extreme habitats and to produce compounds that might not be produced by their terrestrial counterparts. In the last few decades, the systematic investigation of marine/marine-derived microorganisms as sources of novel biologically active agents has exponentially increased. Overall, this Special Issue contains eight articles, including six research articles on different topics related to the microbial natural products derived from marine microbes and two comprehensive review articles. In the following sections, we provide a brief overview of what the reader will find in this Special Issue.

Shaaban et al. reported the isolation and structural identification of three new isoquinolinequinone derivatives (mansouramycins E–G) in addition to the previously reported known compounds mansouramycins A and D from the ethyl acetate extract of the marine-derived *Streptomyces* sp. isolate B1848. The chemical structures of these compounds were elucidated by NMR (1D, 2D), HRMS, comparison with related compounds and computer-assisted methods. The cytotoxic activity of the isolated mansouramycins has been evaluated in a panel of up to 36 tumor cell lines, indicating significant cytotoxicity and good tumor selectivity for the new isolated compound mansouramycin F [2].

Four new cytotoxic indole-diterpenoids (penerpenes K-N), along with twelve other known compounds, have been discovered by Dai et al. from the fermentation broth produced by adding L-tryptophan to the culture medium of *Penicillium* sp. KFD28. The structures of the new compounds were elucidated extensively by NMR, HRMS data analyses and ECD calculations. Penerpene N represents the second example of paxilline-type indole diterpene bearing a 1,3-dioxepane ring. Three compounds (penerpene N, epipaxilline, emindole SB) were found to be cytotoxic to cancer cell lines, of which the known compound, epipaxilline, was the most active and showed cytotoxic activity against the human liver cancer cell line BeL-7402 with an IC_{50} value of 5.3 μM. Moreover, six compounds, namely paxilline, 7-hydroxyl-13-dehydroxypaxilline, 7-hydroxypaxilline-13-ene, 4a-demethylpaspaline-4a-carboxylic acid, PC-M6 and emindole SB, showed antibacterial activities against *Staphylococcus aureus* ATCC 6538 and *Bacillus subtilis* ATCC 6633 [3].

Shu et al. discovered viridicatol as a lead derivative for allergic diseases treatments. Viridicatol is a new quinoline alkaloid derivative purified from the deep-sea-derived fungus *Penicillium griseofulvum*. The structure of viridicatol was established by NMR and X-ray diffraction analysis. In the in vivo biological investigation of allergic reaction treatments in a mouse model study, viridicatol was found to amilorate the inflammatory mediator, stabalized the mast cell elivation of anaphylaxis and repaired the intestinal barrier in mice by suppressing mast cell activation [4].

The study by Lu et al. explored an example of how new strategies could accelerate the discovery of new antibiotics from highly productive natural sources. This article focused on an integrated strategy of combining phylogenetic data and bioactivity tests with a metabolomics-based dereplication approach to fast track the selection process of Mangrove actinomycetia for the discovery of novel biologically active natural products. Metabolomics technologies have been utilized to considerably aid traditional antibiotic discovery approaches in strain prioritization, resulting in increased efficiency in the discovery of new antibiotics from these highly productive and diverse ecosystems. It is a great example of using non-traditional bioactivity and/or taxonomy-based dereplication methods for the selection of new microbial strains for future natural product discoveries. In this study, a total of 521 actinomycetial strains affiliated to 40 genera in 23 families were isolated from 13 different mangrove soil samples by a culture-dependent method. A total of 179 strains affiliated to 40 different genera with unique colony morphology were selected to evaluate antibacterial activity against 12 indicator bacteria. Out of the 179 tested isolates, 47 showed activity against at least one of the tested pathogens. An analysis of 23 out of 47 active isolates using UPLC-HRMS-PCA revealed 6 outliers. Further analysis using the OPLS-DA model identified five compounds from two outliers contributing to bioactivities against drug-sensitive *A. baumannii*. Two *Streptomyces* strains (M22 and H37) were rapidly prioritized for producing potentially new compounds. The scale-up fermentation of *Streptomyces* sp. M22 afforded two new trioxacarcins with keto-reduced trioxacarcinose B, gutingimycin B and trioxacarcin G, together with the known gutingimycin [5].

SARS-CoV-2 (severe acute respiratory syndrome coronavirus-2) is a novel coronavirus strain that emerged at the end of 2019 and has resulted in millions of deaths so far. Marine sulfated polysaccharides (MSPs) are a group of natural products that have been reported from a variety of marine sources. They have recently gained significant attention and are widely examined against a variety of viral diseases. Salih et al. presented a comprehensive report and modeling analysis on marine sulfated polysaccharides from different marine sources as potential antiviral medicines, with an emphasis on SARS CoV-2. They aimed to compile a thorough report on MSPs and their antiviral activities against diverse virus species based on studies published over the last 25 years. The reported MSPs were subjected to molecular docking and dynamic simulation experiments, and it was found that nine of the investigated MSPs candidates exhibited promising results, taking into consideration the newly emerged SARS CoV-2 variants, of which five were not previously reported to exert antiviral activity against SARS CoV-2, including sulfated galactofucan, sulfated polymannuroguluronate (SPMG), sulfated mannan, sulfated heterorhamnan and chondroitin sulfate E (CS-E). These promising results shed light on the importance of sulfated polysaccharides as potential SARS-CoV-2 inhibitors [6].

An original article by Ben Hlima et al. focused on the discovery of new-lipolytic enzymes of biotechnological interest from microalgae with the aid of genomic mining by combining bioinformatics analysis and functional screening to find novel lipases biocatalysts. The in silico characterization of 14 putative *Chlorella vulagaris* lipases with different cellular localizations has been reported in the current study. Membrane-associated lipases were also detected and described in the article for the first time in this species. The 14 lipases display an acyl hydrolase motif (GXSXG) and belong to the α/β hydrolase lipase 3 family and the GX class. These putative lipases could be potential candidates for metabolic engineering to improve microalgae lipid productivity. Finally, the authors of this manuscript have also reported, for the first time, a putative lysosomal acid lipase produced by a green microalgae [7].

Lei Chen et al. provided a comprehensive review of natural products from microorganisms associated with sea cucumbers. Sea cucumbers are a class of marine invertebrates that are extensively used as a source of food in Asian cuisines and have reported pharmacological activities. Numerous microorganisms have been associated with sea cucumbers. Seventy-eight genera of bacteria belonging to forty-seven families in four phyla and twenty-nine genera of fungi belonging to twenty four families in the phylum Ascomycota have been cultured from sea cucumbers. Sea-cucumber-associated microorganisms produce diverse secondary metabolites with various biological activities, including cytotoxic, antimicrobial, enzyme-inhibiting and antiangiogenic activities. In this review, the authors have summarized the list of 145 natural products isolated from microorganisms associated with sea cucumbers between 2000 and 2021, which include polyketides, alkaloids and terpenoids as well as their reported biological activities [8].

Co-cultivation is one of the strategies used for drug discovery and has been known as an effective approach for the enhancement of the production of natural products from microorganisms. Jianwei Chen at al. provided a comprehensive review on the structural diversity of marine microbial metabolites based on the co-culture strategy. As reported by the authors, co-culturing of two or more marine microorganisms together in a solid or liquid medium in a certain environment can activate silent biosynthetic genes to produce cryptic natural products that do not exist in monocultures of the partner microbes based on either their competition or synergetic relationship. This review article by Chen et al. focuses on the significant and excellent examples covering sources, types, structures and the bioactivities of secondary metabolites based on the co-cultures of marine-derived microorganisms from 2009 to 2019. They have summarized 154 phytomolecules reported to be produced by the marine microorganism co-culture with examples of novel and bioactive natural products. They have also provided a detailed discussion on the prospects and current challenges in the field of co-culture approaches [9].

In summary, the research articles on the topic presented in this Special Issue reveal the potential of marine microorganisms and microalgae as the treasure house of new drug development in the future. The research articles illustrate the diversity of marine microbial natural products and their biological activities and highlight the importance of developing new methods to encourage the discovery of new compounds.

Funding: This research received no external funding.

Acknowledgments: The Editor would like to thank all authors that participated in this Research Topic titled "Marine Microbial Diversity as Source of Bioactive Compounds". Special acknowledgment is given to each reviewer who has contributed and whose valuable support is fundamental to the success of the journal of *Marine Drugs*.

References

1. Newman, D.J.; Cragg, G.M. Natural products as sources of new drugs over the nearly four decades from 01/1981 to 09/2019. *J. Nat. Prod.* **2020**, *83*, 770–803. [CrossRef] [PubMed]
2. Shaaban, M.; Shaaban, K.A.; Kelter, G.; Fiebig, H.H.; Laatsch, H. Mansouramycins E–G, Cytotoxic Isoquinolinequinones from Marine Streptomycetes. *Mar. Drugs* **2021**, *19*, 715. [CrossRef] [PubMed]
3. Dai, L.-T.; Yang, L.; Kong, F.-D.; Ma, Q.-Y.; Xie, Q.-Y.; Dai, H.-F.; Yu, Z.-F.; Zhao, Y.-X. Cytotoxic Indole-Diterpenoids from the Marine-Derived Fungus *Penicillium* sp. KFD28. *Mar. Drugs* **2021**, *19*, 613. [CrossRef] [PubMed]
4. Shu, Z.; Liu, Q.; Xing, C.; Zhang, Y.; Zhou, Y.; Zhang, J.; Liu, H.; Cao, M.; Yang, X.; Liu, G. Viridicatol Isolated from Deep-Sea *Penicillium Griseofulvum* Alleviates Anaphylaxis and Repairs the Intestinal Barrier in Mice by Suppressing Mast Cell Activation. *Mar. Drugs* **2020**, *18*, 517. [CrossRef]
5. Lu, Q.-P.; Huang, Y.-M.; Liu, S.-W.; Wu, G.; Yang, Q.; Liu, L.-F.; Zhang, H.-T.; Qi, Y.; Wang, T.; Jiang, Z.-K.; et al. Metabolomics Tools Assisting Classic Screening Methods in Discovering New Antibiotics from Mangrove Actinomycetia in Leizhou Peninsula. *Mar. Drugs* **2021**, *19*, 688. [CrossRef]
6. Salih, A.; Thissera, B.; Yaseen, M.; Hassane, A.; El-Seedi, H.; Sayed, A.; Rateb, M. Marine Sulfated Polysaccharides as Promising Antiviral Agents: A Comprehensive Report and Modeling Study Focusing on SARS CoV-2. *Mar. Drugs* **2021**, *19*, 406. [CrossRef]
7. Ben Hlima, H.; Dammak, M.; Karray, A.; Drira, M.; Michaud, P.; Fendri, I.; Abdelkafi, S. Molecular and Structural Characterizations of Lipases from *Chlorella* by Functional Genomics. *Mar. Drugs* **2021**, *19*, 70. [CrossRef]
8. Chen, L.; Wang, X.-Y.; Liu, R.-Z.; Wang, G.-Y. Culturable Microorganisms Associated with Sea Cucumbers and Microbial Natural Products. *Mar. Drugs* **2021**, *19*, 461. [CrossRef] [PubMed]
9. Chen, J.; Zhang, P.; Ye, X.; Wei, B.; Emam, M.; Zhang, H.; Wang, H. The Structural Diversity of Marine Microbial Secondary Metabolites Based on Co-Culture Strategy: 2009–2019. *Mar. Drugs* **2020**, *18*, 449. [CrossRef] [PubMed]

Review

The Structural Diversity of Marine Microbial Secondary Metabolites Based on Co-Culture Strategy: 2009–2019

Jianwei Chen [1], Panqiao Zhang [1], Xinyi Ye [1], Bin Wei [1], Mahmoud Emam [1,2], Huawei Zhang [1] and Hong Wang [1,*]

1   College of Pharmaceutical Science & Collaborative Innovation Center of Yangtze River Delta Region Green Pharmaceuticals, Zhejiang University of Technology, Hangzhou 310014, China; cjw983617@zjut.edu.cn (J.C.); 18668358826@163.com (P.Z.); xinyiye1020@zjut.edu.cn (X.Y.); binwei@zjut.edu.cn (B.W.); mahmoudemamhegazy2020@gmail.com (M.E.); hwzhang@zjut.edu.cn (H.Z.)

2   Phytochemistry and Plant Systematics Department, National Research Centre, 33 El Bohouth St., Dokki, Giza 12622, Egypt

*   Correspondence: hongw@zjut.edu.cn

Received: 28 July 2020; Accepted: 25 August 2020; Published: 27 August 2020

Abstract: Marine microorganisms have drawn great attention as novel bioactive natural product sources, particularly in the drug discovery area. Using different strategies, marine microbes have the ability to produce a wide variety of molecules. One of these strategies is the co-culturing of marine microbes; if two or more microorganisms are aseptically cultured together in a solid or liquid medium in a certain environment, their competition or synergetic relationship can activate the silent biosynthetic genes to produce cryptic natural products which do not exist in monocultures of the partner microbes. In recent years, the co-cultivation strategy of marine microbes has made more novel natural products with various biological activities. This review focuses on the significant and excellent examples covering sources, types, structures and bioactivities of secondary metabolites based on co-cultures of marine-derived microorganisms from 2009 to 2019. A detailed discussion on future prospects and current challenges in the field of co-culture is also provided on behalf of the authors' own views of development tendencies.

Keywords: co-culture; marine microbes; natural products; structural diversity; biological activities

---

1. Introduction

Although many industrial sectors have stopped their dependence on natural product (NP) drug discovery programs, NPs are still of great interest to many pharmaceutical communities and are important sources of bioactive compounds [1,2]. Marine microbes, as an important source of bioactive NPs, have elicited widespread attention [3–5]. However, the discovery of novel marine microbial NPs is becoming more difficult and the rate of rediscovery of known NPs is being gradually increased. On the other hand, recent genomic sequencing has revealed the presence of numerous biosynthetic gene clusters in some microbes that may be responsible for the biosynthesis of NPs which are not found under classical cultivation conditions [6,7]. Therefore, many alternative strategies have been explored to activate these silent and cryptic biosynthetic genes. The co-culturing of marine microbes involves the culturing of two or more marine microbes together on/in certain conditions; microorganisms can communicate with each other through direct or indirect contact, thereby stimulating the silent gene clusters to produce special NPs [2,8] (Figure 1). This strategy can promote the production of complex and novel skeletons with numerous stereocenters [9–11]. Hence, the co-culturing of marine microbes draws widespread attention in the scientific community as a potential source of unknown bioactive substances classified as alkaloids, polyketides, anthraquinone, flavonoids, cyclopeptides, etc. To exploit the NPs from the co-cultures of marine microbes and understand their medicinal significance, this review summarizes successful examples involved in NPs of marine microbes based on co-cultures from 2009 to 2019 (Table 1).

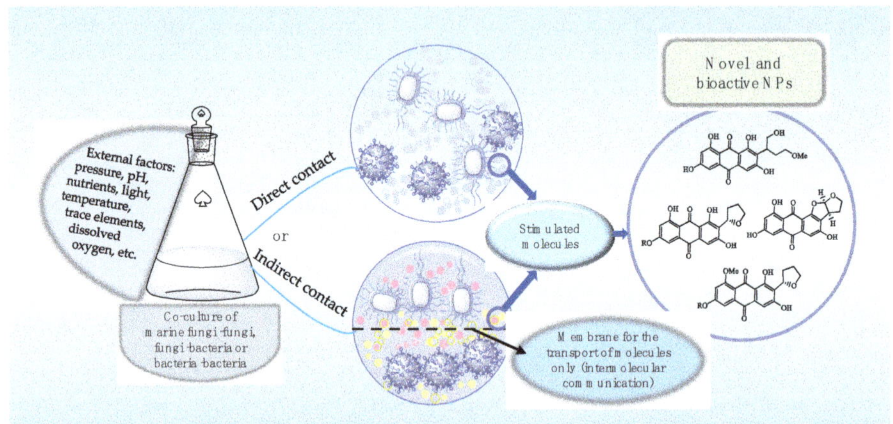

Figure 1. The schematic diagram of novel and bioactive natural products (NPs) using co-cultures of marine fungi−fungi, fungi−bacteria and bacteria−bacteria in direct or indirect contact.

Table 1. Summarized NPs identified from the co-culture of marine microbes: 2009–2019.

Classes	The Number of NPs	Identified Date	Bioactivities	Co-Culture of Marine Microorganisms
				Fungi and fungi
				A. sulphureus KMM 4640 and *I. felina* KMM 4639 *Aspergillus.* sp. FSY-01 and FSW-02 *P. citrinum* SCSGAF 0052 and *A. sclerotiorum* SCSGAF 0053 *Phomopsis* sp. K38 and *Alternaria* sp. E33
				Fungi and bacteria
Alkaloids	80 isolates (**1–80**)	2010 and 2014–2019	Cytotoxicity, enzyme Inhibitors, antimicrobial activities	*Penicillium* sp. DT-F29 and *Bacillus* sp. B31 *A. flavipes* fungus and *S.* sp. CGMCC4.7185 *A. fumigatus* MR2012 and *S. leeuwenhoekii* C34 *A. versicolor* and *B. subtilis*,
				Bacteria and bacteria
				Streptomyces sp. CGMCC4.7185 and *B. mycoides* *Saccharomonospora* sp. UR22 and *Dietzia* sp. UR66
				Fungi and fungi
				Asexual morph and sclerotial morph of *A. alliaceus*
Anthraquinones	13 isolates (**81–93**)	2017–2019	Cytotoxicity and antimicrobial activities	**Fungi and bacteria**
				A. versicolor and *B. subtilis*
				Bacteria and bacteria
				Micromonospora sp. WMMB-235 and *Rhodococcus* sp. WMMA-185

Table 1. *Cont.*

Classes	The Number of NPs	Identified Date	Bioactivities	Co-Culture of Marine Microorganisms
Cyclopeptides	6 isolates (**94–99**)	2014 and 2019	Antifungal and anti-proliferative activities	**Fungi and fungi** *Phomopsis* sp. K38 and *Alternaria* sp. E33 *Aspergillus* sp. BM and 05-BM-05ML **Fungi and bacteria** *A. versicolor* and *B. subtilis*
Macrolides	1 isolate (**100**)	2018	Antitumor and antibacterial activity	**Bacteria and bacteria** *Saccharomonospora* sp. UR22 and *Dietzia* sp. UR66
Phenylpropanoids	23 isolates (**101–123**)	2011, 2015 and 2019	Cytotoxic, antifungal, antibacterial and anti-influenza activities	**Fungi and fungi** *Phomopsis* sp. K38 and *Alternaria* sp. E33 *A. sydowii* EN-534 and *P. citrinum* EN-535 **Fungi and bacteria** *A. versicolor* and *B. subtilis*
Polyketides	12 isolates (**124–135**)	2013, 2014 and 2018	Anti-proliferative, cytotoxicity and antifungal activities	**Fungi and fungi** *Aspergillus* sp. BM and 05 and BM-05ML *Penicillium* sp. Ma(M3)V and *Trichoderma* sp. Gc(M2)1 **Fungi and bacteria** *Penicillium* sp. WC-29-5 and *S. fradiae* 007 **Bacteria and bacteria** *Janthinobacterium* spp. ZZ145 and ZZ148
Steroids	5 isolates (**136–140**)	2009, 2010 and 2014	Antiproliferative activity	**Fungi and fungi** *Aspergillus* sp. FSY-01 and FSW-02 **Fungi and bacteria** *Aspergillus* sp. BM05 and an unknown bacteria (BM05BL)
Terpenoids	2 isolates (**141–142**)	2012 and 2017	Inhibition of diatom *N. annexa* and macroalga *U. pertusa*	**Fungi and bacteria** *A. fumigatus* MR2012 and *S. leeuwenhoekii* C58 **Bacteria and bacteria** *S. cinnabarinus* PK209 and *Alteromonas* sp. KNS-16
Others	12 isolates (**143–154**)	2013, 2016, 2017 and 2019	Antimicrobial, toxicity, cytotoxicity, Hemolytic activities	**Fungi and fungi** *Phomopsis* sp. K38 and *Alternaria* sp. E33 *P. citrinum* SCSGAF 0052 and *A. sclerotiorum* SCSGAF 0053 *A. sulphureus* KMM 4640 and *I. felina* KMM 4639 **Fungi and bacteria** *A. versicolor* and *B. subtilis*

2. Compounds Derived from the Co-Cultures of Marine Microorganisms

Co-culturing or mixed fermentation is considered an important technique of inducing secondary metabolites hidden in the genomes of marine microbes by using appropriate physiological conditions, chemical communication and competition of microbes. Consequently, it is considered an easy, cheap and effective method [12,13]. This finding also explains the chemical communication and antagonism between different marine microorganisms, such as the interactions between marine fungi–fungi, fungi–bacteria and bacteria–bacteria, in which they act as signaling molecules, competitors or defense agents [14]. Herein, the metabolites based on co-cultures of marine microbes were classified according to their skeletons as alkaloids, anthraquinones, cyclopeptides, flavonoids, macrolides, phenylpropanoids, polyketides, steroids, terpenoids and others from 2009–2019. These excellent examples were found from

SciFinder, Science Direct, PubMed, Springer and other databases. Among them, the interactions between marine fungi and bacteria were found to induce the most metabolites (Figure 2A), and the alkaloids played a significant role in co-cultures of marine microbes (Figure 2B), no matter whether the mixed cultivation was of marine fungi–fungi (Figure 2C), fungi–bacteria (Figure 2D) or bacteria–bacteria (Figure 2E).

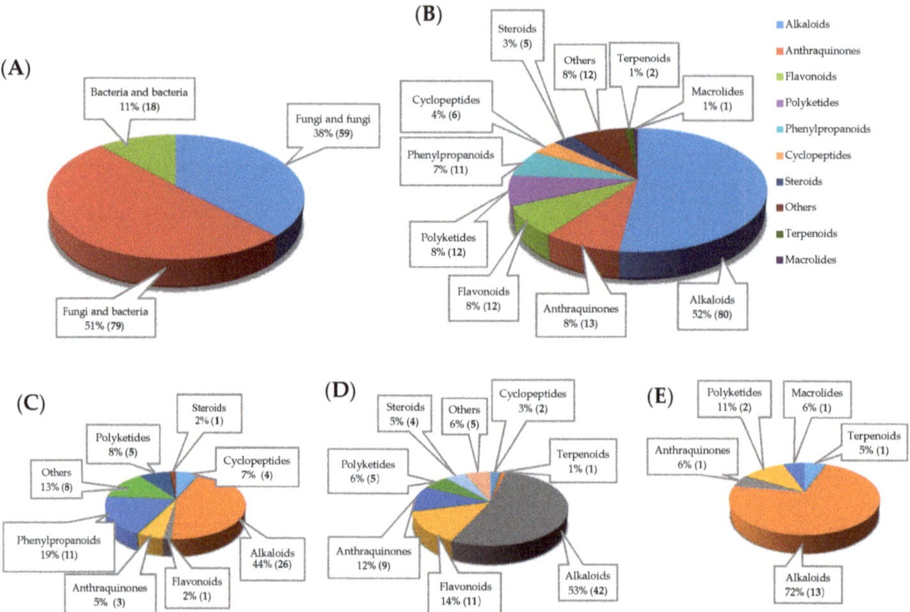

Figure 2. Numbers and the percentage of (**A**) isolates from the co-cultures of different marine microbes; (**B**) different classes of NPs from the co-cultures of marine microbes. The classes, numbers and proportions of NPs isolated from the co-cultures of marine (**C**) fungi and fungi, (**D**) fungi and bacteria, (**E**) bacteria and bacteria.

2.1. Alkaloids

The nitrogenous alkaloids represented the most abundant class of compounds that were produced by the co-cultures of marine microorganisms with diverse skeletons and biological activities [15,16]. Eighty alkaloidal metabolites were isolated and identified from different microbial environments (Figure 2B), and the co-cultures of marine fungi–bacteria represented 51% of the total isolates (Figure 3).

Alkaloids

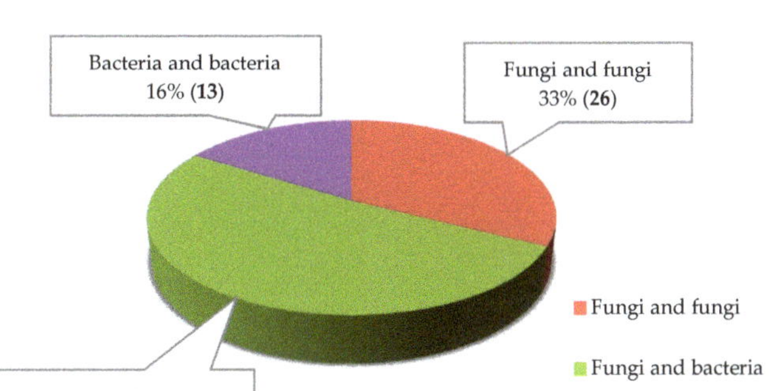

Figure 3. Alkaloids isolated from the co-cultures of marine fungi–fungi, fungi–bacteria and bacteria–bacteria.

2.1.1. Alkaloids Derived from the Co-Cultures of Different Marine Fungi

Several studies of co-cultures of fungal–fungal interactions from different marine sources were summarized as follows; 26 alkaloids were isolated and identified (Figures 2C and 3). The mixed fermentation of marine-derived fungi *Aspergillus sulphureus* KMM 4640 from muddy sand of the eastern Sakhalin shelf (Sea of Okhotsk, 26 m depth) and *Isaria felina* KMM 4639 from sediments (South China Sea, Vietnam shores, 10 m depth), led to the production of five novel prenylated indole alkaloids, 17-hydroxynotoamide D (**1**), 17-O-ethylnotoamide M (**2**), 10-O-acetylsclerotiamide (**3**), 10-O-ethylsclerotiamide (**4**) and 10-O-ethylnotoamide R (**5**) together with known compounds (-)-notoamide B (**6**), notoamide C (**7**), dehydronotoamide C (**8**), notoamide D (**9**), notoamide F (**10**), notoamide Q (**11**), 17-epi-notoamide Q (**12**), notoamide M (**13**) and sclerotiamide (**14**) (Figure 4) [17]. Among them, compounds **1–5** were only produced in the co-culturingprocess.

Compounds **2**, **6**, **8**, **13** and **14** inhibited the proliferation of the human prostate cancer cells 22Rv1 at 100 µM. Notably, **2** and **13** drastically reduced the viability of 22Rv1 prostate cancer cells at 10 µM by 25% and 55%, respectively. 22Rv1 cancer cell lines were resistant to hormone therapy at conventional chemotherapy including two new 2nd generation drugs enzalutamide and abiraterone owing to the presence of the androgen receptor splice variant-7 (AR-V7). Therefore, the active NPs drugs in these cells might be further investigated in the treatment of different human drug-resistant prostate cancer. **6** and **7** displayed weak cytotoxicity against HeLa and L1210 cell lines with half maximal inhibitory concentration (IC$_{50}$) in the range of 22–52 µg/mL [18]. Although **6** and **7** had the similar structure with **9**, compound **9** did not display the similar cytotoxic activity against HeLa and L1210 cell lines. The significant difference in cytotoxicity might be attributed to the possible existence of pyrroloindole system in **9** rather than the dihydroxypyrano-2-oxindole ring system of **6** and **7** [18,19]. In addition, compounds **1**, **2**, **5**, **9**, **13** and **14** did not exhibit any cytotoxicity against human non-malignant (HEK 293 T and MRC-9) or malignant (PC-3, LNCaP, and 22Rv1) cell lines at concentrations up to 100 µM for 48 h [17].

The co-fermentation of marine mangrove epiphytic fungi *Aspergillus* sp. FSY-01 and FSW-02 collected from a rotten fruit of mangrove *Avicennia marina* in Zhanjiang, Guangdong Province, China, yielded a new alkaloid, aspergicin (**15**), together with two known secondary metabolites, neoaspergillic acid (**16**) and aspergicine (**17**) (Figure 5) [20,21]. Notably, compounds **17** and **15** are chemically isomers, and consequently aspergicine (**17**) may be the precursor of aspergicin (**15**) through a proton 1, 2-shift [22].

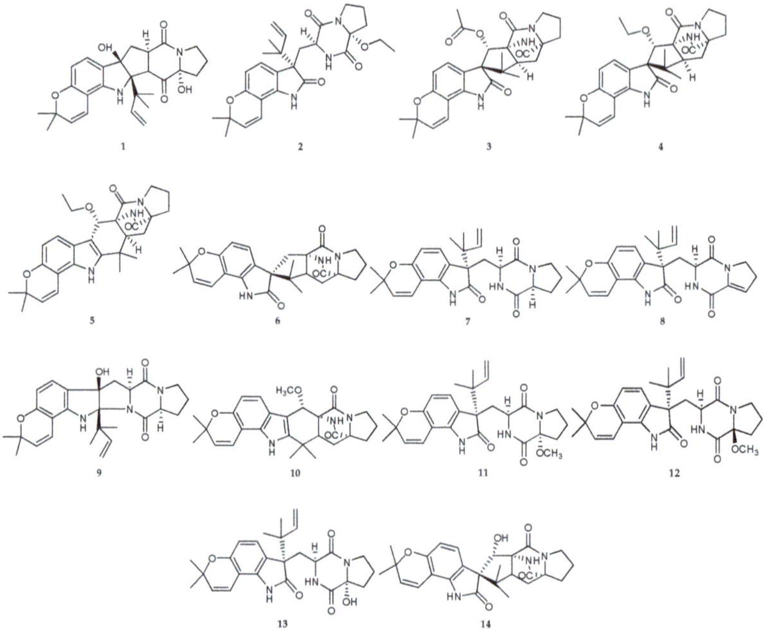

Figure 4. Chemical structures of **1–14**.

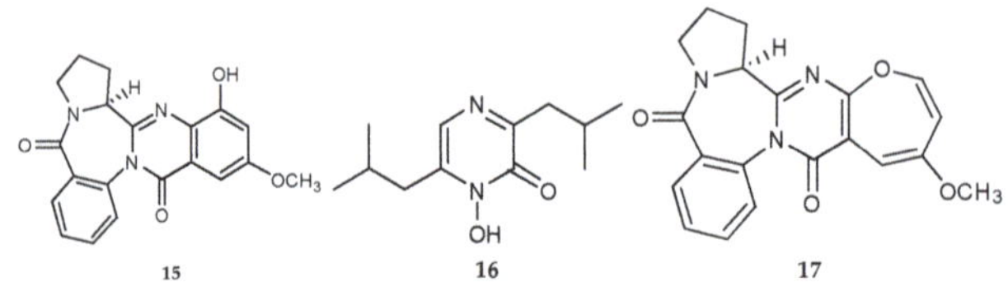

Figure 5. Chemical structures of **15–17**.

Compounds **15** and **16** showed potent inhibitory activities against three Gram-positive bacteria, *Bacillus subtilis* (MIC, minimum inhibitory concentration that inhibits the growth of microbes by 80%, 15.62 and 1.95 µg/mL), *Staphylococcus epidermidis* (MIC 31.25 and 0.49 µg/mL) and *Staphylococcus aureus* (MIC 62.50 and 0.98 µg/mL), and three Gram-negative bacteria, *Escherichia coli* (MIC 31.25 and 15.62 µg/mL), *Bacillus proteus* (MIC 62.50 and 7.80 µg/mL) and *Bacillus dysenteriae* (MIC 15.62 and 7.80 µg/mL), respectively [22].

Marine fungi *Aspergillus sclerotiorum* SCSGAF 0053 and *Penicillium citrinum* SCSGAF 0052 were isolated from the gorgonian corals *Muricella flexuosa* collected from South China Sea, Sanya (18°11′ N, 109°25′ E), Hainan Province, China [23]. Due to the mixed fermentation of marine fungi, a red pigment appeared in the mixed fermentation broth could not be observed in any strain cultured separately. This special phenomenon suggested that a novel biosynthesis route was activated. Four novel alkaloids were obtained, including one oxadiazin derivative sclerotiorumin C (**18**), a pyrrole derivative 1-(4-benzyl-1*H*-pyrrol-3-yl) ethanone (**19**), aluminumneohydroxyaspergillin (**20**) and ferrineohydroxyaspergillin (**21**), together with one known compound ferrineoaspergillin (**22**) (Figure 6) [23]. Compounds **18–21** were only produced in the co-culture process.

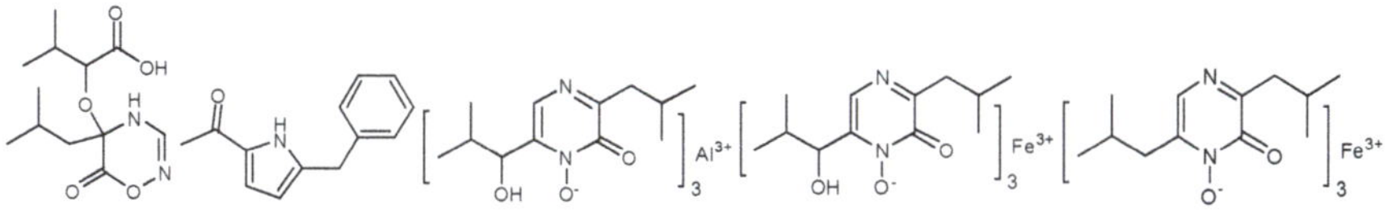

Figure 6. Chemical structures of **18–22**.

Compound **20** exhibited potent toxicity towards brine shrimp with medium lethal concentration (LC_{50}) value of 6.1 µM and high selective cytotoxicity towards histiocytic lymphoma U937 cell line with an IC_{50} value of 4.2 µM. **19**, **21**, and **22** showed moderate toxicity against brine shrimp with LC_{50} values of 46.2, 11.5 and 27.8 µM, respectively. **21** and **22** possessed mild cytotoxicity against U937 with IC_{50} values of 42.0 and 48.0 µM, respectively. These results suggested that the aluminum complex skeletons of compounds showed more potent toxicity and cytotoxicity than ferricomplex structures of compounds [23–26]. Moreover, aspergillic acid and **16** also showed more potent inhibitory activities than neohydroxyaspergillic acid and hydroxyaspergillic acid against *B. subtilis*, *E. coli*, *S. aureus* and *Candida albicans* [27].

The co-culture of mangrove fungi *Phomopsis* sp. K38 and *Alternaria* sp. E33 led to the identification of one new diimide derivative, (-)-byssochlamic acid bisdiimide (**23**) and a novel nonadride derivative, (-)-byssochlamic acid imide (**24**) (Figure 7) [28,29]. Ebada et al. (2014) investigated the mycelial extract of a co-cultivation of marine fungal strains *Aspergillus*. BM-05 and BM-05ML, and identified two alkaloids, protuboxepin A (**25**) and oxepinamide E (**26**) (Figure 7) [30]. **23–24** were only found in the co-culture process.

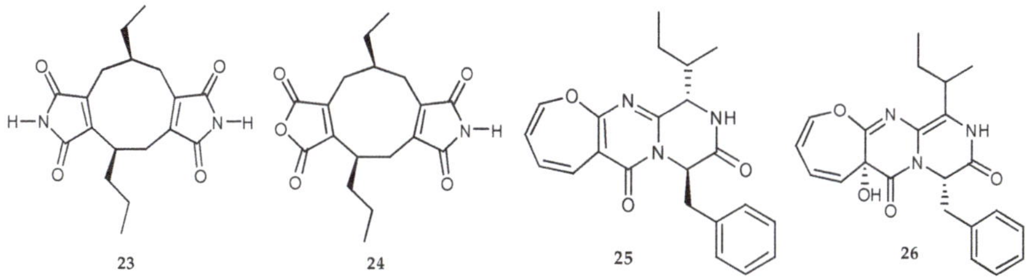

Figure 7. Chemical structures of **23–26**.

Compound **23** exhibited moderate inhibitory activity against HepG2 and Hep-2 with IC_{50} values of 51 µg/mL and 45 µg/mL, respectively. **24** had moderate antifungal activities against *Fusarium oxysporum* and *Fusarium graminearum* with MIC values of 60 µg/mL and 50 µg/mL, respectively [28,29,31]. **25** possessed anti-proliferative activity against human breast cancer adenocarcinoma MDA-MB-231, human acute promyelocytic leukemia HL-60, hepatocellular carcinoma Hep3B, chronic myelogenous leukemia K562 and rat fibroblast 3Y1 cell lines with IC_{50} values of 130, 75, 150, 250 and 180 µM, respectively [32,33]. **26** showed transcriptional activation on liver X receptor α (LXRα) with a half maximal effective concentration (EC_{50}) value of 12.8 µM. It was known that LXR was an important target in drug discovery; LXR agonists had been proven to exhibit remarkable therapeutic effects on diabetes, atherosclerosis, Alzheimer's disease and anti-inflammation. Therefore, **26** was worthy of consideration as a potential lead compound for drug discovery [34].

2.1.2. Alkaloids Derived from the Co-Cultures of Marine Fungi and Bacteria

The alkaloids derived from the co-culture of marine fungi and bacteria were tallied to be 41 isolates (Figures 2D and 3) and can be described as follows; prenylated 2,5-diketopiperazines (2,5-DKPs) were isolated from the co-culture of marine *Penicillium* sp. DT-F29 isolated from marine sediments of Dongtou country, China, and *Bacillus* sp. B31 collected from marine sediments of Changzhi Island, China [35], including ten novel metabolites, 12-β-hydroxy-13-butoxyethoxyfumitremorgin B (**27**), diprostatin A (**28**), 12-hydroxy-13α-ethoxyverruculogen TR-2 (**29**), hydrocycloprostatin A (**30**), 12-β-hydroxy-13α-butoxyethoxyverruculogen TR-2 (**31**), hydrocycloprostatin B (**32**), 26-α-hydroxyfumitremorgin A (**33**), 25-hydroxyfumitremorgin B (**34**), 12-β-hydroxy-13α-methoxyverruculogen (**35**), 25-hydroxyfumitremorgin A (**36**) and thirteen known isolates, verruculogen TR-2 (**37**), 12-α-hydroxy-13-α-prenylverruculogen TR-2 (**38**), 12-hydroxyverruculogen TR-2 (**39**), 13-prenyl fumitremorgin B (**40**), 12-β-hydroxy-13-α-methoxyverruculogen TR-2 (**41**), cycloprostatin C (**42**), cyclotryprostatin B (**43**), spirotryprostatin C (**44**), 12,13-dihydroxyfumitremorgin C (**45**), neofipiperzine C (**46**), prenylcycloprostatin B (**47**), fumitremorgin B (**48**) and fumitremorgin A (**49**) (Figure 8)

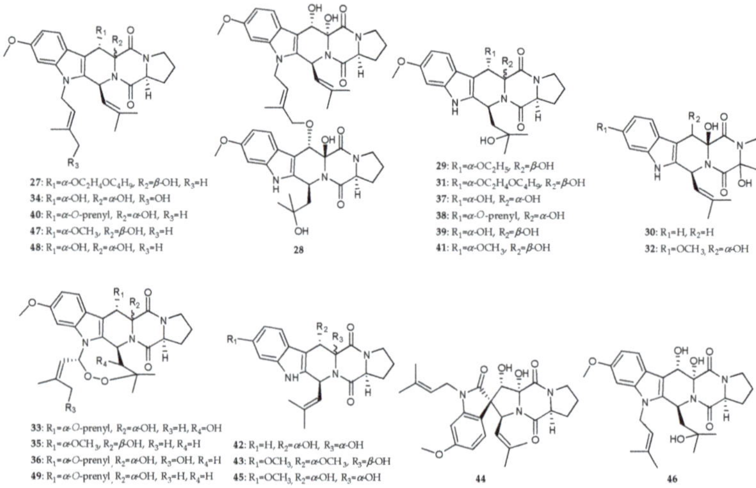

Figure 8. Chemical structures of **27–49**.

The secondary metabolites profile of the co-culture of *Streptomyces* sp. and *Aspergillus flavipes*, obtained from marine sediments of the Nanji Islands of the same habitat, showed an induced biosynthesis of a series of known cytochalasans, including rosellichalasin (**50**), and five aspochalasins (aspochalasin E **51**, aspochalasin P **52**, aspochalasin H **53**, aspochalasin M **54** and 19,20-dihydro-aspochalasin D **55**) (Figure 9) [36]. The chromatographic purification of the combination culture extract from marine-derived *Aspergillus fumigatus* MR2012 and *Streptomyces leeuwenhoekii* C34 led to the isolation of two novel compounds, luteoride D (**56**) and pseurotin G (**57**), along with the known isolates, nocardamine (**58**), terezine D (**59**), 11-*O*-methylpseurotin A (**60**) and lasso peptide chaxapeptin (**61**) [37]. In addition, seven known compounds, notoamide D (**9**), speramide B (**62**), notoamide E (**63**), stephacidin A (**64**), notoamide R (**65**), protuboxepin B (**66**) and 3,10-dehydrocyclopeptine (**67**) (Figure 9) were identified from the mixed-fermentation of the marine-derived fungus *Aspergillus versicolor* isolated from sponge *Agelas oroides* and *B. subtilis* [38].

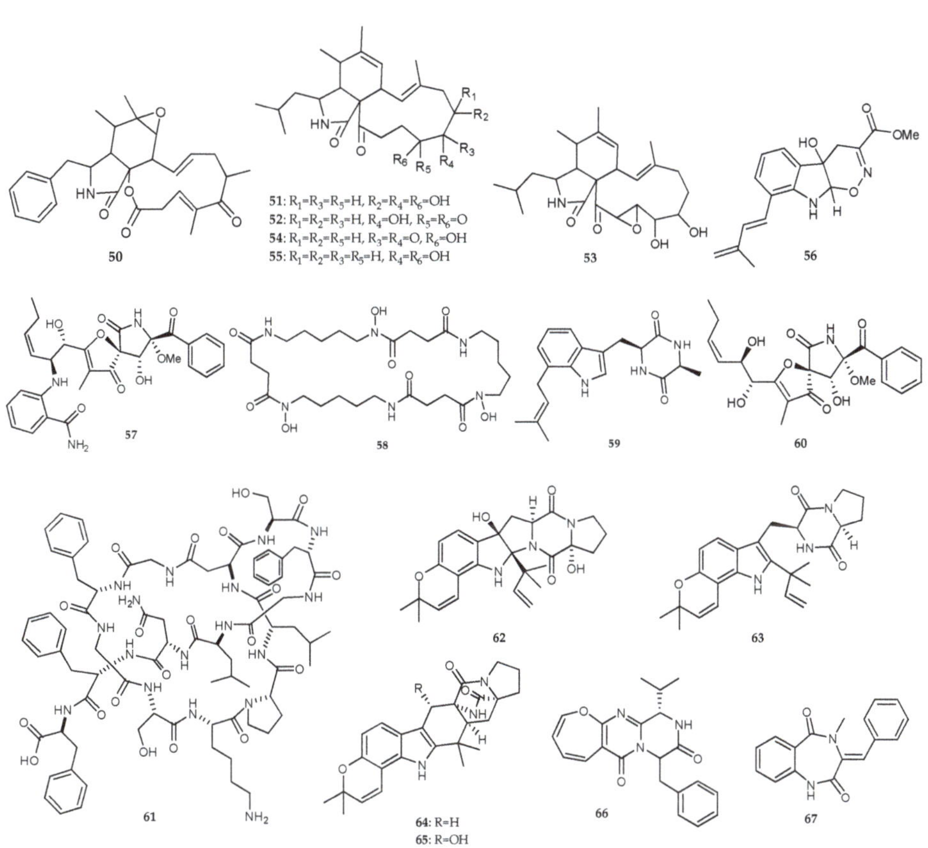

Figure 9. Chemical structures of **50–67**.

Compounds **27**, **28**, **38–40**, **44** and **46–49** displayed strong inhibitory effects on bromodomain-containing protein 4 (BRD4) at 20 μM. Notably, **39** and **48** exhibited the most inhibitory activity with 72.7% and 80.4%, compared with the positive control, BRD4 inhibitor (+)-JQ1 (85.7%) [35]. As reported in the previous study, BRD4 protein was a member of the bromodomain and extra-terminal domain (BET) family that carried two bromodomains and was associated with mitotic chromosomes. Bromodomains targeted genetic and epigenetic alterations and regulated chromatin remodeling, which were important therapeutic targets for major diseases, such as neurological disorders, obesity, cancer and inflammation [39,40]. Thus, these compounds further deserved development and research for the treatment of major diseases. Li et al. (2012) reported that **41** had potent inhibitory activities against *Fusarium oxysporum* f. sp. *Niveum, Alternaria alternate, Fusarium oxysporum* f. sp. *vasinfectum* and *Fusarium solani* with MIC values of 6.25–25 μg/mL and moderate brine shrimp toxicity (LC$_{50}$ 60.7 μg/mL) [41]. The occurrence of **41** could be involved in protecting microbes against invasion by other competing microbes. Therefore, **41** could be considered as a promising lead compound for developing new fungicides. Cui et al. reported **43** could completely inhibit the G2/M phase of tsFT210 cells at concentrations >29.4 μM [42]. Furthermore, Wang et al. (in 2008) showed that **44** had selective cytotoxicity against four cancer cell lines, MOLT-4, HL-60, A-549 and BEL-7402 [43].

Cytochalasans were fungal metabolites that were structurally identified by the presence of a reduced isoindone nucleus connected with a macrocyclic ring [44]. Six cytochalasans (**50–55**) showed strong toxicity against *Streptomyces* sp. with 50–80% inhibition at 2–16 μg/mL, and most of them even exhibited 60% inhibition at 2 µg/mL, but they had no any effect on the fungus *A. flavipes* at the same concentration. This indicated that cytochalasans could help *A. flavipes* to compete with *Streptomyces* sp., which was an important support for their potential ecological role. All cytochalasans also exhibited obvious toxicity against human cell lines, as cytochalasans had the ability to inhibit, specifically, the actin filament elongation by blocking the polymerization sites [45–47]. Thus, all six compounds (**50–55**) exhibited powerful toxicity against *Streptomyces* sp. at 2–16 µg/mL with inhibition rate of 50–80%. Notably, most of these compounds displayed strong inhibitory activity with inhibition rate of 60% even at 2 µg/mL, whereas none of them had antimicrobial activity against the marine-derived producer *A. flavipes* at the same concentration. These findings implied that the co-culture through microbial physical contact could stimulate the expression of silent gene cluster that was responsible for the production of cytochalasans.

The cyclic siderophore, nocardamine (**58**), had inhibitory effects on the proliferation of human tumor cell lines: SK-Mel-5 with an IC_{50} value of 18 µM, T-47D with an IC_{50} value of 6 µM, PRMI-7951 with an IC_{50} value of 14 µM and SK-Mel-28 with an IC_{50} value of 12 µM [48]. Compared with the pure cultures, some novel metabolites were observed in the mixed culture. Two fungal prenylated indole metabolites, **56** and **59**, which were not traced before in *A. fumigatus*, were induced. Both of them had an oxazino [6,5-*b*]indole nucleus which was not previously found in nature. Additionally, the yield of compound **61** was obviously higher than that of the monoculture of *Streptomyces leeuwenhoekii* C58. It was the first time that a bi-lateral cross talk was proved, which resulted in dual induction of both fungal and bacterial metabolites in the same culture conditions. **64** displayed cytotoxic activities toward mouse lymphoma cell line L5178Y with an IC_{50} value of 16.7 µM and in vitro toward testosterone-dependent prostate LNCaP cells with an IC_{50} value of 2.1 µM [49].

2.1.3. Alkaloids Derived from the Co-Cultures of Different Marine Bacteria

Thirteen alkaloids were isolated from the co-culture of different marine bacteria (Figures 2E and 3); the structures of these isolates were listed in Figure 10. The average yields of five known tryptamine derivatives, N-acetyltryptamine (**68**), N-propanoyltryptamine (**69**), bacillamide C (**70**), bacillamide B (**71**) and bacillamide A (**72**) using the co-fermentation of marine strain *Streptomyces* sp. CGMCC4.7185 and *Bacillus mycoides* isolated from marine sediments of the Nanji Island (China, 27°42′ N, 121°08′ E), were 14.9, 2.8, 9.6, 13.7 and 3.0 mg/L, respectively, which were all undetectable under simple culture conditions [50]. This was the first report of applying a microorganism co-culture system to enhance the yields of known compounds [50].

In 2018, El-Hawary et al. identified four indole alkaloids—a novel brominated oxindole alkaloid saccharomonosporine A (**73**), a novel convolutamydine F (**74**) and two known compounds, (*S*) 6-bromo-3-hydroxy-3-(1H-indol-3-yl) indolin-2-one (**75**) and vibrindole (**76**)—from the mixed fermentation culture of two sponge-associated actinomycetes, *Saccharomonospora* sp. UR22 and *Dietzia* sp. UR66 collected from the Red Sea sponge *Callyspongia siphonella* [51].

Two sponge-associated actinomycetes, *Actinokineospora* sp. EG49 isolated from the Red Sea sponge, *Spheciospongia vagabunda*, and *Nocardiopsis* sp. RV163 derived from the Mediterranean sponge, *Dysidea avara*, were co-cultivated together and yielded a novel 5a,6,11a,12-tetrahydro-5a,11a-dimethyl-1,4-benzoxazino[3,2-*b*][1,4]benzoxazine (**77**) and three known metabolites, N-(2-hydroxyphenyl)-acetamide (**78**), 1,6-dihydroxyphenazine (**79**) and 2,2′,3,3′-tetrahydro-2,2′-dimethyl-2,2′-bibenzoxazole (**80**) [52].

Pim-1 kinase is a well-established oncoprotein in several tumor entities, such as prostate cancer, pancreatic cancer, colorectal cancer and myeloid leukemia. Inhibition of Pim-1 kinase would prevent the growth of tumor cells. Compounds **73** and **75** exhibited potent Pim-1 kinase inhibitors with IC_{50} values of 0.3 µM and 0.946 µM, respectively. Docking studies showed the binding model of **73** and **75** in the ATP pocket of Pim-1 kinase. They also exhibited obvious antiproliferative activity against human promyelocytic leukemia HL-60 (IC_{50} 2.8 and 4.9 µM) and human colon adenocarcinoma HT-29

(IC$_{50}$ 3.6 and 3.7 μM). This indicated that **73** and **75** could act as potential Pim-1 kinase inhibitors that mediated the inhibitory effects on the growth of tumor cells [51].

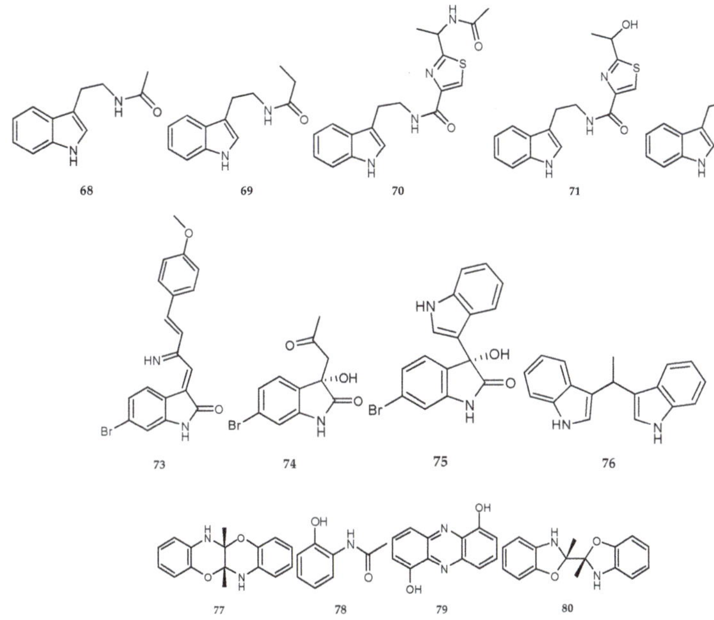

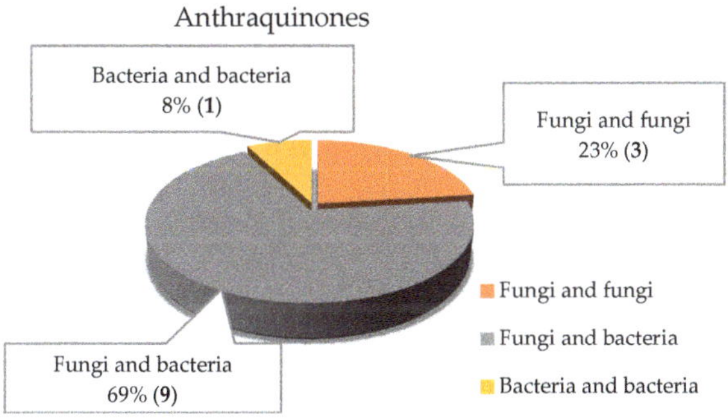

Figure 10. Chemical structures of **68–80**.

In addition, only compound **79** was documented against *Trypanosoma brucei* (IC$_{50}$ 19 μM), *Bacillus* sp. (11 mm inhibition zone diameter) and *Actinokineospora* sp. EG49 (15 mm inhibition zone diameter) [52]. The yield of **79** was very high in the co-culture process. However, it was not detected in the single microbial culture. Co-culture strategy not only enhanced the chemical diversity of the metabolites but also increased the production of metabolites undetected in the single microbial culture.

2.2. Anthraquinones

Thirteen different anthraquinone isolates were obtained from different marine microbial co-cultures; the co-cultures of marine fungi–bacteria represented the majority, 69% (9/13 isolates; Figures 2B and 11).

Figure 11. Anthraquinones isolated from the co-cultures of marine fungi–fungi, fungi–bacteria and bacteria–bacteria.

2.2.1. Anthraquinones Derived from the Co-Cultures of Different Marine Fungi

In the recent study, the combination of cultures from two different developmental stages of marine alga-derived *Aspergillus alliaceus* (teleomorph: *Petromyces alliaceus*) drastically changed the metabolite profile and resulted in the production of allianthrone A (**81**) and two diastereomers, allianthrones B (**82**) and C (**83**) (Figure 12) [53]. **81–83** exhibited cytotoxic activity against SK-Mel-5 melanoma cell lines with IC_{50} (11.0, 12.2, and 19.7 µM) and HCT-116 colon carcinoma cells with IC_{50} (9.0, 10.5 and 13.7 µM), respectively. This study presented the first example of elicitation of novel fungal chemical diversity by a co-existing strategy of two different developmental phenotypes of *Aspergillus* species. For several Aspergilli, e.g., *A. alliaceus*, asexual and sexual life developmental stages were known. However, rarely did they co-cultivate at the same time. Even more surprising was the presence of novel bianthrones when the sclerotial and asexual morphs of the same species co-existed. There were only a few examples that showed differences in secondary metabolites in fungi based on their distinct developmental stages or chemical profiles for the two mating types of heterothallic fungi. However, none of these compounds displayed any activity against *P. aeruginosa*, *E. faecium*, *S. aureus*, *E. coli*, *C. albicans* and *B. subtilis*. Furthermore, non-significant results were obtained against lung (A549), prostate (PC3) and breast (MCF-7) human cancer cells compared with the positive control, etoposide [53].

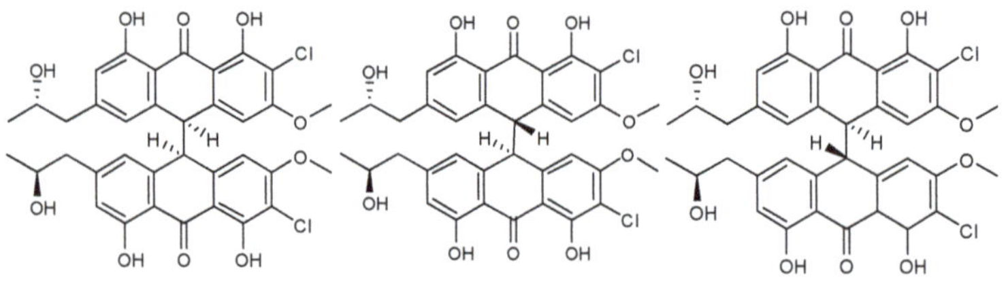

Figure 12. Chemical structures of **81–83**.

2.2.2. Anthraquinones Derived from the Co-Cultures of Marine Fungi and Bacteria

Two novel anthraquinones, (z)(11S,12R)-versicolorin B (**84**) and 6,8-O-dimethylbipolarin (**85**), along with seven known substances bipolarin (**86**), versiconol (**87**), versiconol acetate (**88**), versicolorin B (**89**), 8-O-Methylversicolorin B (**90**), averufin (**91**) and endocrocin (**92**) (Figure 13) were isolated and identified from the mixed fermentation broth of the marine fungus *A. versicolor* and *B. subtilis* [38].

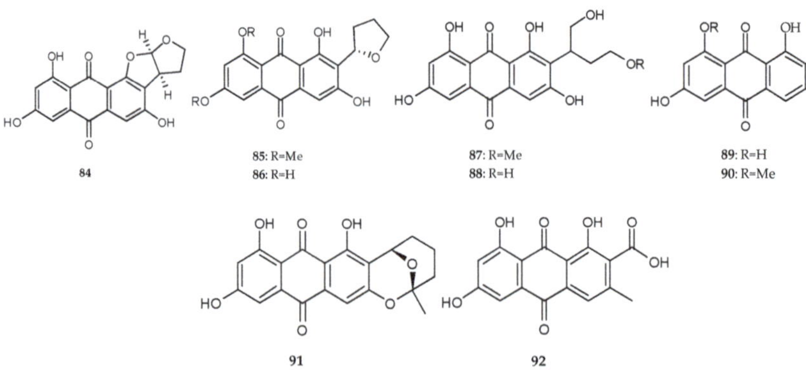

Figure 13. Chemical structures of **84–92**.

Versiconol (**87**) was characterized as an inhibitor of protein tyrosine kinases against EGF-R and v-abl protein tyrosine kinases that were responsible for catalyzing phosphorylation of tyrosine residues of protein substrates, and suppression of MK-cells [54]. **89** displayed inhibitory activity against the Gram-positive *S. aureus* with MIC value of 50 μM and antifungal activity against *Fusarium solani* with MIC values of 16–32 μg/mL [38,55]. The cytotoxic bioassay of **90** was recorded against mouse lymphoma cell line L5178Y with an IC_{50} value of 21.2 μM. Moreover, **91** displayed antibacterial activity against *B. subtilis* (MIC = 8–16 μg/mL) and the Gram-positive *S. aureus* (MIC = 25 μM) and four Gram-positive microbes, including two *E. faecalis* and two *E. faecium* (MIC = 12.5–25 μM) [38,55]. Neither **89** nor **91** had cytotoxicity against L5178Y cell line, which implied that their antimicrobial activities were not associated with their respective general toxicities. Besides, **90** also displayed mild cytotoxic activity against human lung cancer cells H460 and the human prostate cancer cells PC-3 with IC_{50} values of 27.2 and 19.5 μM, respectively [56]. Other compounds did not exhibit distinct cytotoxic activity against L5178Y cell line and antibacterial activity against five Gram-positive microbes, including one *S. aureus*, two *E. faecalis* and two *E. faecium*.

2.2.3. Anthraquinones Derived from the Co-Cultures of Different Marine Bacteria

A new antibiotic, keyicin (**93**) (Figure 14), was purified and identified from a co-culture of two marine invertebrate-associated bacteria *Micromonospora* sp. WMMB-235 and *Rhodococcus* sp. WMMA-185 [57]. It showed selective inhibitory activity against Gram-positive bacteria and could inhibit the growth of *B. subtilis* and Methicillin Sensitive *Staphylococcus aureus* (MSSA) with MIC values of 9.9 μM and 2.5 μM, respectively. In contrast to many anthracyclines, **93** might modulate fatty acid metabolism and exhibit antibacterial activity without nucleic acid damage that is explained by keyicin's mechanism of action (MOA) based on *E. coli* chemical genomics studies [57].

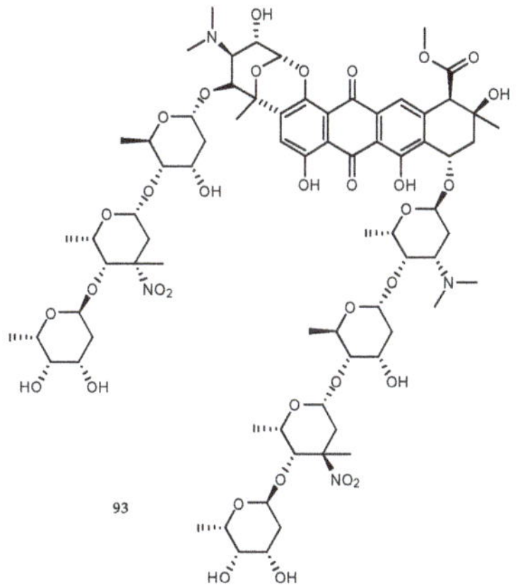

Figure 14. Chemical structures of **93**.

2.3. Cyclopeptides

Cyclopeptides are cyclic compounds mainly formed by the amide bonds of proteinogenic or non-proteinogenic amino acids bound together. Several fungal cyclic peptides have been developed as pharmaceuticals, such as the echinocandins, pneumocandins and cyclosporin A [58]. Six cyclopeptides were produced by the co-cultures of marine fungi–fungi (four isolates, 67%) and fungi–bacteria (two isolates, 33%) from different marine sources. However, marine bacteria–bacteria did not yield these structures in this period of investigation.

2.3.1. Cyclopeptides Derived from the Co-Cultures of Different Marine Fungi

Three new cyclic tetrapeptides, named cyclo-(L-leucyl-*trans*-4-hydroxy-L-prolyl-D-leucyl-*trans*-4-hydroxy-L-proline) (**94**) [59] cyclo (D-Pro-L-Tyr-L-Pro-L-Tyr) (**95**) and cyclo (Gly-L-Phe-L-Pro-L-Tyr)

(96) (Figure 15) [60] were identified from the co-culture of two mangrove fungi *Phomopsis* sp. K38 and *Alternaria* sp. E33 isolated from the South China Sea. Meanwhile, the co-cultivation of two marine alga-derived fungi *Aspergillus* sp. BM-05 and BM-05ML isolated from a brown algal species collected off Helgoland, North Sea, Germany, yielded a new cyclotripeptide, psychrophilin E **(97)** (Figure 15) [30].

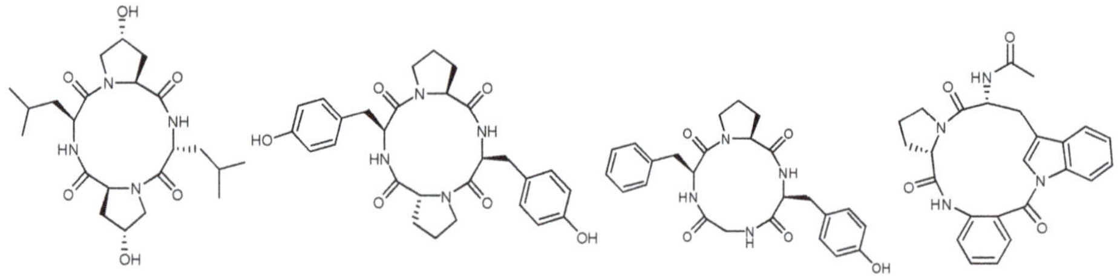

Figure 15. Chemical structures of **94–97**.

Compound **94** exhibited in vitro moderate to high inhibitory activity towards four crop-threatening fungi, *Helminthosporium sativum*, *Gaeumannomyces graminis*, *F. graminearum* and *Rhizoctonia cereals* with MIC values of 130, 220, 250 and 160 µg/mL, respectively [59]. **95** and **96** showed high in vitro antifungal activity against human fungus (*Candida albicans*) with MIC values of 35 µg/mL and 25 µg/mL, respectively [60]. **97** exhibited anti-proliferative activities against four human cancer cells, human cisplatin-resistant ovarian cancer A2780CisR, colon carcinoma HCT116, ovarian cancer A2780 and chronic myelogenous leukemia K562 with IC_{50} values of 49.4, 28.5, 27.3 and 67.8 µM, respectively. The inhibition of HCT116 cells by **97** was more potent than that of the positive control, cisplatin (IC_{50} 33.4 µM) [30].

2.3.2. Cyclopeptides Derived from the Co-Cultures of Marine Fungi and Bacteria

Recently, the chemical investigation of the mixed-fermentation of a marine fungus *Aspergillus versicolor* isolated from the sponge *Agelas oroides* and *B. subtilis* yielded two cyclic pentapeptides, one new cotteslosin C **(98)** and a known cotteslosin A **(99)** (Figure 16) [38]. Both of them did not show significant cytotoxic activity towards mouse lymphoma cell line L5178Y, or even antibacterial activity against five Gram-positive microbes, including one *S. aureus*, two *E. faecalis* and two *E. faecium* [38]. **99** displayed weak cytotoxicity against another three human cancer cell lines, prostate DU145, melanoma MM418c5 and breast T47D, with EC_{50} values of 90, 66 and 94 µg/mL, respectively [61].

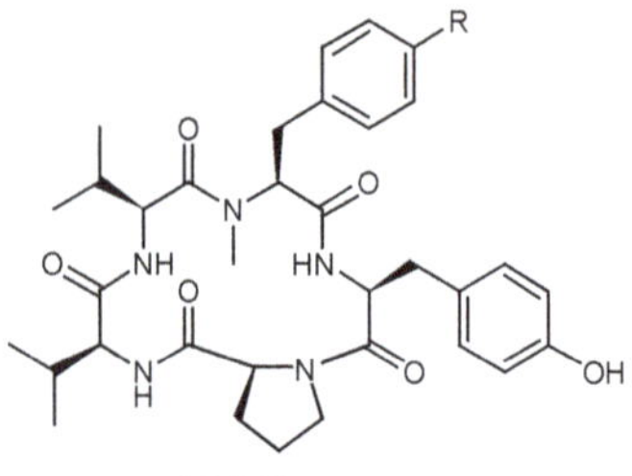

98: R=H
99: R=OH

Figure 16. Chemical structures of **98–99**.

2.4. Macrolide

There were no reported macrolides from the co-cultures of marine fungi–fungi and fungi–bacteria. Only one isolate was identified from a co-culture of marine bacteria–bacteria.

Macrolides Derived from the Co-Cultures of Different Marine Bacteria

A known compound, nonactin (**100**) (Figure 17) was isolated from the co-culture of two marine bacteria, *Saccharomonospora* sp. UR22 and *Dietzia* sp. UR66 [51]. It possessed a macrotetrolide structure integrated from nonactic acid, and exhibited antitumor and antibacterial activity, especially its inhibitory effects on the P170 glycoprotein-mediated efflux of chemotherapeutic agents in multiple-drug-resistant cancer cells [62–65].

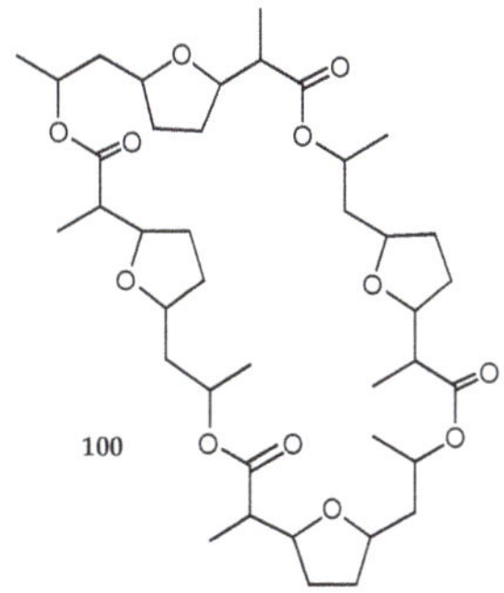

Figure 17. Chemical structures of **100**.

2.5. Phenylpropanoids

Phenylpropanoids are a big and structurally diverse group of secondary metabolites, which bear a C_6–C_3 phenolic scaffold that play crucial roles in a wide spectrum of biological and pharmacological activities [66]. Twenty-three phenylpropanoids were isolated from co-culture of marine fungi–fungi (12 isolates, 52%) and fungi–bacteria (11 isolates, 48%), while there are no reported phenylpropanoids from the co-culture of different marine bacteria.

2.5.1. Phenylpropanoids Derived from the Co-Cultures of Different Marine Fungi

A xanthone derivative known as 8-hydroxy-3-methyl-9-oxo-9*H*-xanthene-1-carboxylic acid methyl ether (**101**) (Figure 18) was discovered from the mixed culture of two mangrove fungi, *Phomopsis* sp. K38 and *Alternaria* sp. E33 [67] from the South China Sea coast. It showed a broad spectrum of antifungal activities against plant pathogens, *Blumeria graminearum*, *Gloeasporium musae*, *F. oxysporum*, *Colletotrichum glocosporioides* and *Peronophthora cichoralearum*.

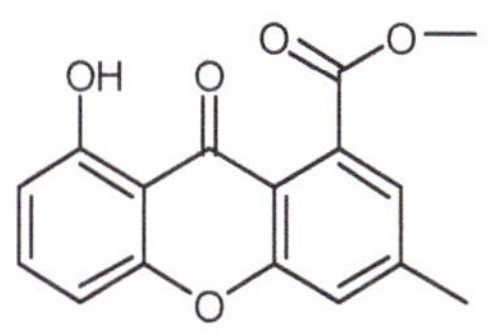

101

Figure 18. Chemical structures of **101**.

Ten citrinin analogues were isolated and identified from the co-culture of two marine algal-derived endophytic fungal strains, *Aspergillus sydowii* EN-534 and *Penicillium citrinum* EN-535 collected from marine red alga *Laurencia okamurai*, including two novel compounds, citrinin dimer *seco*-penicitrinol A (**102**) and citrinin monomer penicitrinol L(**103**), and the known penicitrinone A (**104**), penicitrinone F (**105**), penicitrinol A (**106**), citrinin (**107**), dihydrocitrinone (**108**), decarboxydihydrocitrinone (**109**) phenol A acid (**110**) and phenol A (**111**) (Figure 19) [68]. In addition, one novel coumarin named 7-(γ,γ-dimethylallyloxy)-6-hydroxy-4-methylcoumarin (**112**) (Figure 19) was detected and characterized from the co-culture of the two mangrove fungi, *Phomopsis* sp. K38 and *Alternaria* sp. E33 [69].

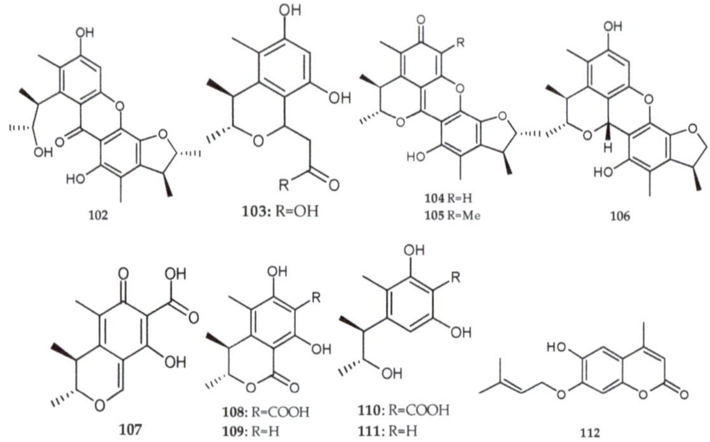

Figure 19. Chemical structures of **102–112**.

Compounds **104**, **106** and **107** exhibited inhibitory activities against two human pathogens *Micrococcus luteus* and *E. coli*, and three aquatic bacteria *Vibrio parahaemolyticus*, *Vibrio alginolyticus* and *Edwardsiella ictaluri* with MIC values of 4–64 μg/mL. **102**, **103** and **105** inhibited *V. alginolyticus* and *E. ictaluri* with MIC values of 32–64 μg/mL. **103** and **105** inhibited *V. parahaemolyticus* and *E. coli* with MIC values of 32 and 64 μg/mL, respectively. Moreover, **102–107** were further evaluated for anti-influenza neuraminidase (homologous protein of H_5N_1) activity. **104** and **105** exhibited significant inhibitory activities with IC_{50} values of 12.9 and 18.5 nM, respectively [68]. Thus, these bioactive substances could be further optimized for the development of antibacterial and anti-influenza agents. In addition to the anti-influenza activity, the activated metabolite penicitrinone A (**104**) also exerted an inhibitory effect on four human cancer cell lines, HL-60, K562, BGC-823 and HeLa cells with IC_{50} values of 43.2, 50.8, 54.2 and 65.6 μM, respectively [70].

2.5.2. Phenylpropanoids Derived from the Co-Cultures of Marine Fungi and Bacteria

The chemical investigation of the mixed culture of the marine fungus *A. versicolor* and *B. subtilis* resulted in the isolation of one novel aflaquinolone, 22-epi-aflaquinolone B (**113**); and ten known metabolites, aflaquinolone A, F and G (**114–116**), 3-*O*-methylviridicatin (**117**), 9-hydroxy-3-methoxyviridicatin (**118**), *O*-demethylsterigmatocystin (**119**), sterigmatocystin (**120**), sterigmatin (**121**), AGI-B4 (**122**) and sydowinin B (**123**) (Figure 20) [38].

The metabolite 3-*O*-methylviridicatin (**117**) was reported to possess inhibitory activity against human immunodeficiency virus (HIV) (Heguy et al., 1998). It could prevent cytokine tumor necrosis factor α (TNF-α), induce the HIV expression with long terminal repeat in HeLa cells (IC_{50}, 5μM) and block the viral replication in the model of chronic infection in OM-10.1 cell lines which directed at the induction of TNF-α [71]. **119** exhibited cytotoxic activities towards mouse lymphoma cell line L5178Y with an IC_{50} value of 5.8 μM. Three xanthone derivatives (**120–122**) showed potent cytotoxic activities towards the mouse lymphoma cell lines with IC_{50} values of 2.3, 2.2 and 2.0 μM, respectively, compared with a positive control, kahalalide F (IC_{50} = 4.3 μM). Sterigmatocystin (**120**) also exhibited strong cytotoxicity towards human hepatoma cells (HepG2) at 3 μM [72]. Its mechanism suggested that it could stimulate a biotransformation process, increase the population of reactive oxygen species and promote the imbalance in the antioxidant defense system caused by the process of lipid peroxidation [73]. Recently, Zingales et al. (2020) displayed the significant role of mitochondria in sterigmatocystin-induced toxicity in SH-SY5Y cells [74]. The reduced viability of SH-SY5Y cells displayed time- and dose-dependence with mitochondrial dysfunction when exposed to **120** in response to the forced dependency of the cells on mitochondrial oxidative phosphorylation [74]. Thus, these findings provided us a valuable direction for the application of neuroprotective mitochondria-target functional peptides. Moreover, compound **122** inhibited human umbilical vein endothelial cells (VEGF-induced proliferation of HUVECs) with an IC_{50} value of 1.4 µM [75]. It is considered as a novel inhibitor of vascular endothelial cell growth factor, which is one of the main stimulants of angiogenesis.

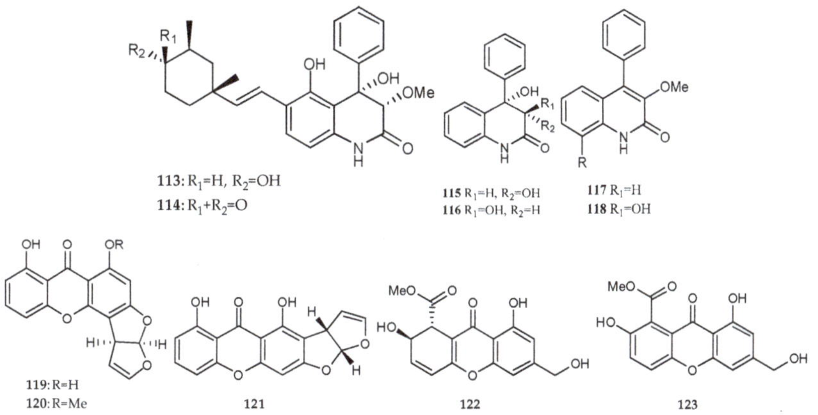

Figure 20. Chemical structures of **113–123**.

2.6. Polyketides

Twelve polyketides were isolated and characterized from the marine microbial co-cultures in recent years (Figures 2B and 21).

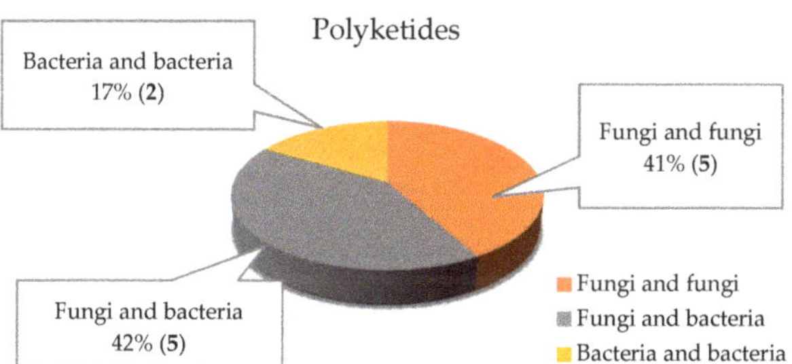

Figure 21. Polyketides isolated from the co-culture of marine fungi–fungi, fungi–bacteria and bacteria–bacteria.

2.6.1. Polyketides Derived from the Co-Cultures of Different Marine Fungi

In 2014, Ebada et al. identified three previously reported polyketide derivatives, sterigmatocystin (**124**), 5-methoxysterigmatocystin (**125**) and aversin (**126**) (Figure 22) from the ethyl acetate extract of two marine alga-derived fungi, *Aspergillus* sp. BM-05 and BM-05ML [30]. Kossuga et al. isolated two new and unusual polyketides: (Z)-2-ethylhex-2-enedioic acid (**127**) and (E)-4-oxo-2-propylideneoct-7-enoic acid (**128**) (Figure 22) from the marine-derived fungi *Penicillium* sp. Ma(M3)V isolated from the marine sponge *Mycale angulosa* co-cultivated with *Trichoderma* sp. Gc(M2)1 isolated from the marine sponge *Geodia corticostylifera* [76]. Two unprecedented polyketides (**127–128**) had a common feature—a conjugated carboxylic acid group that could be biogenetically generated from the methyl group of an acetate rather than a methionine precursor in **127**, and the same group could be derived from C-1 position of an acetate or C-2 position of a propionate in **128** based on the precursor of the ethyl group connected to a double bond. It was an excellent case of a truly novel carbon skeleton induced by the powerful and underexplored method, marine microbial co-cultivation.

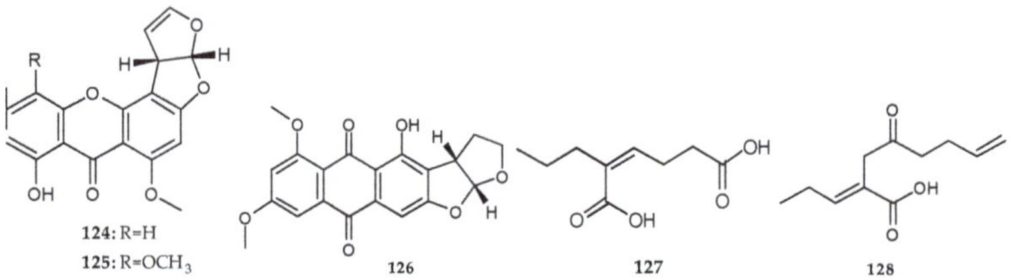

Figure 22. Chemical structures of **124–128**.

Compound **124** showed in vitro anti-proliferative activities towards human cisplatin-resistant ovarian cancer A2780CisR, ovarian cancer A2780 and chronic myelogenous leukemia K562 cell lines with IC_{50} values of 95.5, 30.6 and 57.0 μM, respectively. Moreover, it also displayed more significant anti-proliferative activities against human colon carcinoma HCT116 cells with an IC_{50} value of 10.3 μM (cf. to cisplatin's IC_{50} 33.4 μM). Compound **125** exhibited potent in vitro anti-proliferative activities towards three human cancer cell lines, HCT116, A2780 and human chronic myelogenous leukemia (K562) with IC_{50} values of 4.4, 51.0 and 13.4 μM, respectively [30].

2.6.2. Polyketides Derived from the Co-Cultures of Marine Fungi and Bacteria

A pair of enantiomers (9R,14S)-epoxy-11-deoxyfunicone (**129**) and (9S,14R)-epoxy-11-deoxyfunicone (**130**), along with deoxyfunicone (**131**), alternariol (**132**) and vermistatin (**133**) (Figure 23) were isolated from the co-culture of *Penicillium* sp. WC-29-5 isolated from the mangrove soil around the roots of *Aegiceras corniculatum* and *Streptomyces fradiae* 007 isolated from a sediment sample in the Jiaozhou Bay, Shandong Province, China [77].

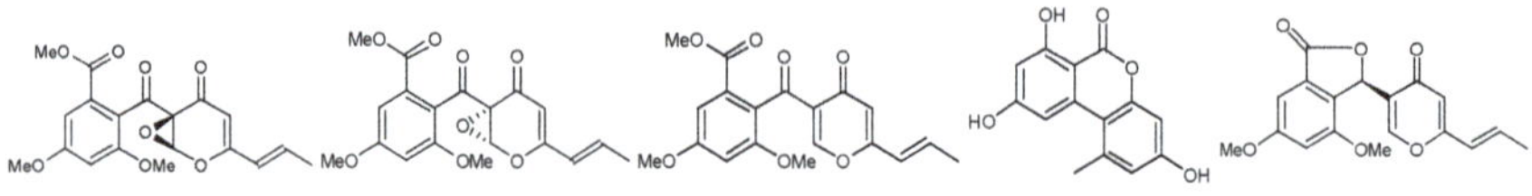

Figure 23. Chemical structures of **129–133**.

Both **129** and **130** exhibited moderate inhibitory activity against H1975 tumor cell lines with IC_{50} values of 3.97 and 5.73 μM, respectively. Deoxyfunicone (**131**) was found to exert anti-inflammatory activity, exhibiting the inhibition effect on overproduction of nitric oxide (NO) and the prostaglandin E_2 in both lipopolysaccharide-provoked BV2 microglial and lipopolysaccharide-stimulated RAW264.7 macrophage cells (IC_{50} = 10.6 and 40.1 μM, respectively) [78]. **132** was known as a cytotoxic, genotoxic, mutagenic and fetotoxic mycotoxin [79,80]. However, in the IL-1β-stimulated Caco-2 cells, the metabolite **132** increased the transcription of TNF-α; inversely reduced the transcription of IL-1β and IL-6; and decreased the transcription and secretion of IL-8, suggesting that **132** possessed immunomodulatory activities on both lipopolysaccharide- and IL-1 β-related pathways in non-immune intestinal epithelial cells [79].

2.6.3. Polyketides Derived from the Co-Cultures of Different Marine Bacteria

Recently, two unusual polyketides, janthinopolyenemycins A (**134**) and B (**135**) (Figure 24) were purified and identified from the co-cultivation broth of two marine bacteria *Janthino bacterium* spp. ZZ145 and ZZ148 isolated from marine soil sample [81]. Both **134** and **135** displayed the same antifungal activity against *C. albicans* with a minimum bactericidal concentration (MBC) value of 31.25 µg/mL and an MIC value of 15.6 µg/mL. However, none of them could suppress the growth of methicillin-resistant *S. aureus* or *E. coli* (MIC > 100 µg/mL) [81].

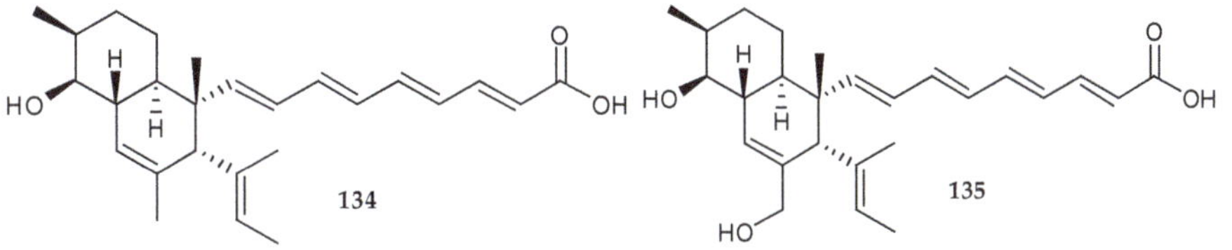

Figure 24. Chemical structures of **134–135**.

2.7. Steroids

Steroids contain a characteristic arrangement of four cycloalkane rings that are joined together. They represent a large family of compounds that play important roles as chemical messengers, and the scaffold is present in many FDA-approved drugs [82–84]. A total of five steroidal metabolites were reported; four of them were isolated from the co-culture of marine fungi–bacteria (80%); only one isolate was identified from the co-culture of marine fungi–fungi (20%). No isolates were obtained from the co-culture of marine bacteria–bacteria.

2.7.1. Steroids Derived from the Co-Cultures of Different Marine Fungi

To the best of our knowledge, the only one steroid, ergosterol (**136**), was found from the co-culture broth of two marine mangrove epiphytic fungi, *Aspergillus* sp. FSY-01 and FSW-02 (Figure 25) [21,85]. It was an essential component of fungal cell membrane with strong specificity and stable structure. Therefore, **136** was widely applied to detecting fungal containment as an indicator of fungal biomass [86].

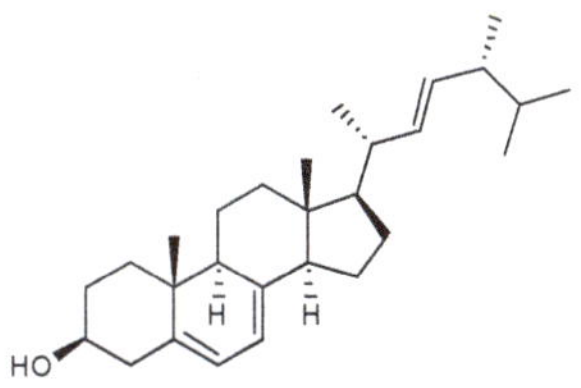

Figure 25. Chemical structures of **136**.

2.7.2. Steroids Derived from the Co-Cultures of Marine Fungi and Bacteria

An unprecedented steroid, 7β-hydroxycholesterol-1β-carboxylic acid (**137**), together with three known steroidal metabolites, 7β-hydroxycholesterol (**138**), 7α-hydroxycholesterol (**139**) and ergosterol-5α,8α-peroxide (**140**) (Figure 26), have been confirmed from the co-culture of two marine alga-derived microbes, *Aspergillus* sp. BM05, and an unidentified bacterium (BM05BL), isolated from the brown alga of the genus *Sargassum* collected off Helgoland, North Sea, Germany [87].

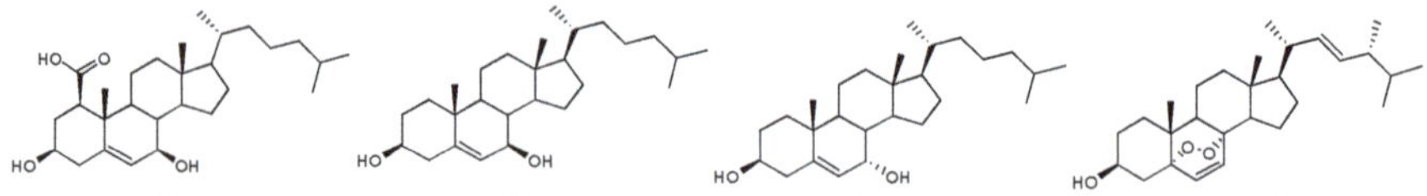

Figure 26. Chemical structures of **137–140**.

Compounds **137–140** showed moderate activities against four human tumor cell lines, A2780, HCT116, K562 and A2780 CisR with the IC_{50} values of 10.0–100.0 µM. At the same time, the total extract of co-culture of *Aspergillus* sp. BM05 and BM05BL showed obvious antiproliferative activity compared with its single steroidal compounds. This implied a synergistic role of these steroidal metabolites in the extract. Furthermore, **140** was reported as a promising new candidate that could overcome the drug-resistant property of malignant cancer cells through abolishing miR-378, a microRNA involved in new tumor initiation, unlimited self-renewal and recurrence of tumor cells after chemotherapy [88].

2.8. Terpenoids

Terpenoids known as isoprenoids are structurally diverse metabolites found in many natural sources. This class of compounds displays a wide sector of important pharmacological entities that confirmed by several preclinical and clinical studies [89,90]. Only two terpenoidals were isolated from the co-cultures of marine fungi–bacteria (one compound, 50%) and bacteria–bacteria (one compound, 50%).

2.8.1. Terpenoids Derived from the Co-Cultures of Marine Fungi and Bacteria

The production of the bacterial sesquiterpene pentalenic acid (**141**) (Figure 27) might be attributed to the competition relationship between marine fungus *A. fumigatus* MR2012 isolated from a Red Sea sediment in Hurghada, Egypt and terrestrial bacterium *S. leeuwenhoekii* C58 collected from the hyper-arid soil of Laguna de Chaxa Salar de Atacama, Chile, in which *S. leeuwenhoekii* C58 suppressed the production of *A. fumigatus* MR2012 and enhanced the production of **141** [37]. This suggested that *S. leeuwenhoekii* C58 appeared to activate the cryptic biosynthetic gene clusters to construct a defense mechanism based on the chemical signals generated by the competitive fungus, *A. fumigatus* MR2012. Thus, the bacterial strain was capable of suppressing the biosynthesis of the fungus metabolites that were present in the axenic cultures.

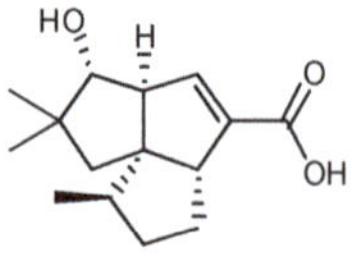

Figure 27. Chemical structures of **141**.

2.8.2. Terpenoids Derived from the Co-Cultures of Different Marine Bacteria

A diterpene lobocompactol (**142**) (Figure 28) was isolated from the co-culture of marine actinomycete *Streptomyces cinnabarinus* PK209 collected from the seaweed rhizosphere, obtained at a depth of 10 m along the coast of Korea and its competitor *Alteromonas* sp. KNS-16. Its productivity was increased 10.4-fold higher than that of the pure culture of PK209 [91]. Moreover, its antifouling activities were recently confirmed against primary fouling organisms, including diatoms, bacteria, and macroalgae zoospores. In order to further determine whether **142** was a non-toxic antifoulant, the therapeutic rate (LC_{50}/EC_{50}) was used to evaluate its toxicity, the LC_{50}/EC_{50} of **142** was more than that of **15**, indicating that the metabolite **142** was a non-toxic antifoulant. Thus, this compound could be valuable as an antifouling agent in both antifouling coating industry and marine ecology.

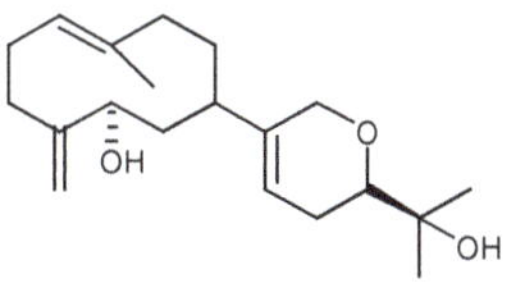

Figure 28. Chemical structures of **142**.

2.9. Others

Twelve compounds with other structures were obtained by co-culture of marine fungi–bacteria (4 compounds, 33%) and fungi–fungi (8 compounds, 67%).

2.9.1. Other Compounds Derived from the Co-Cultures of Different Marine Fungi

A novel polysubstituted benzaldehyde derivative, ethyl-5-ethoxy-2-formyl-3-hydroxy-4-methylbenzoate (**143**) (Figure 29) was identified from the mixed fermentation of the two mangrove fungi, *Phomopsis* sp. K38 and *Alternaria* sp. E33 that were collected from the South China Sea [92]. Another two novel furanone derivatives were identified as sclerotiorumins A and B (**144**, **145**) (Figure 29) from the co-culture of the two marine fungi, *sclerotiorum* SCSGAF 0053 and *P. citrinum* SCSGAF 0052 isolated from gorgonian *Muricella flexuosa* collected from the South China Sea, Sanya (18°11′ N, 109°25′ E), Hainan Province, China [23]. Five diorcinols, including one novel diorcinol J (**146**) and four known diorcinols B-E (**147–150**) (Figure 29), were characterized during the co-culturing of two marine-derived fungi, *A. sulphureus* KMM 4640 and *I. felina* KMM 4639 [93].

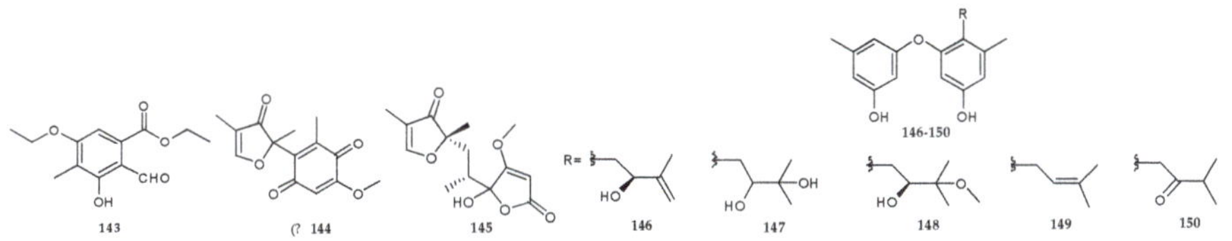

Figure 29. Chemical structures of **143–150**.

Compound **143** showed in vitro inhibitory activity against *G. musae*, *F. graminearum*, *P. sojae* (*Kaufmann* and *Gerdemann*) and *Rhizoctonia solani* Kuhn at 0.25 mM with inhibition zone diameters of 11.57, 12.06, 8.5 and 10.21 mm, respectively. This suggested that **143** had broad inhibitory activity against these microbes [92]. **144** and **145** exhibited weak toxicity against brine shrimp (LC$_{50}$ > 100 µM) and none of them displayed cytotoxicity against the liver hepatocellular carcinoma Huh7 and HepG2 (LC$_{50}$ > 100 µM) and obvious inhibitory activities towards three marine-derived bacteria, *Bacillus stearothermophilus*, *Pseudoalteromonas nigrifaciens* and *Bacillus amyloliquefaciens*, and two common pathogens, *P. aeruginosa* and *S. aureus* [23].

Among the five diorcinols, only **146** showed apparent cytotoxicity against murine Ehrlich carcinoma cells and hemolytic activity against mouse erythrocytes. The significant hemolytic activity of **146** suggested that its cytotoxic activity against murine Ehrlich carcinoma cells was due to a membranolytic mechanism. It is well known that the heat shock protein 70 (HSP70) was frequently overexpressed in tumor cell lines as an ATP-dependent molecular chaperone and played a significant role in refolding misfolded proteins and promoting cell survival under stress [94]. Thus, compounds that could inhibit HSP70 had great potential in tumor therapy. **147** could decrease the expression of HSP70 in the Ehrlich carcinoma cells, which made it possible to develop as a new antitumor drug/lead. Diorcinol D (**149**) was studied for its combined therapy against planktonic *Candida albicans* with a broad-spectrum antifungal agent fluconazole [95]. The combined therapy exhibited considerable antifungal activity against ten clinical isolates of *C. albicans* containing five fluconazole-resistant isolates and five fluconazole-sensitive isolates, whereas fluconazole alone did not display antifungal activity. This suggested that diorcinol D (**149**) restored the susceptibility of fluconazole to *C. albicans*.

Moreover, the efficiencies of fluconazole inhibiting mature biofilms were also drastically boosted by the addition of **149** [95]. The fractional inhibitory concentration index (FICI) model and ΔE model unclosed that the synergistic actions indeed existed in combination of diorcinol D (**149**) and fluconazole [95]. Two resistance mechanisms of azoles were overexpression of efflux pumps genes and alterations of genes (point mutations). **149** mainly suppressed the activity of efflux pump in cells partly by decreasing the expression of Cdr1 (one mediator of azole efflux pumps) in *Candida albicans* CASA1. On the other hand, **149** also inhibited ergosterol synthesis and CYP51 (the target of fluconazole) expression [95]. Thus, the significant synergistic interaction and drug-resistant reversion of fluconazole combined with diorcinol D (**149**) were caused by the two latent mechanisms, the block of efflux pump and ergosterol biosynthesis. Notably, **149** was still needed to further in vivo study in the combination therapy field to settle rock-ribbed clinical fungal infection in response to the azole resistance.

2.9.2. Other Compounds Derived from the Co-Cultures of Marine Fungi and Bacteria

Five known metabolites, diorcinol D (**149**), penicillanone (**151**), diorcinol G (**152**), diorcinol I (**153**) and radiclonic acid (**154**) (Figure 30) were obtained from the co-culture of the sponge-derived fungi *A. versicolor* and *B. subtilis* [38].

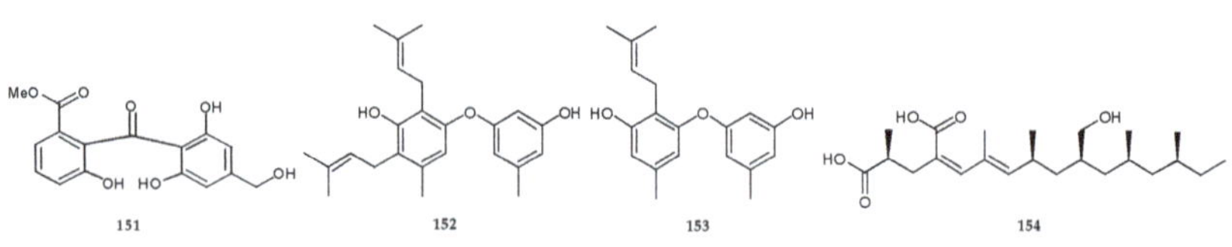

Figure 30. Chemical structures of **151–154**.

Compounds **149**, **152** and **153** displayed antibacterial activities against five Gram-positive microbes, including one *S. aureus*, two *E. faecalis* and two *E. faecium* with the MIC values of 12.5–50 μM. In addition, **152** displayed potent inhibitory activities against all tested bacteria with an MIC value of 12.5 μM. **149** displayed inhibitory activity against *E. coli* with an MIC value of 8 μg/mL; and **153** showed significant antibacterial activity against *S. aureus* with an MIC value of 6.25 μg/mL [96,97]. In contrast, **149**, **152** and **153** did not display any obvious activity against L5178Y cell lines, which suggested that the antimicrobial activities of these products were not associated with their respective general toxicities [38].

3. Conclusions

Marine microorganisms have attracted more attention as natural producers of lead compounds. Marine microbes especially are considered as a renewable and reproducible source that can be easily cultured [98,99]. However, the speed of new lead compound discovery is slowing down. Thus, marine microbial co-culturing represents a powerful strategy for the production of novel bio-substances. The strategy can induce the biosynthesis of novel compounds and various NPs coded by corresponding genomes through the activation of the silent gene clusters or previously unexpressed biosynthetic routes.

In the last ten years, the overall statistical studies showed that 156 metabolites were discovered from the co-culture of different marine microbes. Figure 2 and Table 1 illustrated that 59 compounds were isolated from the co-culturing of different marine fungi; 79 compounds were isolated from marine fungi and bacteria; and only 18 compounds were disclosed from co-culturing different marine bacteria. The metabolites by co-culture of marine fungi and bacteria accounted for the largest proportion (51% of all metabolites of marine microbial co-culture). Alkaloids were the largest group with ≥51.9%, whereas macrolides were the lowest group with <0.65%. Just only one macrolide was identified from the co-cultures of different marine bacteria. Furthermore, co-cultures of different marine bacteria did not produce cyclopeptides, phenylpropanoids and steroids, and co-cultures of different marine fungi did not induce the biosynthesis of terpenoids.

Several studies suggest that *Aspergillus* spp. are the most common fungi that co-fermented with other microbes and produce numerous novel skeletons. The majority of these NPs have antimicrobial or/and antitumor activities. However, some significant restrictions obstruct the development of the co-culture technology; e.g., cryptic and undefined biosynthesis routes and the producers of NPs from the co-cultivation of two or more microorganisms, the particularities of strains and environmental and nutritional requirements, the instability of the ecological relationship, the uncertainty of the interaction relationship and the high contamination probability. Therefore, new technology and equipment need to be created, such as metabolomics analysis and molecular network technology. The new mechanisms of chemical communication of microbes (through direct/mediate contact) also need to be further investigated. In conclusion, co-culture is still shrouded in mystery as a prospective experimental tool for novel bioactive NPs. This article embodies the value and diversity of NPs from the co-cultivation of marine-derived microorganisms and it is considered as a guided reference for studying NPs.

Author Contributions: J.C. and P.Z. conceived and wrote the review; X.Y. and B.W. offered important advice to improve the review; M.E., H.Z. and H.W. revised the paper. All authors have read and agreed to the published version of the manuscript.

Funding: This research work was supported by National Natural Science Foundation of China (number 81773628 and 41776139), and the 111-National Overseas Expertise Introduction Center for Green Pharmaceutical Discipline Innovation (number D17012).

Conflicts of Interest: The authors declare no conflict of interest.

References

1. Newman, D.J.; Cragg, G.M. Natural products as sources of new drugs over the 30 years from 1981 to 2010. *J. Nat. Prod.* **2012**, *75*, 311–335. [CrossRef]
2. Bertrand, S.; Bohni, N.; Schnee, S.; Schumpp, O.; Gindro, K.; Wolfender, J.L. Metabolite induction via microorganism co-culture: A potential way to enhance chemical diversity for drug discovery. *Biotechnol. Adv.* **2014**, *32*, 1180–1204. [CrossRef] [PubMed]
3. Chen, J.W.; Wu, Q.H.; Rowley, D.C.; Al-Kareef, A.M.; Wang, H. Anticancer agent-based marine natural products and related compounds. *J. Aisan. Nat. Prod. Res.* **2015**, *17*, 199–216. [CrossRef]
4. Carroll, A.R.; Copp, B.R.; Davis, R.A.; Keyzers, R.A.; Prinsep, M.R. Marine natural products. *Nat. Prod. Rep.* **2019**, *36*, 122–173. [CrossRef] [PubMed]
5. Chen, J.W.; Wang, B.X.; Lu, Y.J.; Guo, Y.Q.; Sun, J.D.; Wei, B.; Zhang, H.W.; Wang, H. Quorum sensing inhibitors from marine microorganisms and their synthetic derivatives. *Mar. Drugs* **2019**, *17*, 80. [CrossRef] [PubMed]
6. Brakhage, A.A.; Schroeckh, V. Fungal secondary metabolites–Strategies to activate silent gene clusters. *Fungal Genet. Biol.* **2011**, *48*, 15–22. [CrossRef] [PubMed]
7. Mao, D.; Okada, B.K.; Wu, Y.; Xu, F.; Seyedsayamdost, M.R. Recent advances in activating silent biosynthetic gene clusters in bacteria. *Curr. Opin. Microbiol.* **2018**, *45*, 156–163. [CrossRef]
8. Romano, S.; Jackson, S.A.; Patry, S.; Dobson, A.D.W. Extending the "one strain many compounds" (OSMAC) principle to marine microorganisms. *Mar. Drugs* **2018**, *16*, 224. [CrossRef]
9. Dalisay, D.S.; Rogers, E.W.; Edison, A.S.; Molinski, T.F. Structure elucidation at the nanomole scale. 1. Trisoxazole macrolides and thiazole-containing cyclic peptides from the nudibranch *Hexabranchus sanguineus*. *J. Nat. Prod.* **2009**, *72*, 732–738. [CrossRef]
10. Wolfender, J.L.; Queiroz, E.F. Chemical diversity of natural resources and the bioactivity of their constituents. *CHIMIA Int. J. Chem.* **2012**, *66*, 324–329. [CrossRef]
11. Bohni, N.; Cordero-Maldonado, M.L.; Maes, J.; Siverio-Mota, D.; Marcourt, L.; Munck, S.; Kamuhabwa, A.R.; Moshi, M.J.; Esguerra, C.V.; de Witte, P.A.M.; et al. Integration of microfractionation, qNMR and zebrafish screening for the in vivo bioassay-guided isolation and quantitative bioactivity analysis of natural products. *PLoS ONE* **2013**, *8*, e64006. [CrossRef]

12. Scherlach, K.; Hertweck, C. Triggering cryptic natural product biosynthesis in microorganisms. *Org. Biomol. Chem.* **2009**, *7*, 1753–1760. [CrossRef] [PubMed]

13. Abdalla, M.; Sulieman, S.; McGaw, L. Microbial communication: A significant approach for new leads. *S. Afr. J. Bot.* **2017**, *113*, 461–470. [CrossRef]

14. Ola, A.R.B.; Thomy, D.; Lai, D.; Brötz-Oesterhelt, H.; Proksch, P. Inducing secondary metabolite production by the endophytic fungus *Fusarium tricinctum* through coculture with *Bacillus subtilis*. *J. Nat. Prod.* **2013**, *76*, 2094–2099. [CrossRef] [PubMed]

15. Kochanowska-Karamyan, A.J.; Hamann, M.T. Marine indole alkaloids: Potential new drug leads for the control of depression and anxiety. *Chem. Rev.* **2010**, *110*, 4489–4497. [CrossRef]

16. Singh, K.S.; Majik, M.S. Bioactive alkaloids from marine sponges. In *Marine Sponges: Chemicobiological and Biomedical Applications*; Ehrlich, R.P.H., Ed.; Springer: New Delhi, India, 2016; pp. 257–286.

17. Afiyatullov, S.S.; Zhuravleva, O.I.; Antonov, A.S.; Berdyshev, D.V.; Pivkin, M.V.; Denisenko, V.A.; Popov, R.S.; Gerasimenko, A.V.; von Amsberg, G.; Dyshlovoy, S.A.; et al. Prenylated indole alkaloids from co-culture of marine-derived fungi *Aspergillus sulphureus* and *Isaria felina*. *J. Antibiot.* **2018**, *71*, 846–853. [CrossRef] [PubMed]

18. Kato, H.; Yoshida, T.; Tokue, T.; Nojiri, Y.; Hirota, H.; Ohta, T.; Williams, R.M.; Tsukamoto, S. Notoamides A-D: Prenylated indole alkaloids isolated from a marine-derived fungus, *Aspergillus* sp. *Angew. Chem. Int. Ed. Engl.* **2007**, *46*, 2254–2256. [CrossRef] [PubMed]

19. Tsukamoto, S.; Kato, H.; Samizo, M.; Nojiri, Y.; Onuki, H.; Hirota, H.; Ohta, T. Notoamides F-K, prenylated indole alkaloids isolated from a marine-derived *Aspergillus* sp. *J. Nat. Prod.* **2008**, *71*, 2064–2067. [CrossRef]

20. Assante, G.; Camarda, L.; Locci, R.; Merlini, L.; Nasini, G.; Papadopoulos, E. Isolation and structure of red pigments from *Aspergillus flavus* and related species, grown on a differential medium. *J. Agric. Food Chem.* **1981**, *29*, 785–787. [CrossRef]

21. Zhu, F.; Chen, G.Y.; Chen, X.; Huang, M.Z.; Wan, X. Aspergicin, a new antibacterial alkaloid produced by mixed fermentation of two marine-derived mangrove epiphytic fungi. *Chem. Nat. Compd.* **2011**, *47*, 767–769. [CrossRef]

22. Zhu, F.; Li, J.S.; Xie, W.C.; Shi, J.J.; Xu, F.; Song, Z.F.; Liu, Y.L. Structure revision of aspergicin by the crystal structure of aspergicine, a co-occurring isomer produced by co-culture of two mangrove epiphytic fungi. *Nat. Prod. Res.* **2017**, *31*, 2268–2272. [CrossRef] [PubMed]

23. Bao, J.; Wang, J.; Zhang, X.Y.; Nong, X.H.; Qi, S.H. New furanone derivatives and alkaloids from the co-culture of marine-derived fungi *Aspergillus sclerotiorum* and *Penicillium citrinum*. *Chem. Biodivers.* **2017**, *14*, e1600327. [CrossRef] [PubMed]

24. Weiss, U.; Strelitz, F.; Flon, H.; Asheshovl, I.N. Antibiotic compounds with action against bacterial viruses: Neohydroxyaspergillic acid. *Arch. Biochem. Biophys.* **1958**, *74*, 150–157. [CrossRef]

25. Wan, X.Q.; Zhu, F.; Chen, G.Y.; Li, H.M.; Tan, S.Y.; Pan, Y.Q.; Hong, Y. Biological evaluation of neoaspergillic acid, a pyrazine hydroxamic acid produced by mixed cultures of two marine-derived mangrove epiphytic fungi. *J. Biomed. Inform.* **2010**, *5*, 1932–1935.

26. Xu, X.; He, F.; Zhang, X.; Bao, J.; Qi, S. New mycotoxins from marine-derived fungus *Aspergillus* sp. SCSGAF0093. *Food Chem. Toxicol.* **2013**, *53*, 46–51. [CrossRef] [PubMed]

27. MacDonald, J.C.; Micetich, R.G.; Haskins, R.H. Antibiotic activity of neoaspergillic acid. *Can. J. Microbiol.* **1964**, *10*, 90–92. [CrossRef]

28. Li, C.Y.; Ding, W.J.; Shao, C.L.; She, Z.G.; Lin, Y.C. A new diimide derivative from the co-culture broth of two mangrove fungi (strain no. E33 and K38). *J. Asian Nat. Prod. Res.* **2010**, *12*, 809–813. [CrossRef]

29. Ding, W.J.; Lu, Y.C.; Feng, Z.H.; Luo, S.H.; Li, C.Y. A new nonadride derivative from the co-culture broth of two mangrove fungi. *Chem. Nat. Compd.* **2017**, *53*, 691–693. [CrossRef]

30. Ebada, S.S.; Fischer, T.; Hamacher, A.; Du, F.Y.; Roth, Y.O.; Kassack, M.U.; Wang, B.G.; Roth, E.H. Psychrophilin E, a new cyclotripeptide, from co-fermentation of two marine alga-derived fungi of the genus *Aspergillus*. *Nat. Prod. Res.* **2014**, *28*, 776–781. [CrossRef]

31. Wang, J.H.; Ding, W.J.; Wang, R.M.; Du, Y.P.; Liu, H.L.; Kong, X.H.; Li, C.Y. Identification and bioactivity of compounds from the mangrove endophytic fungus *Alternaria* sp. *Mar. Drugs* **2015**, *13*, 4492–4504. [CrossRef]

32. Lee, S.U.; Asami, Y.; Lee, D.; Jang, J.-H.; Ahn, J.S.; Oh, H. Protuboxepins A and B and protubonines A and B from the marine-derived fungus *Aspergillus* sp. SF-5044. *J. Nat. Prod.* **2011**, *74*, 1284–1287. [CrossRef] [PubMed]

33. Asami, Y.; Jang, J.H.; Soung, N.K.; He, L.; Moon, D.O.; Kim, J.W.; Oh, H.; Muroi, M.; Osada, H.; Kim, B.Y.; et al. Protuboxepin A, a marine fungal metabolite, inducing metaphase arrest and chromosomal misalignment in tumor cells. *Bioorg. Med. Chem.* **2012**, *20*, 3799–3806. [CrossRef] [PubMed]

34. Lu, X.H.; Shi, Q.W.; Zheng, Z.H.; Ke, A.B.; Zhang, H.; Huo, C.H.; Ma, Y.; Ren, X.; Li, Y.Y.; Lin, J.; et al. Oxepinamides: Novel liver X receptor agonists from *Aspergillus puniceus*. *Eur. J. Org. Chem.* **2011**, *2011*, 802–807. [CrossRef]

35. Yu, L.Y.; Ding, W.J.; Wang, Q.Q.; Ma, Z.J.; Xu, X.W.; Zhao, X.F.; Chen, Z. Induction of cryptic bioactive 2,5-diketopiperazines in fungus *Penicillium* sp. DT-F29 by microbial co-culture. *Tetrahedron* **2017**, *73*, 907–914. [CrossRef]

36. Yu, L.Y.; Ding, W.J.; Ma, Z.J. Induced production of cytochalasans in co-culture of marine fungus *Aspergillus flavipes* and actinomycete *Streptomyces* sp. *Nat. Prod. Res.* **2016**, *30*, 1718–1723. [CrossRef] [PubMed]

37. Wakefield, J.; Hassan, H.M.; Jaspars, M.; Ebel, R.; Rateb, M.E. Dual induction of new microbial secondary metabolites by fungal bacterial co-cultivation. *Front. Microbiol.* **2017**, *8*, 1284. [CrossRef]

38. Abdel-Wahab, N.M.; Scharf, S.; Özkaya, F.C.; Kurtán, T.; Mándi, A.; Fouad, M.A.; Kamel, M.S.; Müller, W.E.G.; Kalscheuer, R.; Lin, W.H.; et al. Induction of secondary metabolites from the marine-derived fungus *Aspergillus versicolor* through co-cultivation with *Bacillus subtilis*. *Planta Med.* **2019**, *85*, 503–512. [CrossRef]

39. Dey, A.; Chitsaz, F.; Abbasi, A.; Misteli, T.; Ozato, K. The double bromodomain protein Brd4 binds to acetylated chromatin during interphase and mitosis. *Proc. Natl. Acad. Sci. USA* **2003**, *100*, 8758–8763. [CrossRef]

40. Yang, Z.Y.; He, N.H.; Zhou, Q. Brd4 recruits P-TEFb to chromosomes at late mitosis to promote G1 gene expression and cell cycle progression. *Mol. Cell Biol.* **2008**, *28*, 967–976. [CrossRef]

41. Li, X.J.; Zhang, Q.; Zhang, A.L.; Gao, J.M. Metabolites from *Aspergillus fumigatus*, an endophytic fungus associated with *Melia azedarach*, and their antifungal, antifeedant, and toxic activities. *J. Agric. Food Chem.* **2012**, *60*, 3424–3431. [CrossRef]

42. Cui, C.B.; Kakeya, H.; Osada, H. Novel mammalian cell cycle inhibitors, cyclotryprostatins A-D, produced by *Aspergillus fumigatus*, which inhibit mammalian cell cycle at G2/M phase. *Terrahedron* **1997**, *53*, 59–72. [CrossRef]

43. Wang, F.Z.; Fang, Y.C.; Zhu, T.J.; Zhang, M.; Lin, A.Q.; Gu, Q.Q.; Zhu, W.M. Seven new prenylated indole diketopiperazine alkaloids from holothurian-derived fungus *Aspergillus fumigatus*. *Tetrahedron* **2008**, *64*, 7986–7991. [CrossRef]

44. Thomas, E.J. Cytochalasan synthesis: Macrocycle formation via intramolecular Diels-Alder reactions. *Acc. Chem. Res.* **1991**, *24*, 229–235. [CrossRef]

45. Lin, Z.J.; Zhu, T.J.; Wei, H.J.; Zhang, G.J.; Wang, H.; Gu, Q.Q. Spicochalasin A and new aspochalasins from the marine-derived fungus *Spicaria elegans*. *Eur. J. Org. Chem.* **2009**, *2009*, 3045–3051. [CrossRef]

46. Scherlach, K.; Boettger, D.; Remme, N.; Hertweck, C. The chemistry and biology of cytochalasans. *Nat. Prod. Rep.* **2010**, *27*, 869–886. [CrossRef] [PubMed]

47. Xiao, L.; Liu, H.Z.; Wu, N.; Liu, M.; Wei, J.T.; Zhang, Y.Y.; Lin, X.K. Characterization of the high cytochalasin E and rosellichalasin producing-*Aspergillus* sp. nov. F1 isolated from marine solar saltern in China. *World J. Microbiol. Biotechnol.* **2013**, *29*, 11–17. [CrossRef]

48. Kalinovskaya, N.I.; Romanenko, L.A.; Irisawa, T.; Ermakova, S.P.; Kalinovsky, A.I. Marine isolate *Citricoccus* sp. KMM 3890 as a source of a cyclic siderophore nocardamine with antitumor activity. *Microbiol. Res.* **2011**, *166*, 654–661. [CrossRef]

49. Qian-Cutrone, J.F.; Huang, S.; Shu, Y.Z.; Vyas, D.; Fairchild, C.; Menendez, A.; Krampitz, K.; Dalterio, R.; Klohr, S.E.; Gao, Q. Stephacidin A and B: Two structurally novel, selective inhibitors of the testosterone-dependent prostate LNCaP cells. *J. Am. Chem. Soc.* **2002**, *124*, 14556–14557. [CrossRef]

50. Yu, L.Y.; Hu, Z.F.; Ma, Z.J. Production of bioactive tryptamine derivatives by co-culture of marine *Streptomyces* with *Bacillus mycoides*. *Nat. Prod. Res.* **2015**, *29*, 2087–2091. [CrossRef]

51. El-Hawary, S.S.; Sayed, A.M.; Mohammed, R.; Khanfar, M.A.; Rateb, M.E.; Mohammed, T.A.; Hajjar, D.; Hassan, H.M.; Gulder, T.A.M.; Abdelmohsen, U.R. New Pim-1 kinase inhibitor from the co-culture of two sponge-associated *actinomycetes*. *Front. Chem.* **2018**, *6*, 538. [CrossRef]

52. Dashti, Y.; Grkovic, T.; Abdelmohsen, U.R.; Hentschel, U.; Quinn, R.J. Production of induced secondary metabolites by a co-culture of sponge-associated actinomycetes, *Actinokineospora* sp. EG49 and *Nocardiopsis* sp. RV163. *Mar. Drugs* **2014**, *12*, 3046–3059. [CrossRef] [PubMed]

53. Mandelare, P.E.; Adpressa, D.A.; Kaweesa, E.N.; Zakharov, L.N.; Loesgen, S. Coculture of two developmental stages of a marine-derived *Aspergillus alliaceus* results in the production of the cytotoxic bianthrone allianthrone A. *J. Nat. Prod.* **2018**, *81*, 1014–1022. [CrossRef] [PubMed]

54. Petersen, F.; Fredenhagen, A.; Mett, H.; Lydon, N.B.; Delmendo, R.; Jenny, H.B.; Peter, H.H. Paeciloquinones A, B, C, D, E and F: New potent inhibitors of protein tyrosine kinases produced by *Paecilomyces carneus*. I. Taxonomy, fermentation, isolation and biological activity. *J. Antibiot.* **1995**, *48*, 191–198. [CrossRef] [PubMed]

55. Liu, K.; Zheng, Y.K.; Miao, C.P.; Xiong, Z.J.; Xu, L.H.; Guan, H.L.; Yang, Y.B.; Zhao, L.X. The antifungal metabolites obtained from the rhizospheric *Aspergillus* sp. YIM PH30001 against pathogenic fungi of *Panax notoginseng*. *Nat. Prod. Res.* **2014**, *28*, 2334–2337. [CrossRef] [PubMed]

56. Dou, Y.L.; Wang, X.L.; Jiang, D.F.; Wang, H.Y.; Jiao, Y.; Lou, H.X.; Wang, X.N. Metabolites from *Aspergillus versicolor*, an endolichenic fungus from the lichen *Lobaria retigera*. *Drug. Discov. Ther.* **2014**, *8*, 84–88. [CrossRef] [PubMed]

57. Adnani, N.; Chevrette, M.G.; Adibhatla, S.N.; Zhang, F.; Yu, Q.; Braun, D.R.; Nelson, J.; Simpkins, S.W.; McDonald, B.R.; Myers, C.L.; et al. Coculture of marine invertebrate-associated bacteria and interdisciplinary technologies enable biosynthesis and discovery of a new antibiotic, keyicin. *ACS. Chem. Biol.* **2017**, *12*, 3093–3102. [CrossRef] [PubMed]

58. Wang, X.H.; Lin, M.Y.; Xu, D.; Lai, D.W.; Zhou, L.G. Structural diversity and biological activities of fungal cyclic peptides, excluding cyclodipeptides. *Molecules* **2017**, *22*, 2069. [CrossRef]

59. Li, C.Y.; Wang, J.H.; Luo, C.P.; Ding, W.J.; Cox, D.G. A new cyclopeptide with antifungal activity from the co-culture broth of two marine mangrove fungi. *Nat. Prod. Res.* **2014**, *28*, 616–621. [CrossRef]

60. Huang, S.; Ding, W.J.; Li, C.Y.; Cox, D.G. Two new cyclopeptides from the co-culture broth of two marine mangrove fungi and their antifungal activity. *Pharmacogn. Mag.* **2014**, *10*, 410–414.

61. Fremlin, L.J.; Piggott, A.M.; Lacey, E.; Capon, R.J. Cottoquinazoline A and cotteslosins A and B, metabolites from an Australian marine-derived strain of *Aspergillus versicolor*. *J. Nat. Prod.* **2009**, *72*, 666–670. [CrossRef] [PubMed]

62. Meyers, E.; Pansy, F.E.; Perlman, D.; Smith, D.A.; Weisenborn, F.L. The in vitro activity of nonactin and its homologs: Monactin, dinactin and trinactin. *J. Antibiot.* **1965**, *18*, 128–129. [PubMed]

63. Borrel, M.N.; Pereira, E.; Fiallo, M.; Garnier-Suillerot, A. Mobile ionophores are a novel class of P-glycoprotein inhibitors: The effects of ionophores on 4′-O-tetrahydropyranyl-adriamycin incorporation in K562 drug-resistant cells. *Eur. J. Biochem.* **1994**, *223*, 125–133. [CrossRef] [PubMed]

64. Woo, A.J.; Strohl, W.R.; Priestley, N.D. Nonactin biosynthesis: The product of *nonS* catalyzes the formation of the furan ring of nonactic acid. *Antimicrob. Agents Chemother.* **1999**, *43*, 1662–1668. [CrossRef] [PubMed]

65. Kusche, B.R.; Phillips, J.B.; Priestley, N.D. Nonactin biosynthesis: Setting limits on what can be achieved with precursor-directed biosynthesis. *Bioorg. Med. Chem. Lett.* **2009**, *19*, 1233–1235. [CrossRef]

66. Yang, C.H.; Ma, L.; Wei, Z.P.; Han, F.; Gao, J. Advances in isolation and synthesis of xanthone derivatives. *Chin. Herb. Med.* **2012**, *4*, 87–102.

67. Li, C.Y.; Zhang, J.; Shao, C.L.; Ding, W.J.; She, Z.G.; Lin, Y.C. A new xanthone derivative from the co-culture broth of two marine fungi (strain No. E33 and K38). *Chem. Nat. Compd.* **2011**, *47*, 382–384. [CrossRef]

68. Yang, S.Q.; Li, X.M.; Li, X.; Li, H.L.; Meng, L.H.; Wang, B.G. New citrinin analogues produced by coculture of the marine algal-derived endophytic fungal strains *Aspergillus sydowii* EN-534 and *Penicillium citrinum* EN-535. *Phytochem. Lett.* **2018**, *25*, 191–195. [CrossRef]

69. Wang, J.H.; Huang, S.; Li, C.Y.; Ding, W.J.; She, Z.G.; Li, C.L. A new coumarin produced by mixed fermentation of two marine fungi. *Chem. Nat. Compd.* **2015**, *51*, 239–241. [CrossRef]

70. Wu, C.J.; Yi, L.; Cui, C.B.; Li, C.W.; Wang, N.; Han, X. Activation of the silent secondary metabolite production by introducing neomycin-resistance in a marine-derived *Penicillium purpurogenum* G59. *Mar. Drugs* **2015**, *13*, 2465–2487. [CrossRef]

71. Heguy, A.; Cai, P.; Meyn, P.; Houck, D.; Russo, S.; Michitsch, R.; Pearce, C.; Katz, B.; Bringmann, G.; Feineis, D.; et al. Isolation and characterization of the fungal metabolite 3-O-methylviridicatin as an inhibitor of tumour necrosis factor α-induced human immunodeficiency virus replication. *Antivir. Chem. Chemother.* **1998**, *9*, 149–155. [CrossRef]

72. Gao, W.; Jiang, L.P.; Ge, L.; Chen, M.; Geng, C.Y.; Yang, G.; Li, Q.J.; Ji, F.; Yan, Q.; Zou, Y.; et al. Sterigmatocystin-induced oxidative DNA damage in human liver-derived cell line through lysosomal damage. *Toxicol. In Vitro* **2015**, *29*, 1–7. [CrossRef] [PubMed]

73. Sivakumar, V.; Thanislass, J.; Niranjali, S.; Devaraj, H. Lipid peroxidation as a possible secondary mechanism of sterigmatocystin toxicity. *Hum. Exp. Toxicol.* **2001**, *20*, 398–403. [CrossRef] [PubMed]

74. Zingales, V.; Fernandez-Franzon, M.; Ruiz, M.J. The role of mitochondria in sterigmatocystin-induced apoptosis on SH-SY5Y cells. *Food Chem. Toxicol.* **2020**, *142*, 111493. [CrossRef] [PubMed]

75. Kim, H.S.; Park, I.Y.; Park, Y.J.; Lee, J.H.; Hong, Y.S.; Lee, J.J. A novel dihydroxanthenone, AGI-B4 with inhibition of VEGF-induced endothelial cell growth. *J. Antibiot.* **2002**, *55*, 669–672. [CrossRef]

76. Kossuga, M.H.; Ferreira, A.G.; Sette, L.D.; Berlinck, R.G.S. Two polyketides from a co-culture of two marine-derived fungal strains. *Nat. Prod. Commun.* **2013**, *8*, 721–724. [CrossRef]

77. Wang, Y.; Wang, L.P.; Zhuang, Y.B.; Kong, F.D.; Zhang, C.X.; Zhu, W.M. Phenolic polyketides from the co-cultivation of marine-derived *Penicillium* sp. WC-29-5 and *Streptomyces fradiae* 007. *Mar. Drugs* **2014**, *12*, 2079–2088. [CrossRef]

78. Ha, T.M.; Kim, D.C.; Sohn, J.H.; Yim, J.H.; Oh, H. Anti-inflammatory and protein tyrosine phosphatase 1B inhibitory metabolites from the antarctic marine-derived fungal strain *Penicillium glabrum* SF-7123. *Mar. Drugs.* **2020**, *18*, 247. [CrossRef]

79. Schmutz, C.; Cenk, E.; Marko, D. The alternaria mycotoxin alternariol triggers the immune response of IL-1β-stimulated, differentiated Caco-2 cells. *Mol. Nutr. Food Res.* **2019**, *63*, 1900341. [CrossRef]

80. Fliszár-Nyúl, E.; Lemli, B.; Kunsági-Máté, S.; Dellafiora, L.; Dall'Asta, C.; Cruciani, G.; Pethő, G.; Poór, M. Interaction of mycotoxin alternariol with serum albumin. *Int. J. Mol. Sci.* **2019**, *20*, 2352. [CrossRef]

81. Anjum, K.; Sadiq, I.; Chen, L.; Kaleem, S.; Li, X.C.; Zhang, Z.Z.; Lian, X.Y. Novel antifungal janthinopolyenemycins A and B from a co-culture of marine-associated *Janthinobacterium* spp. ZZ145 and ZZ148. *Tetrahedron. Lett.* **2018**, *59*, 3490–3494. [CrossRef]

82. Konoshima, T.; Takasaki, M. Anti-tumor-promoting activities (cancer chemopreventive activities) of natural products. *Stud. Nat. Prod. Chem.* **2000**, *24*, 215–267.

83. Ouellette, R.J.; Rawn, J.D. *Organic Chemistry: Structure, Mechanism, and Synthesis*; Elsevier: Boston, MA, USA, 2014.

84. Roos, G.; Roos, C. *Organic Chemistry Concepts: An EFL Approach*; Academic Press: London, UK, 2014.

85. Dias, D.A.; Urban, S. HPLC and NMR studies of phenoxazone alkaloids from *Pycnoporus cinnabarinus*. *Nat. Prod. Commun.* **2009**, *4*, 489–498. [CrossRef] [PubMed]

86. Ng, H.E.; Raj, S.S.; Wong, S.H.; Tey, D.; Tan, H.M. Estimation of fungal growth using the ergosterol assay: A rapid tool in assessing the microbiological status of grains and feeds. *Lett. Appl. Microbiol.* **2008**, *46*, 113–118. [CrossRef] [PubMed]

87. Ebada, S.S.; Fischer, T.; Klassen, S.; Hamacher, A.; Roth, Y.O.; Kassack, M.U.; Roth, E.H. A new cytotoxic steroid from co-fermentation of two marine alga-derived micro-organisms. *Nat. Prod. Res.* **2014**, *28*, 1241–1245. [CrossRef]

88. Wu, Q.P.; Xie, Y.Z.; Deng, Z.; Li, X.M.; Yang, W.; Jiao, C.W.; Fang, L.; Li, S.Z.; Pan, H.H.; Yee, A.J.; et al. Ergosterol peroxide isolated from Ganoderma lucidum abolishes microRNA miR-378-mediated tumor cells on chemoresistance. *PLoS ONE* **2012**, *7*, e44579. [CrossRef]

89. Ludwiczuk, A.; Skalicka-Woźniak, K.; Georgiev, M.I. Chapter 11–Terpenoids. In *Pharmacognosy Fundamentals, Applications and Strategies*; Badal, S., Delgoda, R., Eds.; Academic Press: London, UK, 2017; pp. 233–266.

90. Teufel, R. Unusual "head-to-torso" coupling of terpene precursors as a new strategy for the structural diversification of natural products. *Methods Enzymol.* **2018**, *604*, 425–439.

91. Cho, J.Y.; Kim, M.S. Induction of antifouling diterpene production by *Streptomyces cinnabarinus* PK209 in co-culture with marine-derived *Alteromonas* sp. KNS-16. *Biosci. Biotechnol. Biochem.* **2012**, *76*, 1849–1854. [CrossRef] [PubMed]

92. Wang, J.H.; Ding, W.J.; Li, C.Y.; Huang, S.P.; She, Z.G.; Lin, Y.C. New polysubstituted benzaldehyde from the co-culture broth of two marine fungi (strains Nos. E33 and K38). *Chem. Nat. Compd.* **2013**, *49*, 799–802. [CrossRef]

93. Zhuravleva, O.I.; Kirichuk, N.N.; Denisenko, V.A.; Dmitrenok, P.S.; Yurchenko, E.A.; Min'ko, E.M.; Ivanets, E.V.; Afiyatullov, S.S. New diorcinol J produced by co-cultivation of marine fungi *Aspergillus sulphureus* and *Isaria felina*. *Chem. Nat. Compd.* **2016**, *52*, 227–230. [CrossRef]

94. Balaburski, G.M.; Leu, J.I.; Beeharry, N.; Hayik, S.; Andrake, M.D.; Zhang, G.; Herlyn, M.; Villanueva, J.; Dunbrack, R.L., Jr.; Yen, T.; et al. A modified HSP70 inhibitor shows broad activity as an anticancer agent. *Mol. Cancer Res.* **2013**, *11*, 219–229. [CrossRef] [PubMed]

95. Li, Y.; Chang, W.; Zhang, M.; Li, X.; Jiao, Y.; Lou, H. Synergistic and drug-resistant reversing effects of diorcinol D combined with fluconazole against *Candida albicans*. *FEMS. Yeast Res.* **2015**, *15*, fov001. [CrossRef] [PubMed]

96. Li, Z.X.; Wang, X.F.; Ren, G.W.; Yuan, X.L.; Deng, N.; Ji, G.X.; Li, W.; Zhang, P. Prenylated diphenyl ethers from the marine algal-derived endophytic fungus *Aspergillus tennesseensis*. *Molecules* **2018**, *23*, 2368. [CrossRef] [PubMed]

97. Xu, X.L.; Yang, H.J.; Xu, H.T.; Yin, L.Y.; Chen, Z.K.; Shen, H.H. Diphenyl ethers from a marine-derived isolate of *Aspergillus* sp. CUGB-F046. *Nat. Prod. Res.* **2018**, *32*, 821–825. [CrossRef]

98. Netzker, T.; Fischer, J.; Weber, J.; Mattern, D.J.; Konig, C.C.; Valiante, V.; Schroeckh, V.; Brakhage, A.A. Microbial communication leading to the activation of silent fungal secondary metabolite gene clusters. *Front. Microbiol.* **2015**, *6*, 299. [CrossRef]

99. Chen, J.W.; Guo, Y.Q.; Lu, Y.J.; Wang, B.X.; Sun, J.D.; Zhang, H.W.; Wang, H. Chemistry and biology of siderophores from marine microbes. *Mar. Drugs* **2019**, *17*, 562. [CrossRef] [PubMed]

 *marine drugs*

Article

Viridicatol Isolated from Deep-Sea *Penicillium Griseofulvum* Alleviates Anaphylaxis and Repairs the Intestinal Barrier in Mice by Suppressing Mast Cell Activation

Zhendan Shu [1], Qingmei Liu [1], Cuiping Xing [2], Yafen Zhang [1], Yu Zhou [1], Jun Zhang [1], Hong Liu [1], Minjie Cao [1], Xianwen Yang [2,*] and Guangming Liu [1,*]

[1] College of Food and Biological Engineering, Xiamen Key Laboratory of Marine Functional Food, Fujian Provincial Engineering Technology Research Center of Marine Functional Food, Fujian Collaborative Innovation Center for Exploitation and Utilization of Marine Biological Resources, Jimei University, 43 Yindou Road, Xiamen 361021, China; zdanshu@163.com (Z.S.); liuqingmei1229@163.com (Q.L.); zyf1697047824@163.com (Y.Z.); carolzy1205@163.com (Y.Z.); 18899201006@163.com (J.Z.); liuhong@jmu.edu.cn (H.L.); mjcao@jmu.edu.cn (M.C.)

[2] Key Laboratory of Marine Biogenetic Resources, Third Institute of Oceanography, Ministry of Natural Resources, 184 Daxue Road, Xiamen 361005, China; xingcuiping123@126.com

* Correspondence: yangxianwen@tio.org.cn (X.Y.); gmliu@jmu.edu.cn (G.L.); Tel.: +86-592-618-0378 (G.L.); Fax: +86-592-618-0470 (G.L.)

Received: 18 September 2020; Accepted: 14 October 2020; Published: 16 October 2020

Abstract: Viridicatol is a quinoline alkaloid isolated from the deep-sea-derived fungus *Penicillium griseofulvum*. The structure of viridicatol was unambiguously established by X-ray diffraction analysis. In this study, a mouse model of ovalbumin-induced food allergy and the rat basophil leukemia (RBL)-2H3 cell model were established to explore the anti-allergic properties of viridicatol. On the basis of the mouse model, we found viridicatol to alleviate the allergy symptoms; decrease the levels of specific immunoglobulin E, mast cell protease-1, histamine, and tumor necrosis factor-α; and promote the production of interleukin-10 in the serum. The treatment of viridicatol also downregulated the population of B cells and mast cells (MCs), as well as upregulated the population of regulatory T cells in the spleen. Moreover, viridicatol alleviated intestinal villi injury and inhibited the degranulation of intestinal MCs to promote intestinal barrier repair in mice. Furthermore, the accumulation of Ca^{2+} in RBL-2H3 cells was significantly suppressed by viridicatol, which could block the activation of MCs. Taken together, these data indicated that deep-sea viridicatol may represent a novel therapeutic for allergic diseases.

Keywords: food allergy; deep-sea-derived viridicatol; X-ray single crystal; intestinal barrier; mast cell; calcium influx

---

1. Introduction

Food allergy is a potentially life-threatening allergic reaction that affects a substantial proportion of the population, being defined as a breakdown of immunological tolerance and clinical symptoms generated in response to ingested food [1]. It is estimated that approximately 8% of children and 5% of adults suffer from food allergies around the world, and the prevalence has also been rising over the past two decades [2]. Food allergy is more common in children than in adults, and it often begins in childhood with the influence of genetic predisposition.

Eggs are indispensable nutrients in children's diet pagodas. However, the prevalence of egg allergy in children ranges from 1.2% to 2.0%. The majority allergens within eggs have been found in egg white, with ovalbumin (OVA) being the most abundant form (at 58%, although it is not the major allergen), which is claimed to frequently elicit allergic reactions [3,4]. Thus, the OVA-induced mice allergy model is often used to evaluate the anti-allergic activity of various materials [5,6].

Food allergy is characterized by a type I hypersensitivity reaction mediated by immunoglobulinE (IgE), and is often accompanied by adverse reactions, including diarrhea, intestinal injury, and hypothermia [7]. Moreover, it is also accompanied by increased intestinal permeability, which is positively correlated with symptom severity, which persists even in a diet without allergens [8]. Food particles and allergens can cross the epithelial barrier and cause allergic reactions induced by IgE production and mast cell (MC) recruitment [9]. In addition, the development of food allergy involves a variety of immune cells, which play different roles in the process. B lymphocytes are mainly responsible for releasing antibodies (e.g., IgE), which are associated with basophil and MC activation [10]. Cellular immunity is primarily characterized by T lymphocytes, including T helper cells (Th) 1 and 2, and regulatory T cells (Tregs). CD4 + Foxp3 + Tregs can secrete transforming growth factor or interleukin (IL)-10, suppress MC activation, and modulate T and B cells, thereby promoting immunosuppression and relieving anaphylaxis [11,12].

As crucial and rapid immune effector cells, MCs play the key role in IgE-mediated allergic responses [13]. Several reports have indicated that targeting MCs may represent a promising strategy for the treatment of food allergies [14]. MCs are activated when IgE binds with high affinity to the IgE receptor (FcεRI) on the surface of MCs [15]. When re-exposed to specific allergens, MCs were activated with a cascade of tyrosine phosphorylation, and then released a series of allergic mediators [16,17]. During the process of MC activation and degranulation, Ca^{2+} is an essential co-factor that regulates the release of various mediators and granule-plasma membrane fusion [18]. It has been reported that glycyrrhizic acid can perform a function as an "MC stabilizer" and inhibit MC degranulation by suppressing Ca^{2+} influx [19]. Therefore, obtaining a stabilizer that can regulate Ca^{2+} influx may be important for inhibiting MC activation and degranulation.

Novel natural compounds from marine resources have attracted increased attention in recent years due to their potential medicinal value [20]. Due to their unique habitat, deep-sea fungi readily produce various secondary metabolites, which have novel structure and distinctive biological properties. Most fully researched secondary metabolites of deep-sea fungi have mainly been from a limited number of genera, including *Penicillium*, *Aspergillus*, *Fusarium*, and *Cladosporium* [21]. Within the facultative marine fungi, species of *Penicillium* and *Talaromyces* are particularly well known for their ability to produce important bioactive compounds [22]. *Penicillium*, an important genus of the phylum Ascomycota, is widely distributed and has a large impact on human life [23]. Moreover, as one of the most studied fungi, *Penicillium* was considered to be a primary source of drug discovery, which produces a wide range of highly active metabolites, including griseofulvin and the blockbuster drug penicillin [24,25]. After the discovery of penicillin, isolating secondary metabolites of *Penicillium* species has received a remarkable amount of interest from researchers due to the interesting structures and possible pharmaceutical applications of the compounds. It has been shown that *Penicillium* metabolites possess a wide range of biological properties such as antibacterial activities, antioxidant properties, and cytotoxic activities against cancer lines [22]. Thus, as the second most common genus of marine fungi, it is of good importance to investigate the properties and metabolites of *Penicillium* [22]. It has been reported that cyclopiane-type diterpenes from the deep sea-derived fungus *Penicillium* possess antimicrobial activity [26]. However, few systematic studies have investigated the anti-food allergy properties of *Penicillium* secondary metabolites, especially for deep-sea *Penicillium*.

In previous studies, the rat basophil leukemia (RBL)-2H3 cell model was established to explore the anti-allergic potential metabolite from deep-sea fungi including *Penicillium*, *Aspergillus*, *Actinomycete*, and so on [27,28]. The metabolites of deep-sea-derived *Aspergillus* and *Actinomycete* relieved allergy by regulating MC function and accelerating MC apoptosis, respectively. Additionally, nine compounds isolated from deep-sea fungus *Penicillium griseofulvum* were tested for their anti-allergic bioactivity using the RBL-2H3 cell model [29]. A further study on the fungus led to the isolation of another potent anti-allergic compound, viridicatol. As reported, viridicatol effectively inhibited the protein activities and expressions of Matrix matalloproteinases (MMP)-2 and -9 against cancer and tumor [30]. However, the investigation of the biological activity of viridicatol is limited at present. This study was aimed at seeking to further validate and explore the associated mechanism of viridicatol using mouse model, flow cytometry, and an RBL-2H3 cell model. The findings of this study are expected to provide a foundation for the development of new marine anti-allergic drugs.

2. Results

2.1. Structural Determination of Viridicatol

The chemical structure of viridicatol (Figure 1A) was determined primarily by a detailed analysis of its 1D and 2D nuclear magnetic resonance (NMR) spectra (Figures S1–S6).

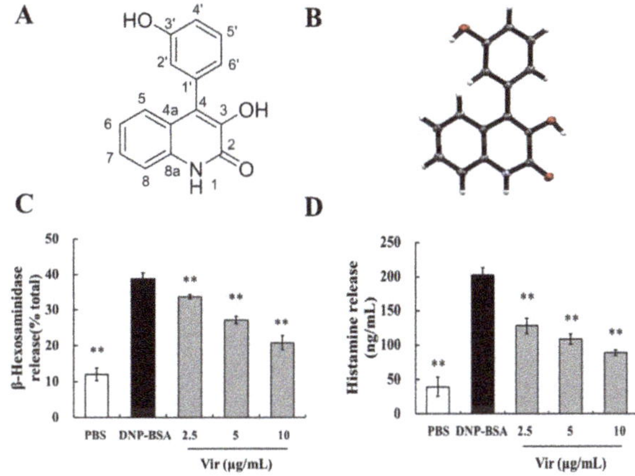

Figure 1. The X-ray crystallography and degranulation efficiency of viridicatol. (**A**) Chemical structure of viridicatol. (**B**) The X-ray crystallography of viridicatol. (**C**) Effects of viridicatol on degranulation. Cells were incubated with 200 ng/mL of dinitrophenyl (DNP)-specific immunoglobulin E (IgE) for 16 h in 48-well plates. The medium was replaced by Tyrode's buffer containing the indicated concentrations of viridicatol (2.5, 5, and 10 µg/mL) followed by stimulation with 500 ng/mL DNP-bovine serum albumin (BSA) for 1 h. The release of β-hexosaminidase was measured. (**D**) Effects of viridicatol on histamine release. The cells were sensitized and treated as described in (**C**), except for stimulation with DNP-BSA for 15 min, and the level of histamine was measured using an ELISA kit. ** $p < 0.01$ compared with the DNP-BSA group. The data represent the mean ± SD of three repeated experiments. Vir: viridicatol.

Viridicatol: ^{1}H-NMR (CD$_3$OD, 400 MHz) δ_H 7.33 (1H, m, H-5), 7.32 (1H, m, H-8), 7.31 (1H, d, $J = 7.6$, H-5′), 7.23 (1H, d, $J = 8.2$, H-7), 7.11 (1H, m, H-6), 6.86 (1H, dd, $J = 9.2, 2.1$, H-4′), 6.81 (1H, d, $J = 8.1$, H-6′), 6.80 (1H, s, H-2′); ^{13}C-NMR (CD$_3$OD, 100 MHz) δ_C 160.5 (s, C-2), 127.3 (s, C-3), 134.4 (s, C-4), 128.0 (d, C-5), 123.8 (d, C-6), 126.3 (d, C-7), 116.5 (d, C-8), 136.2 (s, C-1′), 117.9 (d, C-2′), 158.7 (s, C-3′), 116.0 (d, C-4′), 136.0 (d, C-5′), 122.2 (d, C-6′), 123.0 (s, C-4a), 143.1 (s, C-8a). HRESIMS *m/z* 252.0744 [M − H]$^-$.

2.2. X-ray Crystallography of Viridicatol

The structure of viridicatol was assigned by the single X-ray (Figure 1B). It was obtained as a colorless needle. The crystal data were recorded with an Xcalibur Eos Gemini single-crystal diffractometer with Cu-Kα radiation (λ = 1.54184 Å). Space group P (−1), a = 6.2014 (2) Å, b = 10.2629

(4) Å, c = 10.7415 (5) Å, α = 117.029 (4), β = 96.135 (3), γ = 100.680 (3), V = 583.98 (5) Å^3, Z = 2, D_{calcd} = 1.491 g/cm^3, μ = 0.888 mm^{-1}, F (000) = 274.0. The final *R* indices were [I > 2σ(I)] R_1 = 0.0524, wR_2 = 0.1410. Crystallographic data of viridicatol were deposited in the Cambridge Crystallographic Data Center (deposition number: 2008745).

2.3. Viridicatol Decreased the Release of β-Hexosaminidase and Histamine in RBL-2H3 Cells

In the preliminary screening, we found that viridicatol significantly decreased the release of β-hexosaminidase and histamine in RBL-2H3 cells in a dose-dependent manner (Figure 1C,D). Meanwhile, we found that viridicatol had no measurable cytotoxic effects on RBL-2H3 cells (data not shown), and the IC_{50} value of viridicatol was 6.67 ± 0.6 μg/mL (26.3 μM).

2.4. Viridicatol Relieved OVA-Induced Allergic Symptoms in Mice

An OVA-induced mouse model of food allergy was established to explore the anti-allergic activity of viridicatol in vivo (Figure 2A). As expected, the mice in the OVA group exhibited diarrhea, hypothermia, and a high anaphylactic score after five successive oral challenges [31]. Moreover, after daily treatment with viridicatol, the hypothermia (Figure 2B), loose stool rate (Figure 2C), and anaphylactic score (Figure 2D) were significantly attenuated. Compared with the phosphate-buffered saline (PBS) group, the cecum in the OVA group was significantly increased, which was caused by allergic diarrhea. Following treatments with viridicatol, the onset of diarrhea was suppressed (Figure 2E).

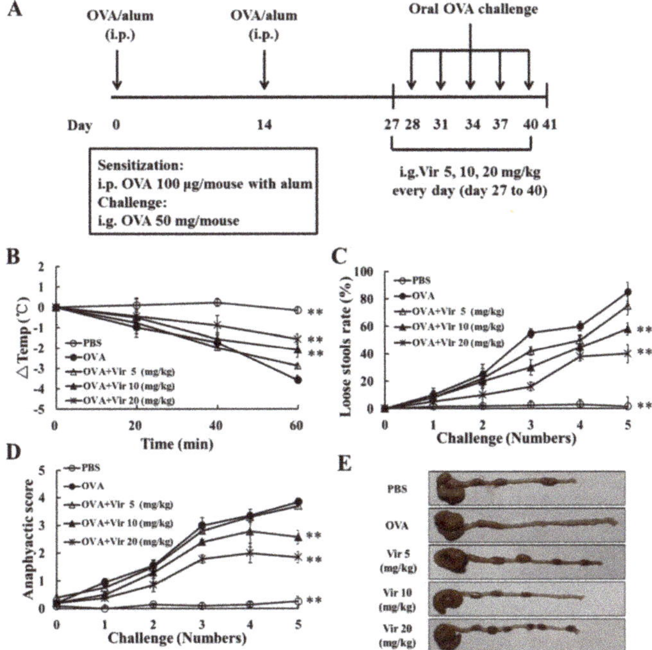

Figure 2. Food allergy model and effects of viridicatol on the allergic symptoms. (**A**) Food allergy model. Mice were sensitized with 100 µg ovalbumin (OVA) and 2 mg alum in 200 µL of phosphate-buffered saline (PBS) by intraperitoneal injection, orally challenged with 50 mg OVA in 200 µL PBS, and the viridicatol concentrates (5, 10, 20 mg/kg) were orally administered. (**B**) The rectal temperature was measured 1 h after the fifth OVA challenge. (**C**) The rate of loose stools was measured 1 h after each OVA challenge. (**D**) The anaphylactic score was measured 1 h after each OVA challenge. (**E**) The representative large macroscopic view of the intestine in each group of mice. ** $p < 0.01$ compared with the OVA group. The data represent the mean ± SD. Vir: viridicatol.

2.5. Effects of Viridicatol on Immunoglobulins, Anaphylactic Mediators, and Cytokines in the Mouse Serum

Severe allergic reactions are associated with the increased secretion of allergic mediators. In this study, viridicatol significantly decreased the level of OVA-specific IgE at a concentration of 20 mg/kg (Figure 3A). However, no effects of viridicatol on OVA-specific IgG1 and IgG2a were observed (Figure 3B,C). As shown in Figure 3D–F, the oral administration of viridicatol significantly decreased the levels of serum histamine, mast cell protease-1 (mMCP-1) , and tumor necrosis factor-α (TNF-α) in a dose-dependent manner compared to that of the OVA group. As shown in Figure 3G, following an oral administration of viridicatol, the level of interleukin (IL)-10 was increased in the serum as a slight trend toward a dose-dependent manner.

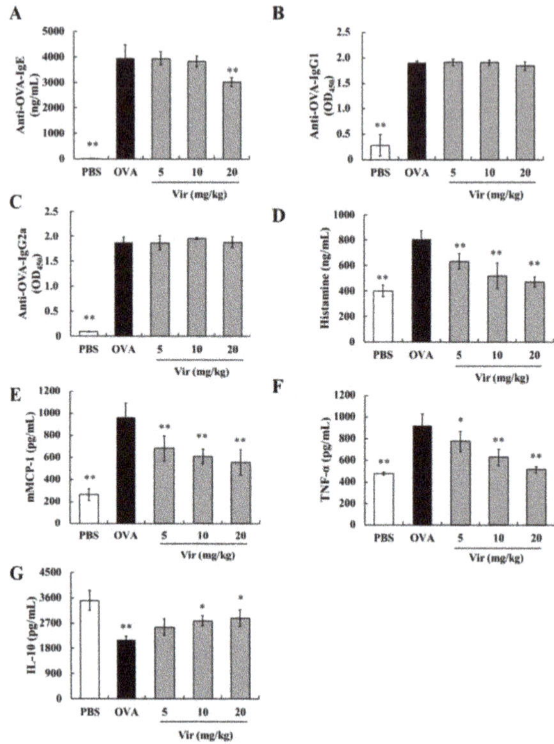

Figure 3. The levels of immunoglobulins, anaphylactic mediators, and cytokines in serum ($n = 6$). (**A**) The level of OVA-specific IgE. (**B**) The level of anti-OVA IgG1. (**C**) The level of anti-OVA IgG2a. (**D**) The level of histamine. (**E**) The level of mMCP-1. (**F**) The level of TNF-α. (**G**) The level of IL-10. In (A,B,C,F,G), serum was from retro-orbital blood collection on day 41. In (D,E), the serum from the mice tail veins was collected within 1 h after the last challenge. $* p < 0.05$, $** p < 0.01$ compared with the OVA group. The data represent the mean $\pm$ SD. Vir: viridicatol.

2.6. Viridicatol Alleviated Intestinal Injury and Inflammation of the Jejunum

While the jejunum villi were regularly arranged without fracture on the surface of the jejunum villi in the PBS group, they were severely damaged in the OVA group. As expected, the damaged jejunum villi were repaired to different degrees following treatments with viridicatol (Figure 4A). In parallel with the clinical assessment, there were obvious signs of inflammation, including serious lymphocytic infiltration using hematoxylin and eosin (H&E). These symptoms were relieved following treatments with viridicatol (Figure 4B).

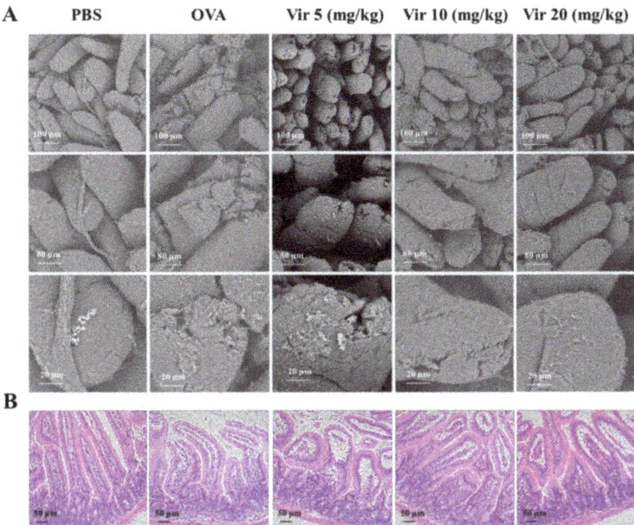

Figure 4. Effects of viridicatol on jejunum tissue injury. (**A**) Injury of the jejunum villi. The scanning electron microscopy (SEM) results shown at 500×, 1000×, and 3000× magnification. (**B**) Representative hematoxylin and eosin (H&E)-stained jejunum sections (magnification: 200×). Vir: viridicatol.

2.7. Effects of Viridicatol on the Subpopulation of B Cells and Tregs in the Spleen

As previously reported, regulating the populations of immune cells involved in the allergic responses is one of the most effective methods to treat food allergies [6,32]. To seek the target cells of viridicatol in vivo, we isolated the splenic lymphocytes from all groups of mice on day 41, and the lymphocyte populations were detected by flow cytometry. Compared to the PBS group (54.04%), the population of CD19$^+$ B cells significantly increased to 64.32% in the OVA group. Moreover, the population of CD19$^+$ B cells in the viridicatol group was significantly decreased at a concentration of 20 mg/kg (Figure 5A).

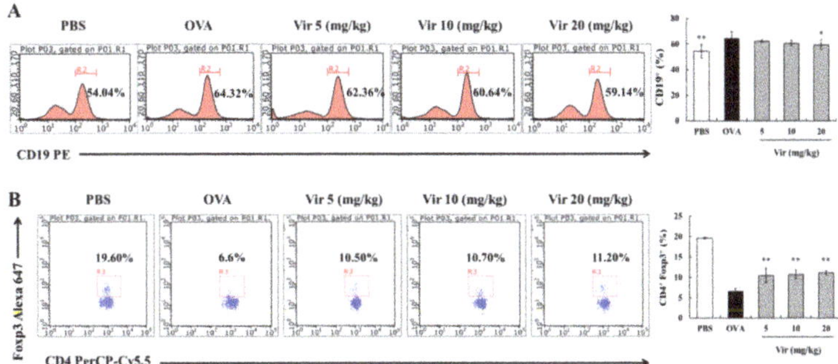

Figure 5. Effects of viridicatol on the population of B cells and regulatory T cells (Tregs) in the spleen. (**A**) Histograms of splenic B cells. Splenic lymphocytes were labeled with anti-CD3 and anti-CD4. (**B**) Scatter diagrams of regulatory T cells (Tregs). Splenic lymphocytes were labeled with anti-CD4 and anti-Foxp3. Spleens were isolated from each group of mice 24 h after the last challenge, labeled with various antibodies, and underwent Fluorescence Activating Cell Sorter (FACS) analysis by flow cytometry. * $p < 0.05$, ** $p < 0.01$ compared with the OVA group. The data represent the mean ± SD. Vir: viridicatol.

Considering the effects of viridicatol on the level of serum IL-10 (Figure 3G), the population of Tregs in the spleen was further investigated. Compared to the PBS group (19.60%), the population of Tregs was significantly decreased in the OVA group (6.6%). As expected, the population of Tregs was significantly upregulated following treatment with viridicatol (Figure 5B). In summary, viridicatol displayed a weak and lack of a dose-dependent effect on the population of B cells and Tregs, respectively.

2.8. Effects of Viridicatol on MC Population and Degranulation

In light of the significant inhibitory effects on histamine and mMCP-1 in the serum, we determined the various populations of FcεRI$^+$c-KIT$^+$ cells (MCs) by flow cytometry. Compared to the PBS group (0.16%), the population of MCs was significantly increased in the OVA group (1.34%). Following treatments with viridicatol, the population of MCs was significantly decreased compared to the OVA group (Figure 6A). The population of MCs was inhibited after treatments with viridicatol using toluidine blue staining (Figure 6B).

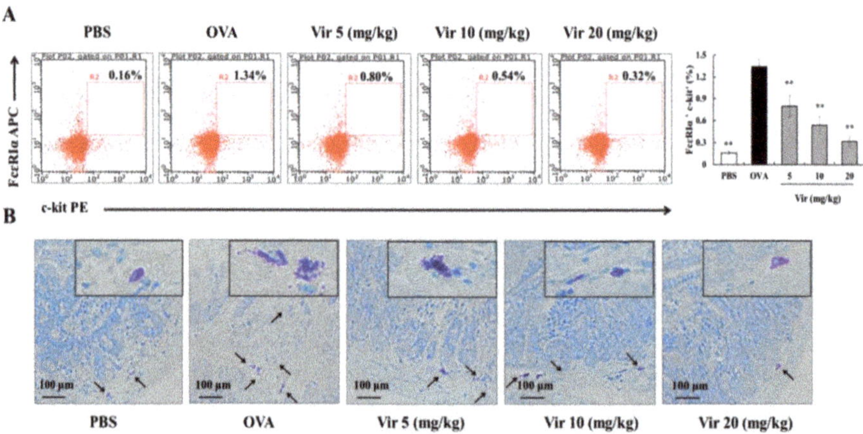

Figure 6. Effects of viridicatol on the mast cell (MC) population and degranulation. (**A**) Scatter diagrams of MCs and FACS analysis by flow cytometry. Splenic lymphocytes were labeled with anti-c-kit and anti-FcεR Iα. (**B**) Representative toluidine blue-stained jejunum sections (magnification: 400×). Jejunum tissues were isolated from each group of mice 24 h after the last challenge and fixed in paraformaldehyde. ** $p < 0.01$ compared with the OVA group. The data represent the mean ± SD. Vir: viridicatol.

2.9. Viridicatol Decreased the Level of Intracellular Calcium

The elevation of Ca^{2+} is a critical process for MC secretory granule translocation [33]. As a calcium ionophore, A23187 is often used to study the status of intracellular Ca^{2+} influx [34]. Moreover, the elevation of intracellular Ca^{2+} stimulated by A23187 was found to be much stronger than that observed in IgE antigen-stimulated cells [16]. In the A32187-induced degranulation model, viridicatol significantly inhibited the release of β-hexosaminidase induced by A23187 in RBL-2H3 cells (Figure 7A). To further study the effect of viridicatol on intracellular Ca^{2+}, we used the Ca^{2+} probe to detect the concentration of intracellular Ca^{2+}. The concentration of intracellular Ca^{2+} was significantly decreased by viridicatol in a dose-dependent manner compared with the A23187 group, as expected (Figure 7B,C). These results indicated that viridicatol may inhibit MC degranulation by suppressing the influx of intracellular Ca^{2+}.

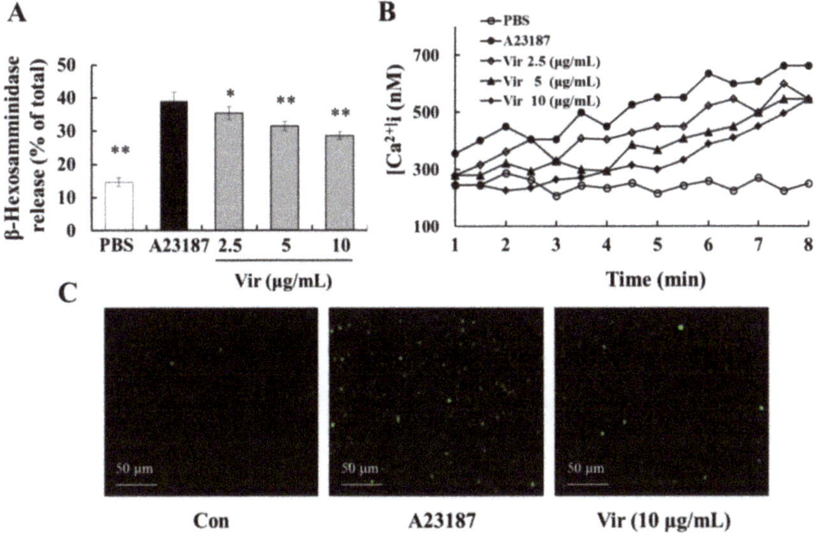

Figure 7. Effects of viridicatol on the degranulation of RBL-2H3 cells induced by A23187 and the concentration of intracellular calcium. (**A**) The release of β-hexosaminidase reduced by the calcium ionophore A23187. RBL-2H3 cells were inoculated in 48-well plates for 12 h and incubated with viridicatol in the presence of 2.5, 5, and 10 µg/mL for 1 h. The cells were then stimulated with 2.5 µM A23187 for 15 min. The release of β-hexosaminidase was measured. (**B**) Effect of Vir on [Ca²⁺]i. (**C**) The accumulation of intracellular calcium (magnification: 200×). * $p < 0.05$, ** $p < 0.01$ compared with the A23187 group. The data represent the mean ± SD of three repeated experiments. Vir: viridicatol.

2.10. Relationship between the Chemical Structure and Inhibitory Effects on Cell Degranulation

The activity of compound is closely related to its chemical structure. Two homologues of viridicatol, viridicatin and 3-*O*-methylviridicatin, were also isolated from *Penicillium griseofulvum*. The chemical structures are presented in Figure 8A–C. In light of the effects of viridicatol on OVA-induced food allergy and the RBL-2H3 cell model, we investigated the effects of viridicatol and its homologous on RBL-2H3 cell degranulation to analyze their inhibitory activity. The effect of these compounds on β-hexosaminidase release in antigen and A23187-induced RBL-2H3 cells was measured at a concentration of 10 µg/mL. In the two models, viridicatol displayed an obvious inhibitory effect on β-hexosaminidase release compared to viridicatin (Figure 8D,E), which may be attributed to the missing conjugation of phenolic hydroxyl. Interestingly, 3-*O*-methylviridicatin hardly inhibited RBL-2H3 cell degranulation, which may have been due to the absence of a phenolic hydroxyl, and the hydroxyl on the mother ring was replaced by a methoxy group.

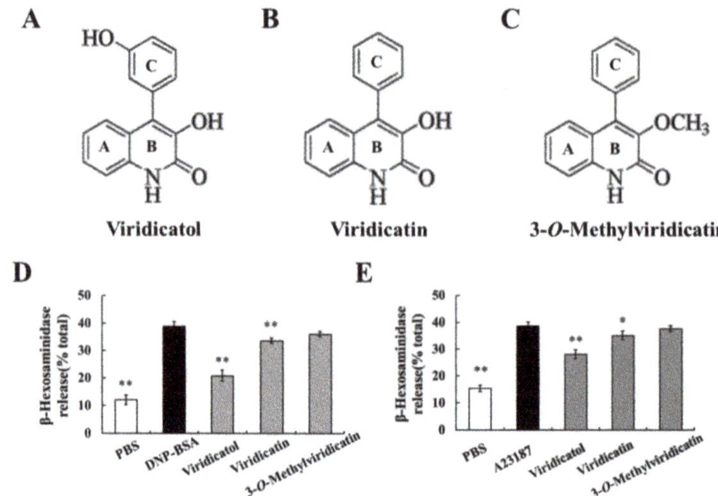

Figure 8. Viridicatol and its homologs for the chemical structural, and the inhibitory effect on RBL-2H3 cells. (**A**) The chemical structure of viridicatol. (**B**) The chemical structure of viridicatin. (**C**) The chemical structure of 3-O-methylviridicatin. (**D**) Effects of viridicatol and its homologs on IgE-induced degranulation. Same as Figure 1C, except that viridicatol was replaced by its homologs at 10 μg/mL. (**E**) Effects of viridicatol and its homologs on A23187-induced degranulation. Same as Figure 7B, except that viridicatol was replaced by its homologs at 10 μg/mL. * $p < 0.05$, ** $p < 0.01$ compared with the DNP-BSA or A23187 groups. The data represent the mean ± SD of three repeated experiments. Vir: viridicatol.

3. Discussion

Due to their extreme environments, deep-sea fungi produce secondary metabolites with novel structures and potent bioactivities [21]. Recently, some secondary metabolites from deep-sea-derived fungi have been discovered to possess biological activities; however, they were poorly studied and seldom utilized. We have found that butenolides, polyketides, and penigrisacids have significant anti-allergic activities [27,29]. Unfortunately, few compounds have been evaluated in mouse models to further elucidate the associated mechanisms in vivo. In this study, viridicatol, a type of quinolone alkaloid, was reported as having anti-tumor activities [30]. However, the anti-allergic activity of viridicatol has not been reported. In this study, viridicatol was found to exhibit significant inhibitory effects on RBL-2H3 cell degranulation. In an OVA-induced mouse model, viridicatol significantly alleviated the allergic symptoms of diarrhea and hypothermia.

Food allergy is accompanied by immune disorders, which involves a variety of immune cells and cytokines [35]. In our study, treatments with viridicatol led to a moderate reduction in the population of splenic B cells, and may contribute to the reduction of serum antigen-specific IgE, which is involved in the process of MC and basophil activation [10]. To some extent, viridicatol was able to upregulate the population of Tregs, which is consistent with increased levels of IL-10. Tregs and the associated cytokine, IL-10, was reported to maintain immune balance and regulate MC activation [36]. MCs are the effector cells involved in inflammation and anaphylaxis, playing a crucial role in allergic reactions [13]. Notably, viridicatol decreased the population of MCs in a dose-dependent manner. Meanwhile, viridicatol could significantly decrease the concentration of serum mMCP-1, TNF-α, and histamine, which was related to a reduction of MC numbers and the inhibition of MC degranulation. On the basis of the results of the OVA-induced model of food allergy, viridicatol was demonstrated to have a stronger effect on MCs than B and Treg cells. Thus, viridicatol may relieve food allergy in mice by decreasing the number and inhibiting the degranulation of MCs.

Previous studies have demonstrated that the occurrence of food allergy is related to damage of intestinal barrier function [37]. The intestinal barrier can be divided into mechanical and immune barriers. Moreover, food allergy can influence intestinal barrier function, which can also enhance the speed of food allergy developments [8]. In this study, viridicatol was found to significantly repair the intestinal barrier by reducing the damage of jejuna tissue villi and alleviating the degree of tissue inflammation. MCs residing in the mucosa of the gastrointestinal tract, as the center of the intestinal immune network, are not only limited to the antigen response, but are also involved in the regulation of the intestinal epithelial barrier and transport, changes in mucosal functionality, and production or amplification of signals to other cells [37]. Following treatments with viridicatol, the length of the cecum was significantly reduced. The activation of MCs resulted in diarrhea, which was likely a part of the mucosal defense response [38]. Bischoff et al. demonstrated that histamine plays a multifunctional role in the intestinal barrier [38]. In addition, Jacobs et al. found that MCs released proteases and TNF-α, which resulted in tight junction protein rearrangement and increased the colic epithelium permeability and intestinal inflammation [39]. The level of mMCP-1 in the blood reflected the extent of degranulation by mast cells in the local tissue [40]. Viridicatol reduced the number of MCs in the jejunum tissue, as well as inhibited MC degranulation and the release of mMCP-1, TNF-α, and histamine, which may be important to the repair of the intestinal barrier following exposure to a food allergen.

At the intersection of many classical signaling pathways, Ca^{2+} converts the received extracellular signals into intracellular signals and regulates lymphocyte activation, differentiation, and various transcriptional processes [41]. Ca^{2+} influx is considered to be a necessary condition for triggering immune function and a key factor for MC activation [42,43]. Ca^{2+} promotes MC degranulation, mediator release, and the activation of gene expression to accelerate a subsequent immune response [44]. Viridicatol was found to suppress Ca^{2+} influx, which was closely associated with MC degranulation. Jacobs et al. suggested that the destruction of the intestinal barrier was accompanied by MC degranulation and increased Ca^{2+} [42]. Therefore, the inhibitory effects of viridicatol on MC degranulation and Ca^{2+} influx may be beneficial for intestinal barrier repair and for relief of food allergy symptoms.

Many anti-allergic drugs have some undesirable adverse effects such as dizziness and drowsiness. Thus, the anti-allergic activities of natural sources have attracted our attention due to their low side effects. As reported, terpenoids, flavenoids, and alkaloids have significant anti-allergic activities and different mechanisms of action [29,45,46]. Alkaloids are nitrogen-containing compounds that have many biological activities, including antiviral, anti-inflammation, and anti-allergy properties. Costa pointed that alkaloids may be associated with the inhibition of eosinophil and mast cell activities [46]. Viridicatol is a quinoline alkaloid that may be absorbed more readily due to the small molecular weight (253.26). Moreover, viridicatol was isolated from the deep-sea fungus *Penicillium griseofulvum*, which produces a variety of anti-allergic metabolites [29]. In this study, viridicatol inhibited the degranulation of RBL-2H3 cells and relieved the allergic reactions in mice, which may be related to its structure and thus it is worth further exploring. The bioactivity of these compounds is closely related to their structures. For example, the number and position of hydroxyl groups have been shown to be associated with anti-allergic activities [45]. Moreover, both in antigen-induced and A23187-induced RBL-2H3 cell degranulation models, the inhibitory effect of viridicatol on β-hexosaminidase were stronger than viridicatin, which may due to the presence of an extra phenolic hydroxyl. Interestingly, no significant effect of 3-*O*-methylviridicatin on the release of β-hexosaminidase were observed, which may have resulted from the different chemical structure of 3-*O*-methylviridicatin compared with viridicatol: one lost hydroxyl in ring C and one hydroxyl was replaced by a methoxy in ring B. Phenolic hydroxyl groups have been reported to contribute to the increased efficiency of these compounds [47].

Taken together, deep-sea-derived viridicatol could relieve OVA-induced allergic symptoms. Treatments with viridicatol had significant effects on decreasing the level of anti-OVA-IgE, histamine, mMCP-1, and TNF-α. Moreover, viridicatol significantly upregulated the level of IL-10 in the serum. Viridicatol treatment was also found to downregulate the populations of B cells and MCs in the spleen, as well as upregulate Tregs in the spleen. Additionally, viridicatol repaired the intestinal barrier and alleviated the degree of tissue infiltration. Notably, viridicatol significantly suppressed MC degranulation in the jejunum of mice, which may be attributable to the decreased intracellular flow of Ca^{2+} in MCs. Compared to many alkaloids, viridicatol has a relatively small molecular weight, allowing it to be absorbed more readily, and it is expected to be stabilizing for MCs, which could possess a higher practical value. In conclusion, this research provided some insight into the prevention of food allergy and the application of marine fungi. Marine fungi have a large impact in terms of application in the pharmaceutical industry, since many of their metabolites have entered the clinical pipeline and have the potential of being exploited as novel drugs. The present findings demonstrated that viridicatol have the potential to be applied in the therapy of food allergies. Further systematic safety assessments and optimal dosage determination are necessary for the clinical application of viridicatol.

4. Materials and Methods

4.1. Reagents an General Experimental Procedures

Fetal bovine serum (FBS) was purchased from Gibco (Grand Island, NY, USA). Roswell Park Memorial Institute (RPMI) 1640, Eagle's minimum essential medium (EMEM), penicillin–streptomycin solution, and trypsin 0.25% (1×) solution were purchased from HyClone Co. (Logan, UT, USA). Imject alum was obtained from Thermo Fisher Scientific Inc. (Waltham, MA, USA). Evans Blue, anti-dinitrophenyl (DNP)-IgE, ovalbumin (OVA), calcium ionophore A23187, and 4-methyl-umbelliferyl-N-acetyl-β-D-glucosaminide were obtained from Sigma-Aldrich (St Louis, MO, USA). DNP-bovine serum albumin (BSA) was purchased from Biosearch (Petaluma, CA, USA). Goat anti-mouse IgG1 (ab98693) and IgG2a (ab98698) antibodies were purchased from Abcam (Cambridge, United Kingdom). Calcium kit-Fluo 3-AM was purchased from Dojindo Laboratories (Mashiki, Japan). ELISA (enzyme-linked immunosorbent assay) kits for histamine (EHP184), anti-OVA-IgE (500840), and IL-10 (BMS614INST) were obtained from IBL (Hamburg, Germany), Cayman (Ann Arbor, MI, USA), and eBioscience (San Diego, CA, USA), respectively. Antibodies for flow cytometry analysis were obtained from BioLegend (San Diego, CA, USA) and BD (New York, NY, USA). P-phycoerythrin (PE) anti-mouse CD117(c-Kit) (105807), Allophycocyanin (APC) anti-mouse FcϵRIα (134316), PE anti-mouse CD19 (115508), Mouse Th/Treg Phenotyping Kit (560767) were also used. All other chemicals were of analytical grade. The HRESIMS spectrum was recorded on a Waters Xevo G2 Q-TOF mass spectrometer (Waters Corporation, Milford, MA, USA). The NMR spectra were recorded on a Bruker 400 MHz spectrometer (Bruker, Fällanden, Switzerland). The single X-ray crystal data were recorded with an XtaLAB AFC12 Kappa single diffractometer using Cu Kα radiation (Rigaku, Japan). Copies of the data can be obtained, free of charge, on application to Cambridge Crystallographic Data Center (CCDC), 12 Union Road, Cambridge CB21EZ, United Kingdom, (fax: +44(0)-1233-336033; email: deposit@ccdc.cam.ac.uk.

4.2. Cell Culture

The RBL-2H3 cell line was purchased from the American Type Culture Collection (ATCC, Manassas, VA, USA). Cells were cultured in EMEM supplemented with 10% FBS, 1% penicillin, and 1% streptomycin at 37 °C in an atmosphere containing 5% CO_2.

4.3. Animals

Female BALB/c mice (aged 6–7 weeks) were purchased from Charles River Laboratories (Beijing, China). Six mice were randomly assigned to each experimental group and all mice were maintained in an automatic light/dark cycle (light periods of 12 h) environment and were acclimatized for 1 week before the experiment.

This experiment was performed in strict accordance with the National Institutes of Health (NIH) Guidelines for the Care and Use of Laboratory Animals. All protocols were approved by the Animal Research Ethics Board of Jimei University (Xiamen, China, no. SCXK 2016-0006).

4.4. Viridicatol Extraction, Isolation, and Purification

On the basis of a previous method of fermentation, we defatted the crude extract with petroleum ether and CH_2Cl_2 by column chromatography over silica gel[29]. The MeOH layer was evaporated under reduced pressure to provide a defatted extract (55.4 g). The extract was then subjected to column chromatography on silica gel using a CH_2Cl_2-MeOH gradient ($0\rightarrow100\%$, 49 mm × 460 mm) to yield six fractions (Fr. 1–Fr.6). Fr. 5 (40.0 g) was separated by column chromatography over ODS (H_2O-MeOH, $5\rightarrow80\%$, 49 mm × 460 mm) to obtain 15 subfractions (sfrs. 5.1–sfrs. 5.15). Sfr. 5.15 was subjected to column chromatography on Sephadex LH-20 (MeOH) to obtain viridicatol (443.0 mg).

4.5. Determination of Histamine and β-Hexosaminidase Release in RBL-2H3 Cells

It has previously been reported that the level of β-hexosaminidase and histamine from RBL-2H3 cells activated by IgE–FcεRI complex were detected [45]. The cellular cytotoxicity of viridicatol was measured using MTT (3-(4,5-dimethyl-2-thiazolyl)-2, 5-diphenyl-2-H-tetrazolium bromide) assay.

4.6. OVA-Induced Model of Food Allergy

Groups of mice (each group of six) were sensitized with 2 mg alum and 100 μg OVA in 200 μL of phosphate-buffered saline (PBS) intraperitoneally (i.p.) on days 0 and 14. The mice were challenged with 50 mg OVA in 200 μL PBS by gavage every 3 days from days 28 to 40 for a total of 5 times. The mice in the treatment groups were treated with viridicatol by gavage every day starting on day 27. The mice in the PBS group were sensitized with 2 mg alum in 200 μL PBS and challenged with PBS, which served as the control group. The anti-allergic effect of viridicatol was determined on the basis of diarrhea, anaphylaxis, and rectal temperature [31]; mice were placed alone in cages, their feces were counted to calculate the proportion of loose stools in the total stools, and symptoms were scored from 0 to 5.

4.7. Measurement of Immunoglobulins, Anaphylactic Mediators, and Cytokines in the Mouse Serum

On day 41, the mouse eyeball sera were collected. OVA-specific IgE, IL-10, and TNF-α were measured by commercial ELISA kits. OVA-specific IgG1 and IgG2a were measured by an indirect ELISA, as previously reported [48]. Serum was obtained from the tail vein blood after 30 min of the final challenge on day 40 to detect the level of histamine and mMCP-1 using a commercial ELISA kit. All sera were stored at −80 °C.

4.8. Morphology and Staining Analysis of the Intestinal Tissue

On day 41, the mice were sacrificed, and the cecum tissue was collected for observation. The jejunum tissue was fixed, dehydrated, and coated with gold-palladium, and then the jejunum villi were observed by Phenom Pro Model desktop scanning electron microscopy (SEM, Phenom-world, Eindhoven, The Netherlands) as previously reported [48]. The jejunum tissue was fixed in 4% paraformaldehyde, and Servicebio was entrusted to conduct hematoxylin and eosin (H&E) staining and toluene blue staining for the jejunal tissues.

4.9. Splenic Lymphocyte Population Analysis by Flow Cytometry

As previously reported, single-cell suspensions were obtained on day 41 by density gradient centrifugation from the mouse spleen tissue suspension [45]. The cells were labeled with anti-CD3, anti-CD4, anti-CD19, anti-Foxp3, anti-FcεRI, and anti-c-KIT on the surface. Intracellular expression markers were stained with Foxp3-PerCP-Cy5.5 after fixing and permeabilizing as described in the protocol. The cells were gated and demonstrated by flow cytometry(Millipore, Billerica, MA, USA), and data were analyzed with a Guava easyCyte 6-2L system using Guava Soft 3.1.1 software.

4.10. Release of β-Hexosaminidase and Ca^{2+} Influx of RBL-2H3 Cells Induced by A23187

To measure the release of β-hexosaminidase induced by A23187, we inoculated RBL-2H3 cells in 48-well plates for 12 h and incubated them with or without viridicatol for 1 h. The cells were then stimulated with 2.5 μM A23187 for 15 min, after which β-hexosaminidase release was measured. The concentration of Ca^{2+} was measured using a Calcium kit-Fluo 3 according to the manufacturer's instructions. RBL-2H3 cells (5×10^4 in 100 μL) were seeded into a black 96-well plate and cultured for 16 h. The cells were then washed and incubated with Fluo-3 AM (10 μM) for 1 h. Next, the cells were washed and treated with viridicatol (2.5, 5, 10 μg/mL) in HBSS for 1 h. After we stimulated it with A23187 (2.5 μM), the fluorescent intensity was immediately monitored using the microplate reader at an excitation wavelength of 490 nm and an emission wavelength of 530 nm every 30 s. $[Ca^{2+}]i$ was calculated as follows: $[Ca^{2+}]i$ (nm) = Kd [(F − Fmin)/(Fmax − F)], Fmin: the background fluorescence of 5 mM Ethylenebis(oxyethylenenitrilo) tetraacetic acid (EGTA), Fmax: the maximum fluorescence with of 0.1% Triton X-100, Kd: the dissociation constant of Ca^{2+} and Fluo-3 (450 nM). Moreover, the cells were photographed with a fluorescence microscope (Echo, San Diego, CA, USA) to observe the variation of intracellular Ca^{2+}.

4.11. Statistical Analysis

All experiments were repeated at least 3 times, and the data were presented as the mean ± standard deviation (SD). Statistical differences were analyzed using a one-way analysis of variance (ANOVA). A threshold *p*-value of 0.05 or 0.01 indicated a significant difference.

Supplementary Materials: The following are available online at http://www.mdpi.com/1660-3397/18/10/517/s1, Figures S1–S6: 1D and 2D NMR spectra of viridicatol. Figure S7: The HR-ESI-MS spectrum of viridicatol.

Author Contributions: Z.S. conducted anti-food allergic experiments. G.L. and X.Y. designed and coordinated the project; J.Z. and Y.Z. (Yu Zhou) assisted Z.S.'s experiments. C.X. conducted fungal fermentation as well as compound isolation and characterization. Q.L. and Y.Z. (Yafen Zhang) designed the experiments. H.L. and M.C. contributed to the coordination of the project. All authors have read and agreed to the published version of the manuscript.

Funding: This research was funded by the grants from the National Natural Scientific Foundation of China (31871720, 32072336, 32001695), the Science and Technology Program of Fujian Province (2018N5009, 2018R0071), and the National Key R&D Program of China (2019YFD0901703).

Conflicts of Interest: The authors declare no conflict of interest.

References

1. Renz, H.; Allen, K.J.; Sicherer, S.H.; Sampson, H.A.; Lack, G.; Beyer, K.; Oettgen, H.C. Food allergy. *Nat. Rev. Dis. Primers.* **2018**, *4*, 17098. [CrossRef] [PubMed]

2. Sicherer, S.H.; Sampson, H.A. Food allergy: Epidemiology, pathogenesis, diagnosis, and treatment. *J. Allergy Clin. Immunol.* **2014**, *133*, 291–307. [CrossRef] [PubMed]

3. Begin, P.; Filion, C.; Graham, F.; Lacombe-Barrios, J.; Paradis, J.; Paradis, L.; Des Roches, A. Consultation with registered dietitian to prevent accidental reactions to food: Insight from an egg allergy influenza vaccination cohort. *Eur. J. Clin. Nutr.* **2017**, *71*, 287–289. [CrossRef] [PubMed]

4. Dhanapala, P.; De Silva, C.; Doran, T.; Suphioglu, C. Cracking the egg: An insight into egg hypersensitivity. *Mol. Immunol.* **2015**, *66*, 375–383. [CrossRef] [PubMed]

5. Zhang, T.T.; Hu, Z.Y.; Cheng, Y.W.; Xu, H.X.; Velickovic, T.C.; He, K.; Sun, F.; He, Z.D.; Liu, Z.G.; Wu, X.L. Changes in Allergenicity of Ovalbumin in Vitro and in Vivo on Conjugation with Quercetin. *J. Agr. Food Chem.* **2020**, *68*, 4027–4035. [CrossRef] [PubMed]

6. Castillo-Courtade, L.; Han, S.; Lee, S.; Mian, F.M.; Buck, R.; Forsythe, P. Attenuation of food allergy symptoms following treatment with human milk oligosaccharides in a mouse model. *Allergy* **2015**, *70*, 1091–1102. [CrossRef]

7. Galli, S.J.; Tsai, M.; Piliponsky, A.M. The development of allergic inflammation. *Nature* **2008**, *454*, 445–454. [CrossRef]

8. Konig, J.; Wells, J.; Cani, P.D.; Garcia-Rodenas, C.L.; MacDonald, T.; Mercenier, A.; Whyte, J.; Troost, F.; Brummer, R.J. Human Intestinal Barrier Function in Health and Disease. *Clin. Transl. Gastroenterol.* **2016**, *7*, e196. [CrossRef]

9. Groschwitz, K.R.; Hogan, S.P. Intestinal barrier function: Molecular regulation and disease pathogenesis. *J. Allergy Clin. Immu.* **2009**, *124*, 3–20. [CrossRef]

10. Takhar, P.; Smurthwaite, L.; Coker, H.A.; Fear, D.J.; Banfield, G.K.; Carr, V.A.; Durham, S.R.; Gould, H.J. Allergen drives class switching to IgE in the nasal mucosa in allergic rhinitis. *J. Immunol.* **2005**, *174*, 5024–5032. [CrossRef]

11. Rivas, M.N.; Chatila, T.A. Regulatory T cells in allergic diseases. *J. Allergy Clin. Immu.* **2016**, *138*, 639–652. [CrossRef]

12. Tordesillas, L.; Mondoulet, L.; Blazquez, A.B.; Benhamou, P.H.; Sampson, H.A.; Berin, M.C. Epicutaneous immunotherapy induces gastrointestinal LAP(+) regulatory T cells and prevents food-induced anaphylaxis. *J. Allergy Clin. Immu.* **2017**, *139*, 189. [CrossRef] [PubMed]

13. Harvima, I.T.; Levi-Schaffer, F.; Draber, P.; Friedman, S.; Polakovicova, I.; Gibbs, B.F.; Blank, U.; Nilsson, G.; Maurer, M. Molecular targets on mast cells and basophils for novel therapies. *J. Allergy Clin. Immunol.* **2014**, *134*, 530–544. [CrossRef] [PubMed]

14. Gilfillan, A.M.; Beaven, M.A. Regulation of mast cell responses in health and disease. *Crit. Rev. Immunol.* **2011**, *31*, 475–529. [CrossRef] [PubMed]

15. Kim, S.H.; Park, S.B.; Kang, S.M.; Jeon, H.; Lim, J.P.; Kwon, T.K.; Park, W.H.; Kim, H.M.; Shin, T.Y. Anti-allergic effects of Teucrium japonicum on mast cell-mediated allergy model. *Food Chem. Toxicol.* **2009**, *47*, 398–403. [CrossRef] [PubMed]

16. Matsuo, N.; Yamada, K.; Shoji, K.; Mori, M.; Sugano, M. Effect of tea polyphenols on histamine release from rat basophilic leukemia (RBL-2H3) cells: The structure-inhibitory activity relationship. *Allergy* **1997**, *52*, 58–64. [CrossRef]

17. Izawa, K.; Yamanishi, Y.; Maehara, A.; Takahashi, M.; Isobe, M.; Ito, S.; Kaitani, A.; Matsukawa, T.; Matsuoka, T.; Nakahara, F.; et al. The receptor LMIR3 negatively regulates mast cell activation and allergic responses by binding to extracellular ceramide. *Immunity* **2012**, *37*, 827–839. [CrossRef]

18. Nishida, K.; Yamasaki, S.; Ito, Y.; Kabu, K.; Hattori, K.; Tezuka, T.; Nishizumi, H.; Kitamura, D.; Goitsuka, R.; Geha, R.S.; et al. FcεRI-mediated mast cell degranulation requires calcium-independent microtubule-dependent translocation of granules to the plasma membrane. *J. Cell Biol.* **2005**, *170*, 115–126. [CrossRef]

19. Han, S.; Sun, L.; He, F.; Che, H. Anti-allergic activity of glycyrrhizic acid on IgE-mediated allergic reaction by regulation of allergy-related immune cells. *Sci. Rep.* **2017**, *7*, 7222. [CrossRef]

20. Blunt, J.W.; Copp, B.R.; Keyzers, R.A.; Munro, M.H.; Prinsep, M.R. Marine natural products. *Nat. Prod. Rep.* **2012**, *29*, 144–222. [CrossRef]

21. Imhoff, J.F. Natural Products from Marine Fungi–Still an Underrepresented Resource. *Mar. Drugs* **2016**, *14*, 19. [CrossRef] [PubMed]

22. Nicoletti, R.; Trincone, A. Bioactive Compounds Produced by Strains of Penicillium and Talaromyces of Marine Origin. *Mar. Drugs* **2016**, *14*, 37. [CrossRef]

23. Zhang, P.; Wei, Q.; Yuan, X.; Xu, K. Newly reported alkaloids produced by marine-derived Penicillium species (covering 2014-2018). *Bioorg. Chem.* **2020**, *99*, 103840. [CrossRef] [PubMed]

24. Paget, G.E.; Walpole, A.L. Some cytological effects of griseofulvin. *Nature* **1958**, *182*, 1320–1321. [CrossRef] [PubMed]

25. Garcia-Leoni, M.E.; Cercenado, E.; Rodeno, P.; Bernaldo de Quiros, J.C.; Martinez-Hernandez, D.; Bouza, E. Susceptibility of Streptococcus pneumoniae to penicillin: A prospective microbiological and clinical study. *Clin. Infect. Dis.* **1992**, *14*, 427–435. [CrossRef]

26. Youssef, D.T.A.; Alahdal, A.M. Cytotoxic and Antimicrobial Compounds from the Marine-Derived Fungus, Penicillium Species. *Molecules* **2018**, *23*, 394. [CrossRef]

27. Liu, Q.M.; Xie, C.L.; Gao, Y.Y.; Liu, B.; Lin, W.X.; Liu, H.; Cao, M.J.; Su, W.J.; Yang, X.W.; Liu, G.M. Deep-Sea-Derived Butyrolactone I Suppresses Ovalbumin-Induced Anaphylaxis by Regulating Mast Cell Function in a Murine Model. *J. Agric. Food Chem.* **2018**, *66*, 5581–5592. [CrossRef]

28. Gao, Y.Y.; Liu, Q.M.; Liu, B.; Xie, C.L.; Cao, M.J.; Yang, X.W.; Liu, G.M. Inhibitory Activities of Compounds from the Marine Actinomycete Williamsia sp MCCC 1A11233 Variant on IgE-Mediated Mast Cells and Passive Cutaneous Anaphylaxis. *J. Agr. Food. Chem.* **2017**, *65*, 10749–10756. [CrossRef]

29. Xing, C.P.; Xie, C.L.; Xia, J.M.; Liu, Q.M.; Lin, W.X.; Ye, D.Z.; Liu, G.M.; Yang, X.W. Penigrisacids A-D, Four New Sesquiterpenes from the Deep-Sea-Derived Penicillium griseofulvum. *Mar. Drugs* **2019**, *17*, 507. [CrossRef]

30. Liang, P.; Zhang, Y.Y.; Yang, P.; Grond, S.; Zhang, Y.; Qian, Z.J. Viridicatol and viridicatin isolated from a shark-gill-derived fungus Penicillium polonicum AP2T1 as MMP-2 and MMP-9 inhibitors in HT1080 cells by MAPKs signaling pathway and docking studies. *Med. Chem. Res.* **2019**, *28*, 1039–1048. [CrossRef]

31. Yamaki, K.; Yoshino, S. Tyrosine kinase inhibitor sunitinib relieves systemic and oral antigen-induced anaphylaxes in mice. *Allergy* **2012**, *67*, 114–122. [CrossRef]

32. Zhang, J.; Chen, Y.; Luo, H.; Sun, L.; Xu, M.; Yu, J.; Zhou, Q.; Meng, G.; Yang, S. Recent Update on the Pharmacological Effects and Mechanisms of Dihydromyricetin. *Front. Pharmacol.* **2018**, *9*, 1204. [CrossRef]

33. Kamei, R.; Fujimura, T.; Matsuda, M.; Kakihara, K.; Hirakawa, N.; Baba, K.; Ono, K.; Arakawa, K.; Kawamoto, S. A flavanone derivative from the Asian medicinal herb (Perilla frutescens) potently suppresses IgE-mediated immediate hypersensitivity reactions. *Biochem. Biophys. Res. Commun.* **2017**, *483*, 674–679. [CrossRef] [PubMed]

34. Barbosa, M.; Lopes, G.; Valentao, P.; Ferreres, F.; Gil-Izquierdo, A.; Pereira, D.M.; Andrade, P.B. Edible seaweeds' phlorotannins in allergy: A natural multi-target approach. *Food Chem.* **2018**, *265*, 233–241. [CrossRef]

35. Lee, D.; Kim, H.S.; Shin, E.; Do, S.G.; Lee, C.K.; Kim, Y.M.; Lee, M.B.; Min, K.Y.; Koo, J.; Kim, S.J.; et al. Polysaccharide isolated from Aloe vera gel suppresses ovalbumin-induced food allergy through inhibition of Th2 immunity in mice. *Biomed. Pharmacother.* **2018**, *101*, 201–210. [CrossRef]

36. Veldhoen, M.; Hocking, R.J.; Atkins, C.J.; Locksley, R.M.; Stockinger, B. TGFbeta in the context of an inflammatory cytokine milieu supports de novo differentiation of IL-17-producing T cells. *Immunity* **2006**, *24*, 179–189. [CrossRef]

37. Forbes, E.E.; Groschwitz, K.; Abonia, J.P.; Brandt, E.B.; Cohen, E.; Blanchard, C.; Ahrens, R.; Seidu, L.; McKenzie, A.; Strait, R.; et al. IL-9- and mast cell-mediated intestinal permeability predisposes to oral antigen hypersensitivity. *J. Exp. Med.* **2008**, *205*, 897–913. [CrossRef]

38. Bischoff, S.C. Physiological and pathophysiological functions of intestinal mast cells. *Semin. Immunopathol.* **2009**, *31*, 185–205. [CrossRef] [PubMed]

39. Jacob, C.; Yang, P.C.; Darmoul, D.; Amadesi, S.; Saito, T.; Cottrell, G.S.; Coelho, A.M.; Singh, P.; Grady, E.F.; Perdue, M.; et al. Mast cell tryptase controls paracellular permeability of the intestine. Role of protease-activated receptor 2 and beta-arrestins. *J. Biol. Chem.* **2005**, *280*, 31936–31948. [CrossRef]

40. Nguyet, T.M.N.; Lomunova, M.; Le, B.V.; Lee, J.S.; Park, S.K.; Kang, J.S.; Kim, Y.H.; Hwang, I. The mast cell stabilizing activity of Chaga mushroom critical for its therapeutic effect on food allergy is derived from inotodiol. *Int. Immunopharmacol.* **2018**, *54*, 286–295. [CrossRef]

41. Trebak, M.; Kinet, J.P. Calcium signalling in T cells. *Nat. Rev. Immunol.* **2019**, *19*, 154–169. [CrossRef] [PubMed]

42. Lewis, R.S. Calcium signaling mechanisms in T lymphocytes. *Annu. Rev. Immunol.* **2001**, *19*, 497–521. [CrossRef] [PubMed]

43. Rao, A.; Hogan, P.G. Calcium signaling in cells of the immune and hematopoietic systems. *Immunol. Rev.* **2009**, *231*, 5–9. [CrossRef]

44. Di Capite, J.; Parekh, A.B. CRAC channels and Ca2+ signaling in mast cells. *Immunol. Rev.* **2009**, *231*, 45–58. [CrossRef]

45. Zhang, Y.F.; Liu, Q.M.; Liu, B.; Shu, Z.D.; Han, J.; Liu, H.; Cao, M.J.; Yang, X.W.; Gu, W.; Liu, G.M. Dihydromyricetin inhibited ovalbumin-induced mice allergic responses by suppressing the activation of mast cells. *Food. Funct.* **2019**, *10*, 7131–7141. [CrossRef] [PubMed]

46. Costa, H.F.; Leite, F.C.; Alves, A.F.; Barbosa, J.M.; dos Santos, C.R.B.; Piuvezam, M.R. Managing murine food allergy with Cissampelos sympodialis Eichl (Menispermaceae) and its alkaloids. *Int. Immunopharmacol.* **2013**, *17*, 300–308. [CrossRef]

47. Xin, M.L.; Ma, Y.J.; Xu, K.; Chen, M.C. Dihydromyricetin An effective non-hindered phenol antioxidant for linear low-density polyethylene stabilisation. *J. Therm. Anal. Calorim.* **2013**, *114*, 1167–1175. [CrossRef]

48. Liu, Q.; Zhang, Y.; Shu, Z.; Liu, M.; Zeng, R.; Wang, Y.; Liu, H.; Cao, M.; Su, W.; Liu, G. Sulfated oligosaccharide of Gracilaria lemaneiformis protect against food allergic response in mice by up-regulating immunosuppression. *Carbohydr. Polym.* **2020**, *230*, 115567. [CrossRef]

marine drugs

Article

Molecular and Structural Characterizations of Lipases from *Chlorella* by Functional Genomics

Hajer Ben Hlima [1], Mouna Dammak [1], Aida Karray [2], Maroua Drira [3], Philippe Michaud [4,*], Imen Fendri [3] and Slim Abdelkafi [1,*]

[1]   Laboratoire de Génie Enzymatique et de Microbiologie, Equipe de Biotechnologie des Algues, Ecole Nationale d'Ingénieurs de Sfax, Université de Sfax, Sfax 3038, Tunisia; hajer.benhlima@enis.tn (H.B.H.); dammakmouna23@yahoo.fr (M.D.)
[2]   Laboratoire de Biochimie et de Génie Enzymatique des Lipases, Ecole Nationale d'Ingénieurs de Sfax, Université de Sfax, Sfax 3038, Tunisia; aida.karray@enis.tn
[3]   Laboratoire de Biotechnologie Végétale Appliquée à l'Amélioration des Cultures, Faculté des Sciences de Sfax, Université de Sfax, Sfax 3038, Tunisia; driramaroua@yahoo.fr (M.D.); imen.fendri@fss.usf.tn (I.F.)
[4]   CNRS, SIGMA Clermont, Institut Pascal, Université Clermont-Auvergne, F-63000 Clermont-Ferrand, France
*   Correspondence: philippe.michaud@uca.fr (P.M.); slim.abdelkafi@enis.tn (S.A.)

Abstract: Microalgae have been poorly investigated for new-lipolytic enzymes of biotechnological interest. In silico study combining analysis of sequences homologies and bioinformatic tools allowed the identification and preliminary characterization of 14 putative lipases expressed by *Chlorella vulagaris*. These proteins have different molecular weights, subcellular localizations, low instability index range and at least 40% of sequence identity with other microalgal lipases. Sequence comparison indicated that the catalytic triad corresponded to residues Ser, Asp and His, with the nucleophilic residue Ser positioned within the consensus GXSXG pentapeptide. 3D models were generated using different approaches and templates and demonstrated that these putative enzymes share a similar core with common α/β hydrolases fold belonging to family 3 lipases and class GX. Six lipases were predicted to have a transmembrane domain and a lysosomal acid lipase was identified. A similar mammalian enzyme plays an important role in breaking down cholesteryl esters and triglycerides and its deficiency causes serious digestive problems in human. More structural insight would provide important information on the enzyme characteristics.

Keywords: *Chlorella*; enzymes; lipases; molecular modeling

Citation: Ben Hlima, H.; Dammak, M.; Karray, A.; Drira, M.; Michaud, P.; Fendri, I.; Abdelkafi, S. Molecular and Structural Characterizations of Lipases from *Chlorella* by Functional Genomics. *Mar. Drugs* **2021**, *19*, 70. https://doi.org/10.3390/md19020070

Academic Editor: Khaled A. Shaaban
Received: 27 December 2020
Accepted: 26 January 2021
Published: 28 January 2021

1. Introduction

The industrial enzymes market is estimated to be valued at USD 5.9 billion in 2020 and is projected to reach USD 8.7 billion by 2026, recording a Compound Annual Growth Rate (CAGR) of 6.5%, in terms of value [1]. The majority of enzymes currently used in industrial processes (more than 75%) are hydrolases [2]. Lipases represent the third most commercialized enzymes, after carbohydrases and proteases [3], and their production has constantly increased, they now account for more than one-fifth of the global enzyme market. The global Lipase Market size is anticipated to develop at a notable CAGR of about 7% over the calculated period from the current value of USD 0.6 billion in 2020. Lipases form an integral part of the industries ranging from biodiesels, food, nutraceuticals and detergents with little utilization in bioremediation, agriculture, cosmetics and leather [4].

Although lipases are produced by a huge number of organisms (bacterial, plant and animal origin), microbial lipases have attracted far more interest from researchers and industries than lipases from other sources, due to both their specific features and ease of production on large scale [5–7]. Notwithstanding current achievements, there is still a quest for lipases with improved and/or novel catalytic features like stability in harsh environments. Marine organisms can be an adequate source for such lipases as marine enzymes have demonstrated their useful for both process improvement and for the development of new process or products. Relevant types of lipases from marine organisms were identified and their novel features were discussed. They display, for example, salt tolerance, calcium independence and thermostable activities; they can also be stable in alkaline environment and were suggested to have antibiofilm action and higher catalytic efficiencies at temperatures lower than those from terrestrial microbial and/or mammal lipases [8]. However, few microalgal lipases and genes encoding lipases have been investigated and compared to bacterial, fungal, animal and plant lipases. In 2010, Demir and Tukel isolated and characterized for the first time a lipase from the photosynthetic cyanobacterium *Arthrospira. platensis* [9]. The lipase was a monomeric protein of 45 kDa with an isoelectric point of 5.9. It was specific for the 3-position in the ester bond. Godet et al. [10] isolated a new gene from the microalgae *Isochrysis galbana* encoding a 49 kDa lipase that shares similarities with fungal known lipase sequences. *Chlorella vulgaris* is a microalga belonging to the order of the *Chlorococcales*, which has a green color. It contains a significant number of intracellular proteins, carbohydrates, lipids, vitamin C, β-carotenes and B vitamins (B1, B2, B6 and B12), which is why it is commonly used for the preparation of food supplements. It is considered as raw materials for chemical compounds that have been affected by its primary and secondary metabolism, such as lipids, whose main application is for the generation of biodiesel [11]. This microalga has one of the highest lipid accumulating abilities in microalgae (50% of its DW), very high volumetric lipid productivity (VLP) of about 80 mg/L.day with a high growth rate in large-scale outdoor cultivation systems. Genetic manipulation technique for this microalga has already been established, showing great promise for improving its oleaginous phenotype by metabolic engineering [12]. Recently, its whole genome sequence was revealed by next-generation sequencing technologies, and the major metabolic pathways were identified [13]. Lipid metabolism has also been analyzed in multi-omics studies, including transcriptomics and proteomics to obtain the mechanistic insight of its lipid biosynthesis [14]. However, the TAG lipases have not been investigated yet. It will be of great importance to estimate the number and characteristics of its lipases, attracting knockdown targets for enhancement of lipid productivity. Here, a bioinformatic screening of a *C. vulgaris* genome was done to explore the presence of genes encoding putative lipases. The potential properties of the candidates are discussed on the basis of their three-dimensional (3D) model structures.

2. Results

2.1. Sequence Retrieval

The results of the amino acid sequence search showed that 14 protein sequences from nine *C. vulgaris* strains of the UTEX259 UTEX259 culture collection (taxid 3077)-scaffolds met determined criteria. The accession numbers of Transcriptome Shotgun Assembly (TSA) and Whole Genome Shotgun (WGS) sequences are given in Table 1 andTable 2, respectively. As can be seen from the Table 1, all found lipase sequences belong to AB_hydrolase family (Interpro number IPR029058) and display Acyl hydrolase motif GXSXG. Nine of them show high sequence identity to Lipase_3 domain-containing protein (*Chlorella variabilis*) from the ESTHER database. Two sequences, namely Lip_5800 and Lip_5999, present high identity with sn1-specific diacylglycerol lipases alpha from *Auxenochlorella protothecoides* and *Micractinium conductrix*, respectively. In addition, 46.6% of sequence identity with chloroplastic Phospholipase A1 from *M. conductrix* was also detected with Lip_3448. Sequence homology analysis with multiple alignments revealed that these 14 sequences could be broadly clustered into two groups; 3 probable sn1-diacylglycerol lipases and 11 other lipase_3 family. Subsequently, gene prediction experiments were carried out with ab initio gene models (Table 2). These predictions showed different scaffold localization of the predicted lipase sequences with an exon number varying from 8 (Lip_5800 and Lip_5462) to 23 (Lip_2999). Lip_4551 and Lip_6297 lipases genes were found to be tandemly arrayed in the genome structure. These two genes have different sequence and size and their adjacent organization could allow faster transcription [15].

Table 1. Putative TAG lipases in *C. vulgaris* with their sequence-based motif and family search.

TSA ID	Family InterPro	Family Pfam	Acyl Hydrolase Motif (GXSXG)	Highest Identity in ESTHER Database	Accession Number ESTHER Database
GHLX01005462.1	IPR029058, AB_hydrolase	PF04083, Abhydro_lipase	GHSQG	62.2% Lipase (*M. conductrix*)	A0A2P6V8F3
GHLX01003448.1	IPR029058, AB_hydrolase IPR002921, Fungal_lipase-like	PF01764, Lipase_3	GHSLG	46.6% Phospholipase A(1) chloroplastic (*M. conductrix*)	A0A2P6VDJ3
GHLX01004364.1	IPR029058, AB_hydrolase IPR002921, Fungal_lipase-like	PF01764, Lipase_3	GHSLG	79.9% Lipase_3 domain-containing protein (*C. variabilis*)	E1ZB31
GHLX01003076.1	IPR029058, AB_hydrolase IPR002921, Fungal_lipase-like	PF01764, Lipase_3	GHSLG	41.5% Lipase_3 domain-containing protein (*C. variabilis*)	E1ZMR0
GHLX01002999.1	IPR029058, AB_hydrolase IPR002921, Fungal_lipase-like	PF01764, Lipase_3	GHSLG	56.8% Lipase_3 domain-containing protein (*C. variabilis*)	E1Z559
GHLX01001704.1	IPR029058, AB_hydrolase IPR002921, Fungal_lipase-like	PF01764, Lipase_3	GHSLG	56.1%Lipase_3 domain-containing protein (*C. variabilis*)	E1ZAU0
GHLX01004551.1	IPR029058, AB_hydrolase IPR002921, Fungal_lipase-like	PF01764, Lipase_3	GHSLG	53.5% Lipase_3 domain-containing protein (*C. variabilis*)	E1Z6D5
GHLX01003928.1	IPR029058, AB_hydrolase IPR002921, Fungal_lipase-like	PF01764, Lipase_3	GHSLG	54.5% Lipase_3 domain-containing protein (*C. variabilis*)	E1ZMR0
GHLX01006297.1	IPR029058, AB_hydrolase IPR002921, Fungal_lipase-like	PF01764, Lipase_3	GHSLG	61.7% Lipase_3 domain-containing protein (*C. variabilis*)	E1Z6D6
GHLX01001795.1	IPR029058, AB_hydrolase IPR002921, Fungal_lipase-like	PF01764, Lipase_3	GFSLG	60.8% Lipase_3 domain-containing protein (*C.a variabilis*)	E1Z814
GHLX01004575.1	IPR029058, AB_hydrolase IPR002921, Fungal_lipase-like	PF01764, Lipase_3	GHSLG	55.3% Lipase_3 domain-containing protein (*C. variabilis*)	E1Z559
GHLX01004232.1	IPR029058, AB_hydrolase IPR002921, Fungal_lipase-like	PF01764, Lipase_3	GHSLG	52.4% Alpha beta-hydrolase (*C. sorokiniana*)	A0A2P6TJS1
GHLX01005999	IPR029058, AB_hydrolase IPR002921, Fungal_lipase-like	PF01764, Lipase_3	GHSLG	62.2% *sn*1-specific diacylglycerol lipase alpha (*Auxenochlorella protothecoides*)	A0A087SMB1
GHLX01005800	IPR029058, AB_hydrolaseIPR002921, Fungal_lipase-like	PF01764, Lipase_3	GHSLG	44.5% *sn*1-specific diacylglycerol lipase alpha (*M. conductrix*)	A0A2P6V840

Table 2. Genes annotation of the putative TAG predicted lipases.

Transcriptome Shotgun Assembly ID	Genome Survey Sequences ID	Start	End	Gene Length	Strand	5′ UTR	3′ UTR	StartCodon	StopCodon	Exon Number
GHLX01005462.1	VATW01000002.1	1,249,709	1,253,360	3651	+	1,249,709	1,253,181	1,249,877	1,253,178	8
GHLX01003448.1	VATW01000019.1	480,480	486,471	5991	+	480,893	485,948	480,896	485,947	15
GHLX01004364.1	VATW01000012.1	444,856	447,981	3125	−	447,981	444,884	447,869	444,885	9
GHLX01003076.1	VATW01000017.1	300,534	306,599	6065	−	306,599	300,680	306,266	300,681	16
GHLX01002999.1	VATW01000004.1	234,154	243,389	9235	−	243,389	235,621	243,213	235,622	23
GHLX01001704.1	VATW01000077.1	53,966	57,537	3571	+	54,324	57,537	54,485	57,503	12
GHLX01004551.1	VATW01000014.1	391,368	400,369	9001	+	391,465	400,369	391,466	399,488	18
GHLX01003928.1	VATW01000021.1	364,412	371,783	7371	+	364,615	371,704	364,616	371,701	17
GHLX01006297.1	VATW01000014.1	387,248	391,356	4108	+	387,638	391,356	387,830	391,200	10
GHLX01001795.1	VATW01000003.1	1,009,232	1,012,963	3731	+	1,009,437	1,012,963	1,009,559	1,012,785	9
GHLX01004575.1	VATW01000004.1	243,414	249,169	5755	−	248,807	243,564	248,803	243,565	19
GHLX01004232.1	VATW01000004.1	387,912	396,833	8921	+	388,099	393,753	388,100	393,622	16
GHLX01005999.1	VATW01000002.1	467,247	474,411	7164	−	474,331	467,332	474,153	467,333	16
GHLX01005800.1	VATW01000021.1	56,136	59,601	3465	+	56,233	59,492	56,234	59,489	8

2.2. Physicochemical Characterization of Protein Sequences

ProtParam parameters shown in Table 3 reveal protein lengths varying from 421 to 1145 amino acids corresponding to diverse molecular masses (from 44.8 to 124.3 kDa). Various theoretical isoelectric points (Ip) were also found (4.09 to 9.34) and all proteins were predicted to have high molar extinction coefficients (46,300 to 193,210). Predicted repeats, motifs and localizations are given in Table 4. Among all predicted lipases, seven proteins have transmembrane motifs, including four predicted as being localized in plasma membrane and three in chloroplastic membrane. The seven other lipases have different cellular localizations (cytoplasmic, mitochondrian, chloroplastic or extracellular space), with five of them possessing a predicted signal peptide sequence. This enhances the possibility of extracellularity prediction however the signal peptides of chloroplasts and mitochondria are also N-terminal cleavable peptides [16]. They are less characterized than the secretory ones, but they are both rare in negatively charged amino acids and able to fold into amphiphilic α-helices [17].

The half-life is a prediction of the time it takes for half of the amount of protein in a cell to disappear after its synthesis in the cell; for all predicted lipases, it was found to be 30 h in mammalian (in vitro), more than 20 h in yeast, (in vivo) and more than 10 h in *Escherichia coli* (in vivo). ProtParam classifies also all studies proteins as stable (Instability index < 40).

Soluble predicted lipases have molecular weights between 44.8 and 102.5 kDa and Ip between 4.09 and 8.5. Concordant results were found by Ursu et al. [18]. The authors demonstrated, using the 2-DE profile of *C. vulgaris* soluble proteins, the presence of two protein groups that have been identified considering their isoelectrical points: a main group, having an Ip range of 4.0–5.5, and a minor group, with an Ip range of 6.0–8.0. However, the majority of separated proteins have apparent molecular weights range between 12 and 75 kDa. The difference observed herein could be explained by the fact that some proteins are not expressed under the culture conditions used by the authors.

Table 3. ProtParam parameters of the predicted lipases.

Protein Name	Length (Amino Acids)	Molecular Mass (Da)	Theoretical Ip	Total Number of Negatively Charged Residues (Asp + Glu)	Total Number of Positively Charged Residues (Arg + Lys)	Molar Extinction $(M^{-1}\ cm^{-1})$	Half-Life	Grand Average of Hydropathicity Index (GRAVY)
Lip_5462	469	50,526.70	6.94	35	34	58,830 58,330	30 h (mammalian reticulocytes, in vitro), >20 h (yeast, in vivo), >10 h (*Escherichia coli*, in vivo).	0.056
Lip_3448	810	87,834.98	4.50	125	55	113,955 113,330	30 h (mammalian reticulocytes, in vitro), >20 h (yeast, in vivo), >10 h (*Escherichia coli*, in vivo).	−0.223
Lip_4364	433	47,169.73	6.08	35	28	105,475 104,850	30 h (mammalian reticulocytes, in vitro), >20 h (yeast, in vivo), >10 h (*Escherichia coli*, in vivo).	0.083
Lip_3076	966	104,984.87	8.78	80	91	164,875 163,750	30 h (mammalian reticulocytes, in vitro), >20 h (yeast, in vivo), >10 h (*Escherichia coli*, in vivo).	0.086
Lip_2999	1104	121,199.69	8.67	98	110	193,210 191,710	30 h (mammalian reticulocytes, in vitro), >20 h (yeast, in vivo), >10 h (*Escherichia coli*, in vivo).	−0.125
Lip_1704	421	44,824.88	8.57	29	35	47,050 46,300	30 h (mammalian reticulocytes, in vitro), >20 h (yeast, in vivo), >10 h (*Escherichia coli*, in vivo).	0.057
Lip_4551	1145	124,302.95	7.43	103	103	157,425 156,300	30 h (mammalian reticulocytes, in vitro), >20 h (yeast, in vivo), >10 h (*Escherichia coli*, in vivo).	−0.057
Lip_3928	934	100,739.17	8.92	80	93	132,905 131,780	30 h (mammalian reticulocytes, in vitro), >20 h (yeast, in vivo), >10 h (*Escherichia coli*, in vivo).	−0.013
Lip_6297	530	56,818.32	6.12	51	47	69,940 69,440	30 h (mammalian reticulocytes, in vitro), >20 h (yeast, in vivo), >10 h (*Escherichia coli*, in vivo).	0.101
Lip_1795	557	59,817.51	4.09	67	23	97,330 96,830	30 h (mammalian reticulocytes, in vitro), >20 h (yeast, in vivo), >10 h (*Escherichia coli*, in vivo).	0.110
Lip_4575	726	81,163.18	9.26	67	28	162,885 162,260	30 h (mammalian reticulocytes, in vitro), >20 h (yeast, in vivo), >10 h (*Escherichia coli*, in vivo).	−0.117
Lip_4232	779	85,389.08	9.34	62	83	107,675 106,800	30 h (mammalian reticulocytes, in vitro), >20 h (yeast, in vivo), >10 h (*Escherichia coli*, in vivo).	0.082
Lip_5999	1003	102,522.28	5.31	112	86	85,425 84,800	30 h (mammalian reticulocytes, in vitro), >20 h (yeast, in vivo), >10 h (*Escherichia coli*, in vivo).	−0.143
Lip_5800	629	67,075.99	4.98	98	66	58,160 57,410	30 h (mammalian reticulocytes, in vitro), >20 h (yeast, in vivo), >10 h (*Escherichia coli*, in vivo).	−0.307

Table 4. Predicted repeats, motifs and localization of putative lipases.

Protein Name	Predicted Localization	Signal Peptide Sequence	Membrane Helix	N Glycosylation Sites
Lip_5462	Ps (extracellular space)	MNVGRVAALFACLLQGACLALAVQ	-	325
Lip_3448	Mito	-	-	170
Lip_4364	Ps (extracellular space)	MRPAITEALLAVLVCLVVGANGA	-	134/180/273/280
Lip_3076	Chloro (membrane)	-	42–64/84–106/129–151/161–183/266–288/314–336/348–370	8/676
Lip_2999	Chloro (membrane)	-	120–142/162–184/203–225/245–267/293–315/341–363/400–422	187/349/383/1030
Lip_1704	Chloro	MKLGLPLLLAALLLAAAAPATAR	-	230/260/305/369
Lip_4551	Chloro (membrane)	-	139–161/176–198/222–244/254–276/313–330/362–384/411–433/453–475/482–504	279/946
Lip_3928	plasma membrane	-	62–84/174–196/220–242/254–276	-
Lip_6297	Ps (extracellular space)	MFIRVQSRVVSAVFTAIIFSLLFMSLVPTLQGN		392
Lip_1795	cyto	-	-	19/53/307
Lip_4575	plasma membrane	MYIANTSVGGVLTLASFAMLAHGL	6–28/48–70/80–102/115–137/170–189/196–218	5
Lip_4232	plasma membrane	-	31–53/66–88/108–130/145–167/202–224/251–273/300–322	475
Lip_5999	chloro	-	-	-
Lip_5800	chloro	-	-	487

2.3. D-Structural Modeling

The programs for 3D structural modeling automatically selected template structures mostly from fungal lipases as shown in Table S1 (PDBs: 6A0w, 6qpr, 4jei, 3o0d, 6unv, 4tgl, 3tgl). All models presented the typical α/β-hydrolase fold, with mostly parallel β-sheets, flanked on both sides by α-helixes. The highly conserved catalytic triad (serine, aspartic/glutamic acid and histidine) and the oxyanionic hole were well orientated in the space. The α/β hydrolase fold is one of the most thriving architectures in proteins across kingdoms, providing the skeleton for diverse enzymes [19] as well as an emerging class of non-catalytic but functionally important receptors [20]. Some of the predicted structures were very similar with the typical lipase motifs and are formed by one domain, but some other possesses an extra-transmembrane domain which could be quite bulky (Lip_4551 displays 9 helices against 4 for Lip_3928). Few membrane-bound lipases over intracellular or extracellular counterparts were studied. Recently the catalytic behavior of a membrane-associated lipolytic enzyme (MBL-Enzyme) from the microalgae *Nannochloropsis oceanica* was investigated by Savvidou et al. [21].

3. Discussion

TAG lipases responsible for the degradation of the lipids accumulated in oil bodies are attractive knockdown targets for the enhancement of the lipid productivity and storage in microalgae. Nonetheless, considering the numerous data available on bacterial, terrestrial plant and animal lipases those from algae and more especially microalgae have been relatively neglected. Therefore, more emphasize has to be given to the characterization of algal lipases, and hence, further work is needed in these aspects. Future approaches to maximize the enzymatic potential of microalgae are likely to focus on three different strategies: (i) the use of ever-increasing amounts of available omics data to optimize microalgal strains for the production of valuable products, through the overexpression of one or more enzymes by the use of genome editing tools; (ii) the identification and subsequent characterization of metabolic pathways involving the production of specific enzymes, such as lipases which are still poorly characterized; (iii) the search for genes with direct biotechnological applications in microalgal genomes and transcriptomes datasets. The feasibility of employing any of the aforementioned approaches or a combination of them will be directly influenced by progress in growth and genetic manipulation of microalgae.

In this study, we have used computational approach to identify lipase genes and to classify the respective lipases from a *C. vulgaris* strain. Lipases operate usually at the interface between lipid and water. An important feature of many lipases that is used for lipase classification is the presence of a mobile subdomain lid or flap located over the active site [22]. Among the 14 putative TAG lipases identified after *C. vulgaris* genome analysis, 10 have high identity in ESTHER database with Lipase_3 domain-containing protein. Family 3 of lipolytic enzymes are widely distributed in animals, plants and prokaryotes and possess the conserved consensus sequence GXSXG. Members of this family were demonstrated to be very closely related and exhibit the canonical α/β-hydrolase fold as well as the typical catalytic triad. Enzymes of this class exhibit also high activities at low temperature (less than 15 °C) believed to originate from a conserved sequence motifs they display [23]. Four lipases out of the 10 aforementioned were predicted to be either cytoplasmic, chloroplastic or extracellular. The six remaining could be anchored to a membrane with a distinct N-terminal transmembrane domain formed by at least four transmembrane helices (Figures 1 and 2). Lip_4551 and Lip_2999 were predicted with quite similar 3D models composed of three domains: a catalytic domain containing the catalytic triad and a one helix lid, an N terminal transmembrane domain formed by long helices and a C terminal domain with mainly α helices (Figure 2). It has been reported that 10 additional modules can be attached to the core domain including lid modules, cap modules, N-terminal or C-terminal domains. Accordingly, superfamilies could be assigned to five groups (core, lid, cap, one additional domain or two additional domains) [24]. Predicted transmembrane domains by bioinformatic tools were already reported for microalga lipases [25]. Some authors characterized and used as a self-immobilized lipase for esterification reactions membrane bound lipase from microalga [21]. The membrane localization could be in intracellular or extracellular counterparts or even in lipid droplets (LD). In eukaryotes, some TAG lipases and their cofactors have been demonstrated to localize to LDs [26]. For example, Diatom Oleosome-Associated Protein 1 (DOAP1) is translocated from the ER to LDs in *Fistulifera solaris* [27].

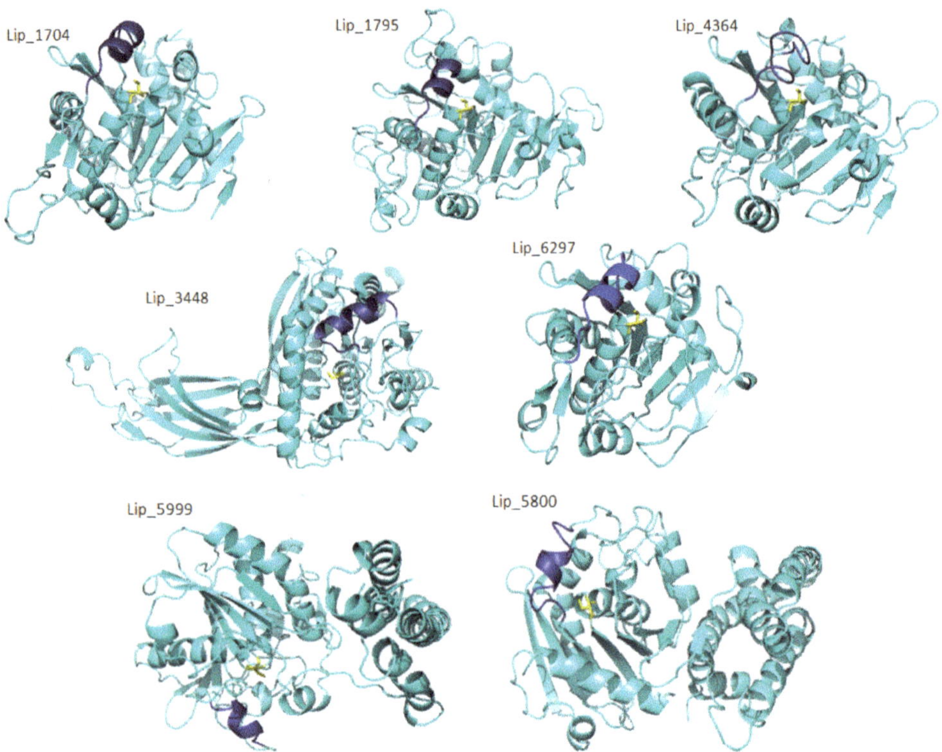

Figure 1. 3D models of seven putative lipases without transmembrane domains. Four of them (Lip_1704, Lip_1795, Lip_4364, Lip_6297) display only the core module with a Rossman Fold architecture. Lip_3448 presents a C2 N-terminal domain, while Lip_5800 and Lip_5999 present a C-terminal alpha helices module. Lids are shown in dark blue and active site serine in yellow sticks.

As for Lip_3448, the N terminal module is a PLAT domain found in a variety of lipid-associated proteins. It forms a β-sandwich composed of two β-sheets of four β-strands each, which is known as a C2 domain in pfam classification. Interestingly, two predicted lipases have a C terminal module only composed of α helices. These two proteins (Lip_5800 and Lip_5999) are shown to be closely related in cladogram of sequence similarity. Hence, the predicted lipases could be classified into a main core with Rossman fold architecture lipases (Lip_1704, Lip_1795, Lip_4364, Lip_6297, Lip_5462, Lip_3928, Lip_4575), two domain lipases (Lip_4232, Lip_3448, Lip_3076, Lip_5800, Lip_5999) and three domain lipases including a transmembrane domain (Lip_2999 and Lip_4551).

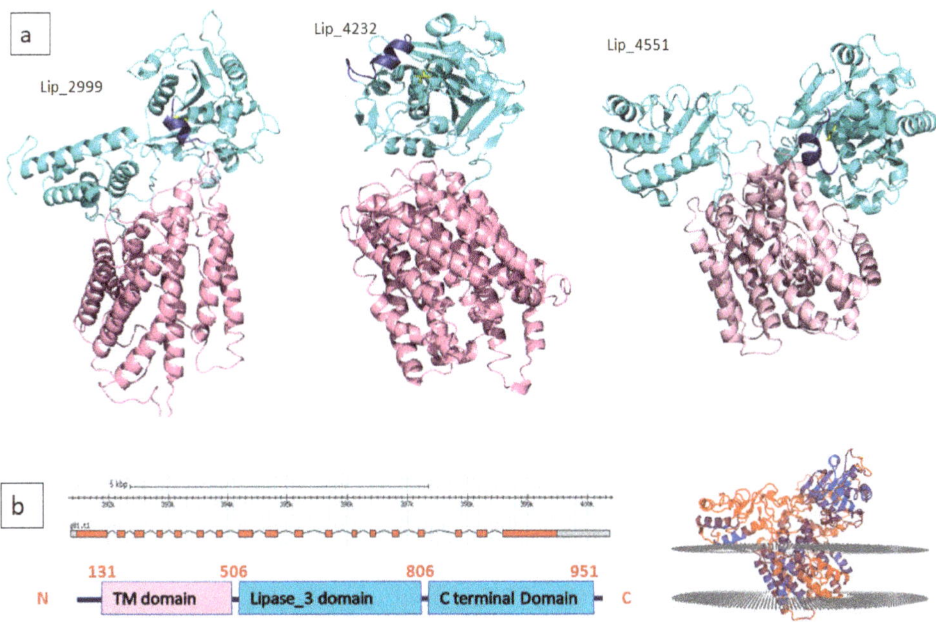

Figure 2. (**a**) Three predicted membrane-associated lipases with a transmembrane module shown in lilac; lids are shown in dark blue and active site serine in yellow sticks. (**b**) Gene annotation and domain boundaries of Lip_4551 (left panel), Qmeanbrane result for transmembrane localization for Lip_4551 (right panel).

Oxyanion holes are crucial for high-energy oxyanion intermediate stabilization. They consist of two residues, which donate their backbone amide protons to stabilize the substrate in the transition state. In fact, during hydrolysis, a negatively charged tetrahedral intermediate is generated and the oxygen ion formed modulates the catalytic efficiency of the enzyme [28]. The first residue is located in the structurally conserved nucleophilic elbow. As a consequence, its backbone amide is positioned identically in all lipases. In contrast, the second oxyanion hole residue is not located in a region with conserved sequence and structure between lipases, but in a loop between the β3-strand and the αA-helix in the core module [29,30]. Consequently, lipases are classified into three classes according to their oxyanion hole type: GX, GGGX and Y [31]. In all lipases, the first oxyanion hole is a conserved glycine which contacts the nucleophilic elbow (highlighted with a star in Figure 3). When the oxyanion hole is formed by the amide backbone of the C-terminal neighbor X of this conserved glycine, it is termed as 'GX type,' with X being the second oxyanion hole residue. In our case, the inspection of the multiple sequence alignment of the 14 lipases demonstrates they belong all to the GX class with the conserved glycine (G) residue followed by an alanine (A), cysteine (C) or serine (S) residue (Figure 3). The lipases with GX oxyanion hole type are widely distributed and diverse, and they usually prefer hydrolyzing medium and long chain substrates [32]. The type of amino acid X is conserved inside the superfamilies; for example, it is hydrophilic in *Candida antarctica* like lipases (T), filamentous fungi lipases and cutinases (S, T), and hydrophobic in *Moraxella* (F), *Mycoplasma* (F, W) and *Pseudomonas* lipases (L, F, M) [29].

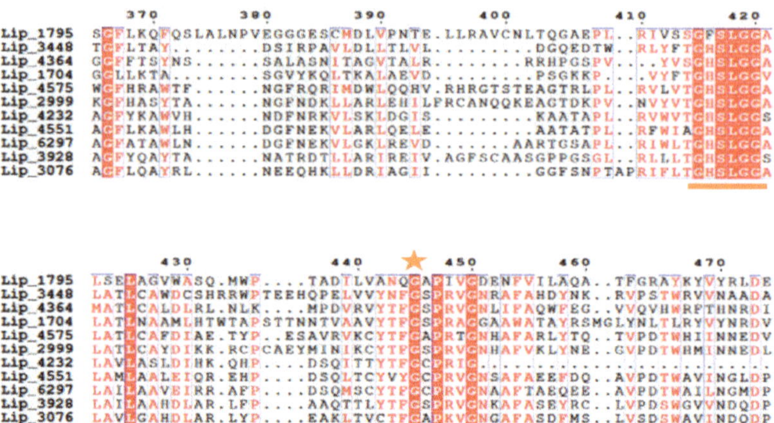

Figure 3. Multiple sequence alignment of putative lipases showing the conserved lipase 3 motif GXSXG and the conserved G residue for GX classification highlighted with orange star.

According to the shape of the binding site cavity, lipases can be divided into three categories: (i) lipases with a funnel-like binding site (lipases from the mammalian pancreas and cutinase), (ii) lipases with tunnel-like binding sites (lipases from *Candida rugosa*, and *Candida antarctica* A) [33] and (iii) lipases with a crevice-like binding site (lipases from *Rhizomucor* sp. and *Rhizopus* sp.) [34]. It should be noted that most of the template structures used for 3D modeling are lipases from *Rhizomucor miehei*. In addition, the inspection of predicted open lid models like Lip_1704 showed a crevice-like cavity shape as shown in Figure 4.

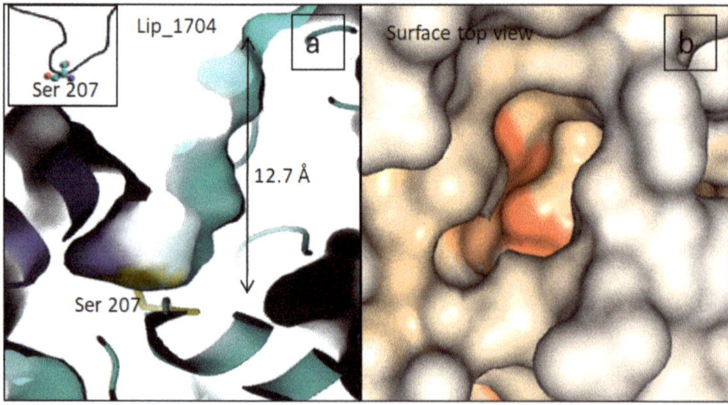

Figure 4. (**a**) Slabbed close up view of the active site cavity for Lip_1704 showing a crevice-like shape (**b**) A surface top view with DEPTH showing the shape of substrate entrance in the same protein.

The amphipathic nature of the lid is crucial for the substrate specificity. It provides new insight into the structural basis of lipase substrate specificity and a way to tune the substrate preference of lipases. Based on the type of lid domain, lipases were also classified into three groups, such as lipases without lids, lipases with one loop or one helix lids and lipases with two or more helix lids. It has been reported that high temperature lipases contain larger lid domains with two or more helices, and that all mono- and diacylglycerol lipases have a small lid with a form of loop or helix [22]. As shown in Figures 1 and 2, almost all lipases found in *C. vulgaris* have small lids with one loop (Lip_4364) or one helix (Lip_1704). However, Lip_5462 displays an entire cap domain with three small helices lid covering a deep cavity of 15.6 Å and shows, surprisingly, 40% of sequence identity with human lysosomal acid lipase (LAL). In fact, it has been demonstrated that, in addition to the direct association of lipases to oil bodies, macro-autophagy (referred to as lipophagy) plays a critical roles in lipid catabolism in eukaryotes [35]. During this type of autophagy, autophagosomes containing a portion of an oil body are merged with lysosomes containing LAL, which could contribute to TAG degradation [36]. Transcriptomic analysis of *Neochloris oleoabundans* (an oleaginous microalga) reveals up regulation of an LAL encoding gene under nitrogen starvation condition [37]. Accordingly, the in silico prediction method used for lipases of *C. vulgaris* allowed the identification of Lip_5264, which could be transported to lysosomes. This enzyme was predicted to have a signal peptide and 40% of sequence identity with the human LAL. It consists of a core domain belonging to the classical α/β hydrolase-fold family with a classical catalytic triad (Ser-161, His-378, Asp-347), an oxyanion hole and a "cap" domain, which probably regulates substrate entry to the catalytic site (Figure 5). LAL breaks down cholesteryl esters (CEs) and TGs into free cholesterol, glycerol and fatty acids (1–3). Defective LAL have been associated with two autosomal recessive diseases in humans: Wolman's disease and CE storage disease [38,39]. The gene of Lip_5264 consists of 8 exons spread over almost 4 kb, while human LAL consists of 10 exons spread over 36 kb. Lip_5264 encodes a 445 amino acid mature protein following the cleavage of 24 signaling peptide residues, with an expected molecular mass of 50 kDa whereas human LAL encodes for 378 residues with a signal peptide of 21 amino acids and a molecular mass of 43 kDa. The two compared proteins are glycosylated and share high structure identity, as shown in Figure 5c with some differences, including the lid helices, which contain a cluster of highly conserved Cys residues C 236 and C 243 (Lip_5264 numbering) (Figure 5d). The lysosomal proteins in microalga have not yet been fully investigated, and it remains unclear how lipophagy contributes to lipid degradation. These should be an attractive research topic in a future work. Microalgae are a good source of nutrients for human nutrition. However, they are also rich in various biomolecules, which may have a potential in promoting human health. Defective or diminished LAL activity of human LAL has been associated with some mutations and the molecular mechanisms of these loss-of-function mutants leading to WD and CESD have yet to be explored. Some study demonstrated that these mutations could be located in the signal peptide or in the lid domain [40]. A complete physicochemical characterization of this *C. vulgaris* LAL combined with a deep structure–function relationship investigation of the probable mutation effect using a structure-based molecular model speculating the loss of function could be of interest. The current treatment options for CESD phenotypes are limited to diets excluding cholesterol and lipid-rich food, cholesterol lowering drugs such as statins and ultimately liver transplantation. Recombinant LAL replacement therapy has been shown to be effective in animal models and human clinical trials and was recently authorized in Europe and the United States [41].

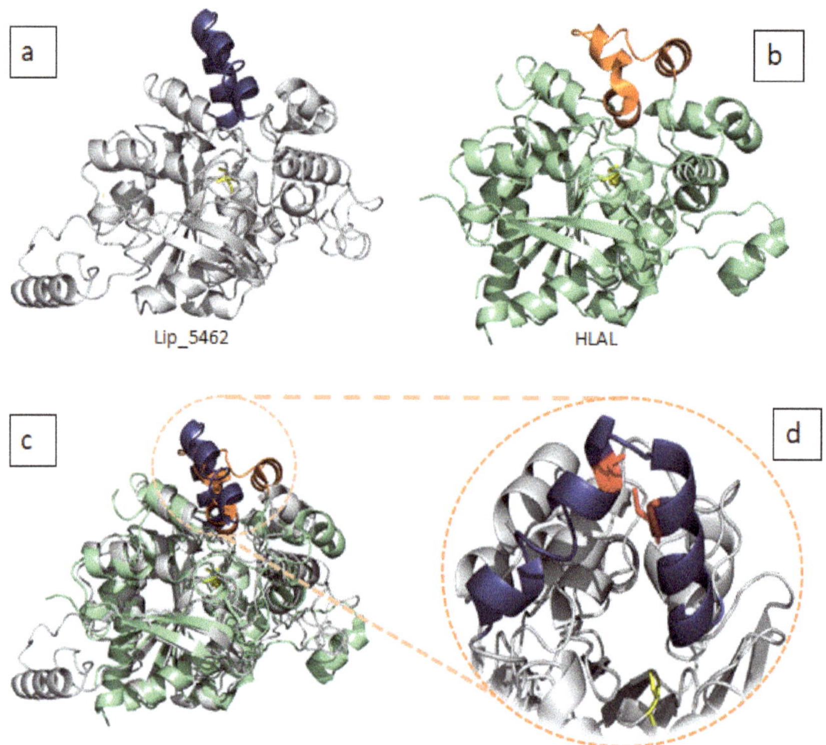

Figure 5. (**a**) 3D model of putative Lysosomal acid lipase Lip_5462 lid is shown in dark blue and catalytic serine in yellow sticks. (**b**) Crystal structure of human Lysosomal acid lipase PDB ID: 6V7N. Lid is shown in orange and catalytic serine in yellow sticks. (**c**) The overlay of the two aforementioned structures showing high structure similarities and lid differences. (**d**) A close up view of the three helices lid of Lip_5462 showing conserved cysteine residues in red sticks.

4. Materials and Methods

4.1. Sequence Retrieval

BlastP search was performed using amino acid sequences of functionally characterized lipases from terrestrial plants (*Trifolium pretense* and *Diplocarpon rosae*), fungi (*Colletotrichum chlorophyti*), microalga (*Scenedesmus* sp. and *Symbiodinium microadriaticum*) and bacteria (*Pseudomonas fluorescens* and *B. subtilis*) available in the NCBI database (http://ncbi.nlm.gov/protein/). The FASTA sequences were searched using tblastn modality against Transcriptome Shotgun Assembly database (TSA) of *C. vulgaris* strain UTEX259 UTEX259 (taxid 3077) and every hit with an E-value $< 10^{-5}$ was identified as putative Lipase transcript. The open reading frames (ORFs) were searched using the ORF finder program [42] and the longer ones were blasted a second time against non-redundant protein database to ensure that the respective TSA corresponds to a putative Lipase ORF. The selected TSA sequences were submitted to a blastn search against the whole Genome Shotgun contigs (WGS) database of the same *C. vulgaris* strain (taxid 3077) and single hits with E-value $< 10^{-100}$ were identified as scaffolds with putative Lipase genes. Gene predictions from the selected WGS scaffolds were performed using ab initio gene models through Augustus [43]. The application was trained on the gene structures of *Chlamydomonas reinhardtii* and the TSA sequences were used in cDNA uploaded option. The final output ORF and protein sequences were saved for further in silico analysis.

4.2. Multiple Sequence Alignment

The multiple sequence alignment and calculation of cladogram illustrating sequence similarity relationships among the 14 putative lipase sequences was executed by MAFFT (v7.310) with G-INS-1 strategy, unalign level 0.8, leave gappy region options for alignment and UPGMA as average linkage method for clustering. Rendering was done using ESPript [44].

4.3. Physicochemical Characterization of Protein Sequences

Basic physicochemical properties such as molecular weight, extinction coefficient, isoelectric point, aliphatic index, grand average of hydropathicity and instability index were estimated by ProtParam tool (http://web.expasy.org/protparam/) [35]. Extinction coefficients were calculated assuming all pairs of Cys residues form cystines or assuming all Cys residues are reduced. Sequence analysis and lipase motifs search were performed with InterPro [45] and the Expasy my hits search tool (https://myhits.isb-sib.ch/cgi-bin/motif_scan), respectively. These sequences were also compared in the ESTHER database to check higher sequence identity [19]. For predicting subcellular localization Deepmito [46], Mitoprot v1.101 [47], HECTAR v1.3 [48] and TMHMM v2.0 [49] were performed. Putative signal peptides in each sequence were predicted using the SignalP 4.0 server [50]. Since N-glycosylation was widly described for lipases prediction of N-glycosylation sites were performed using NetOGlyc 4.0 Server [51].

4.4. Tertiary Structure Prediction, Structure Validation and Quality Prediction

Three-dimensional models of the selected putative enzymes were generated using different approaches. For sequences with acceptable homology in the template of the programs, UCSF Chimera (https://www.rbvi.ucsf.edu/chimera/) and the automated protein homology modeling server SWISS-MODEL (http://swissmodel.expasy.org/) were used. For sequences with low homology with the structures in the database, multiple-threading alignments using the I-TASSER approach (zhanglab.ccmb.med.umich.edu/I-TASSER/) was used. I-TASSER is an automated bioinformatics tool for predicting protein structures from an amino acid sequence followed by iterative structural assembly simulations and atomic-level structure refinement.

The predicted structures were evaluated to ensure correctness of the model stereochemistry, as checked by a Ramachandran plot (http://mordred.bioc.cam.ac.uk/~rapper/rampage.php) (Lovell et al., 2003) and Verify 3D [52]. The Ramachandran plot scores of the predicted structures showed more than 90% of the amino acids were in favorable regions. ProSA-web Z-score plot (https://prosa.services.came.sbg.ac.at/prosa.php) [53] was used to check whether the Z-score of the input structures is within the range of typically found for the native proteins of a similar size. The Z-score values of all protein structures checked in this study were highlighted as a black dot, which indicates being in the range of native conformations. The final modeled structures were further energetically minimized and molecular dynamics simulation was performed with CABS-flex 2.0 (http://212.87.3.12/CABSflex2). The latter program is an efficient simulation engine that allows modeling of the large-scale conformational change related to protein flexibility [54]. The models were comprehensively analyzed using PyMol (http://pymol.org/) to check for the presence of a lid, and the existence and orientation of the catalytic triad. The depth of the putative intramolecular tunnels was calculated with DEPTH [55] taking residues from the oxyanion hole in each candidate as the cavity end point.

5. Conclusions

Genomic mining by combining bioinformatics analysis and functional screening provides opportunities to find out novel biocatalysts, such as lipases. The present study allowed the in silico characterization of 14 putative *C. vulgaris* lipases with different cellular localization. Membrane associated lipases were also detected and described for the first time in this species. The 14 lipases display an acyl hydrolase motif (GXSXG) and belong to the α/β hydrolase lipase 3 family and GX class. These putative lipases could be candidates for metabolic engineering in a future study to improve this microalga lipid productivity. In this study, we also report, for the first time, a putative lysosomal acid lipase produced by a green microalga. Further investigation on the generated 3D models, such as docking studies and MD simulations, will provide important information on the substrate catalytic process and the binding characteristics and could be of interest to understand molecular mechanisms of the loss-of-function mutants leading to WD and CESD in humans.

Supplementary Materials: The following are available online at https://www.mdpi.com/1660-3397/19/2/70/s1, Table S1 Templates used for 3D model generation.

Author Contributions: Conceptualization, H.B.H. and A.K.; methodology, H.B.H. and M.D. (Mouna Dammak); software, H.B.H.; validation, S.A., I.F. and P.M.; formal analysis, H.B.H., A.K., M.D. (Maroua Drira) and M.D. (Mouna Dammak); investigation, H.B.H., M.D. (Mouna Dammak), A.K. and M.D. (Maroua Drira); resources, S.A.; data curation, M.D. (Maroua Drira) and M.D. (Mouna Dammak); writing—original draft preparation, H.B.H.; writing—review and editing, H.B.H., S.A., M.D. (Mouna Dammak) and A.K.; visualization, S.A., P.M. and I.F.; supervision, S.A.; project administration, H.B.H. All authors have read and agreed to the published version of the manuscript.

Funding: This research received no external funding.

Institutional Review Board Statement: Not applicable.

Acknowledgments: The authors are grateful to the Tunisian Ministry of Higher Education and Scientific Research for financial assistance.

Conflicts of Interest: The authors declare no conflict of interest.

References

1. Market, M.N. Enzymes Market. 2020. Available online: https://www.marketsandmarkets.com.Market-Reports/enzymes-market-46202020.html (accessed on 22 June 2020).
2. Prakash, D.; Nawani, N.; Prakash, M.; Bodas, M.; Mandal, A.; Khetmalas, M.; Kapadnis, B. Actinomycetes: A Repertory of Green Catalysts with a Potential Revenue Resource. *BioMed Res. Int.* **2013**, *2013*, 1–8. [CrossRef] [PubMed]
3. Hasan, F.; Shah, A.A.; Hameed, A. Industrial applications of microbial lipases. *Enzym. Microb. Technol.* **2006**, *39*, 235–251. [CrossRef]
4. Andualema, B.; Gessesse, A. Microbial Lipases and Their Industrial Applications: Review. *Biotechnology* **2012**, *11*, 100–118. [CrossRef]
5. Fendri, I.; Chaari, A.; Dhouib, A.; Jlassi, B.; Abousalham, A.; Carriere, F.; Sayadi, S.; Abdelkafi, S. Isolation, identification and characterization of a new lipolytic *Pseudomonas* sp., strain AHD-1, from Tunisian soil. *Environ. Technol.* **2010**, *31*, 87–95. [CrossRef] [PubMed]
6. Barouh, N.; Abdelkafi, S.; Fouquet, B.; Pina, M.; Scheirlinckx, F.; Carrière, F.; Villeneuve, P. Neutral Lipid Characterization of Non-Water-Soluble Fractions of Carica Papaya Latex. *J. Am. Oil Chem. Soc.* **2010**, *87*, 987–995. [CrossRef]
7. Khannous, L.; Mouna, J.; Dammak, M.; Miladi, R.; Chaaben, N.; Khemakhem, B.; Gharsallah, N.; Fendri, I. Isolation of a novel amylase and lipase-producing Pseudomonas luteola strain: Study of amylase production conditions. *Lipids Health Dis.* **2014**, *13*, 9. [CrossRef]
8. Navvabi, A.; Razzaghi, M.; Fernandes, P.; Karami, L.; Homaei, A. Novel lipases discovery specifically from marine organisms for industrial production and practical applications. *Process. Biochem.* **2018**, *70*, 61–70. [CrossRef]
9. Demir, B.S.; Tükel, S.S. Purification and characterization of lipase from Spirulina platensis. *J. Mol. Catal. B Enzym.* **2010**, *64*, 123–128. [CrossRef]
10. Godet, S.; Hérault, J.; Pencreac'H, G.; Ergan, F.; Loiseau, C. Isolation and analysis of a gene from the marine microalga Isochrysis galbana that encodes a lipase-like protein. *Environ. Boil. Fishes* **2012**, *24*, 1547–1553. [CrossRef]
11. Ben Hlima, H.; Dammak, M.; Karkouch, N.; Hentati, F.; Laroche, C.; Michaud, P.; Fendri, I.; Abdelkafi, S. Optimal cultivation towards enhanced biomass and floridean starch production by Porphyridium marinum. *Int. J. Biol. Macromol.* **2019**, *129*, 152–161. [CrossRef]
12. Talebi, A.F.; Tohidfar, M.; Tabatabaei, M.; Bagheri, A.; Mohsenpor, M.; Mohtashami, S.K. Genetic manipulation, a feasible tool to enhance unique characteristic of Chlorella vulgaris as a feedstock for biodiesel production. *Mol. Biol. Rep.* **2013**, *40*, 4421–4428. [CrossRef] [PubMed]
13. Guarnieri, M.T.; Levering, J.; Henard, C.A.; Boore, J.L.; Betenbaugh, M.J.; Zengler, K.; Knoshaug, E.P. Genome Sequence of the Oleaginous Green Alga, Chlorella vulgaris UTEX 395. *Front. Bioeng. Biotechnol.* **2018**, *6*, 37. [CrossRef] [PubMed]

14. Cecchin, M.; Marcolungo, L.; Rossato, M.; Girolomoni, L.; Cosentino, E.; Cuine, S.; Li-Beisson, Y.; Delledonne, M.; Ballottari, M. *Chlorella vulgaris* genome assembly and annotation reveals the molecular basis for metabolic acclimation to high light conditions. *Plant J.* **2019**, *100*, 1289–1305. [CrossRef] [PubMed]

15. Lodish, H.; Berk, A.; Kaiser, C.; Krieger, M.; Bretscher, A.; Ploegh, H.; Amon, A.; Scott, M. Genes, genomics, and chromosomes. In *Molecular Cell Biology*, 7th ed.; W.H. Freeman Company: New York, NY, USA, 2013; pp. 227–230.

16. Schatz, G.; Dobberstein, B. Common Principles of Protein Translocation across Membranes. *Science* **1996**, *271*, 1519–1526. [CrossRef]

17. Bruce, B.D. Chloroplast transit peptides: Structure, function and evolution. *Trends Cell Biol.* **2000**, *10*, 440–447. [CrossRef]

18. Ursu, A.-V.; Marcati, A.; Sayd, T.; Sante-Lhoutellier, V.; Djelveh, G.; Michaud, P. Extraction, fractionation and functional properties of proteins from the microalgae Chlorella vulgaris. *Bioresour. Technol.* **2014**, *157*, 134–139. [CrossRef]

19. Lenfant, N.; Hotelier, T.; Velluet, E.; Bourne, Y.; Marchot, P.; Chatonnet, A. ESTHER, the database of the α/β-hydrolase fold superfamily of proteins: Tools to explore diversity of functions. *Nucleic Acids Res.* **2012**, *41*, D423–D429. [CrossRef]

20. Janssen, B.J.; Snowden, K.C. Strigolactone and karrikin signal perception: Receptors, enzymes, or both? *Front. Plant Sci.* **2012**, *3*, 296. [CrossRef]

21. Savvidou, M.G.; Katsabea, A.; Kotidis, P.; Mamma, D.; Lymperopoulou, T.V.; Kekos, D.; Kolisis, F.N. Studies on the catalytic behavior of a membrane-bound lipolytic enzyme from the microalgae Nannochloropsis oceanica CCMP1779. *Enzym. Microb. Technol.* **2018**, *116*, 64–71. [CrossRef]

22. Khan, F.I.; Lan, D.; Durrani, R.; Huan, W.; Zhao, Z.; Wang, Y. The Lid Domain in Lipases: Structural and Functional Determinant of Enzymatic Properties. *Front. Bioeng. Biotechnol.* **2017**, *5*, 16. [CrossRef]

23. Ramnath, L.; Sithole, B.; Govinden, R. Classification of lipolytic enzymes and their biotechnological applications in the pulping industry. *Can. J. Microbiol.* **2017**, *63*, 179–192. [CrossRef] [PubMed]

24. Bauer, T.L.; Buchholz, P.C.; Pleiss, J. The modular structure of α/β-hydrolases. *FEBS J.* **2020**, *287*, 1035–1053. [CrossRef] [PubMed]

25. Nomaguchi, T.; Maeda, Y.; Liang, Y.; Yoshino, T.; Asahi, T.; Tanaka, T. Comprehensive analysis of triacylglycerol lipases in the oleaginous diatom Fistulifera solaris JPCC DA0580 with transcriptomics under lipid degradation. *J. Biosci. Bioeng.* **2018**, *126*, 258–265. [CrossRef] [PubMed]

26. Zimmermann, R.; Strauss, J.G.; Haemmerle, G.; Schoiswohl, G.; Birner-Gruenberger, R.; Riederer, M.; Lass, A.; Neuberger, G.; Eisenhaber, F.; Hermetter, A.; et al. Fat Mobilization in Adipose Tissue Is Promoted by Adipose Triglyceride Lipase. *Science* **2004**, *306*, 1383–1386. [CrossRef]

27. Leyland, B.; Boussiba, S.; Khozin-Goldberg, I. A Review of Diatom Lipid Droplets. *Biology* **2020**, *9*, 38. [CrossRef]

28. Gupta, R.; Kumari, A.; Syal, P.; Singh, Y. Molecular and functional diversity of yeast and fungal lipases: Their role in biotechnology and cellular physiology. *Prog. Lipid Res.* **2015**, *57*, 40–54. [CrossRef]

29. Pleiss, J.; Fischer, M.; Peiker, M.; Thiele, C.; Schmid, R.D. Lipase engineering database: Understanding and exploiting sequence-structure-function relationships. *J. Mol. Cat. B Enzym.* **2000**, *10*, 491–508. [CrossRef]

30. Abdelkafi, S.; Ogata, H.; Barouh, N.; Fouquet, B.; Lebrun, R.; Pina, M.; Scheirlinckx, F.; Villeneuve, P.; Carrière, F. Identification and biochemical characterization of a GDSL-motif carboxylester hydrolase from Carica papaya latex. *Biochim. Biophys. Acta-(BBA) Mol. Cell Biol. Lipids* **2009**, *1791*, 1048–1056. [CrossRef]

31. Martínez-Corona, R.; Marrufo, G.V.; Penagos, C.C.; Madrigal-Perez, L.A.; González-Hernández, J.C. Bioinformatic characterization of the extracellular lipases from *Kluyveromyces marxianus*. *Yeast* **2020**, *37*, 149–162. [CrossRef]

32. Hitch, T.C.A.; Clavel, T. A proposed update for the classification and description of bacterial lipolytic enzymes. *PeerJ* **2019**, *7*, e7249. [CrossRef]

33. Barriuso, J.; Vaquero, M.E.; Prieto, A.; Martínez, M.J. Structural traits and catalytic versatility of the lipases from the Candida rugosa-like family: A review. *Biotechnol. Adv.* **2016**, *34*, 874–885. [CrossRef] [PubMed]

34. Pleiss, J.; Fischer, M.; Schmid, R.D. Anatomy of lipase binding sites: The scissile fatty acid binding site. *Chem. Phys. Lipids* **1998**, *93*, 67–80. [CrossRef]

35. Settembre, C.; Ballabio, A. Lysosome: Regulator of lipid degradation pathways. *Trends Cell Biol.* **2014**, *24*, 743–750. [CrossRef] [PubMed]

36. Lübke, T.; Lobel, P.; Sleat, D.E. Proteomics of the lysosome. *Biochim. Biophys. Acta (BBA)-Mol. Cell Res.* **2009**, *1793*, 625–635. [CrossRef] [PubMed]

37. Rismani-Yazdi, H.; Haznedaroglu, B.Z.; Hsin, C.; Peccia, J. Transcriptomic analysis of the oleaginous microalga Neochloris oleoabundans reveals metabolic insights into triacylglyceride accumulation. *Biotechnol. Biofuels* **2012**, *5*, 74. [CrossRef]

38. Anderson, R.A.; Bryson, G.M.; Parks, J.S. Lysosomal acid lipase mutations that determine phenotype in Wolman and choles-terol ester storage disease. *Mol. Gene. Metab.* **1999**, *68*, 345–346.

39. Ries, S.; Aslanidis, C.; Fehringer, P.; Carel, J.C.; Gendrel, D.; Schmitz, G. A new mutation in the gene for lysosomal acid li-pase leads to Wolman disease in an African kindred. *J. Lipid Res.* **1996**, *37*, 1761–1765. [CrossRef]

40. Frampton, J.E. Sebelipase Alfa: A Review in Lysosomal Acid Lipase Deficiency. *Am. J. Cardiovasc. Drugs* **2016**, *16*, 461–468. [CrossRef]

41. Gomaraschi, M.; Bonacina, F.; Norata, G.D. Lysosomal Acid Lipase: From Cellular Lipid Handler to Immunometabolic Target. *Trends Pharmacol. Sci.* **2019**, *40*, 104–115. [CrossRef]

42. Rombel, I.T.; Sykes, K.F.; Rayner, S.; Johnston, S.A. ORF-FINDER: A vector for high-throughput gene identification. *Gene* **2002**, *282*, 33–41. [CrossRef]
43. Hoff, K.J.; Stanke, M. WebAUGUSTUS—A web service for training AUGUSTUS and predicting genes in eukaryotes. *Nucleic Acids Res.* **2013**, *41*, W123–W128. [CrossRef] [PubMed]
44. Gouet, P.; Robert, X.; Courcelle, E. ESPript/ENDscript: Extracting and rendering sequence and 3D information from atomic structures of proteins. *Nucleic Acids Res.* **2003**, *31*, 3320–3323. [CrossRef] [PubMed]
45. Finn, R.D.; Attwood, T.K.; Babbitt, P.C.; Bateman, A.; Bork, P.; Bridge, A.J.; Chang, H.-Y.; Dosztányi, Z.; El-Gebali, S.; Fraser, M.; et al. InterPro in 2017—Beyond protein family and domain annotations. *Nucleic Acids Res.* **2017**, *45*, D190–D199. [CrossRef] [PubMed]
46. Savojardo, C.; Bruciaferri, N.; Tartari, G.; Martelli, P.L.; Casadio, R. DeepMito: Accurate prediction of protein sub-mitochondrial localization using convolutional neural networks. *Bioinformatics* **2020**, *36*, 56–64. [CrossRef] [PubMed]
47. Claros, M.G.; Vincens, P. Computational Method to Predict Mitochondrially Imported Proteins and their Targeting Sequences. *JBIC J. Biol. Inorg. Chem.* **1996**, *241*, 779–786. [CrossRef]
48. Gschloessl, B.; Guermeur, Y.; Cock, J.M. HECTAR: A method to predict subcellular targeting in heterokonts. *BMC Bioinform.* **2008**, *9*, 393. [CrossRef]
49. Krogh, A.; Larsson, B.; Von Heijne, G.; Sonnhammer, E.L.L. Predicting transmembrane protein topology with a hidden Markov model: Application to complete genomes. *J. Mol. Biol.* **2001**, *305*, 567–580. [CrossRef]
50. Petersen, T.N.; Brunak, S.; Von Heijne, G.; Nielsen, H. SignalP 4.0: Discriminating signal peptides from transmembrane regions. *Nat. Methods* **2011**, *8*, 785–786. [CrossRef]
51. Wang, S.; Mao, Y.; Narimatsu, Y.; Ye, Z.; Tian, W.; Goth, C.K.; Lira-Navarrete, E.; Pedersen, N.B.; Benito-Vicente, A.; Martin, C.; et al. Site-specific O-glycosylation of members of the low-density lipoprotein receptor superfamily enhances ligand interactions. *EMBO J.* **2013**, *32*, 1478–1488. [CrossRef]
52. Eisenberg, D.; Lüthy, R.; Bowie, J.U. VERIFY3D: Assessment of protein models with three-dimensional profiles. *Methods Enzymol.* **1997**, *277*, 396–404.
53. Wiederstein, M.; Sippl, M.J. ProSA-web: Interactive web servicefor the recognition of errors in three-dimensional structures ofproteins. *Nucleic Acids Res.* **2007**, *35*, 407–410. [CrossRef] [PubMed]
54. Kuriata, A.; Gierut, A.M.; Oleniecki, T.; Ciemny, M.P.; Kolinski, A.; Kurcinski, M.; Kmiecik, S. CABS-flex 2.0: A web server for fast simulations of flexibility of protein structures. *Nucleic Acids Res.* **2018**, *46*, W338–W343. [CrossRef] [PubMed]
55. Tan, K.P.; Nguyen, T.B.; Patel, S.; Varadarajan, R.; Madhusudhan, M.S. Depth: A web server to compute depth, cavity sizes, detect potential small-molecule ligand-binding cavities and predict the pKa of ionizable residues in proteins. *Nucleic Acids Res.* **2013**, *41*, W314–W321. [CrossRef] [PubMed]

marine drugs

Article

Marine Sulfated Polysaccharides as Promising Antiviral Agents: A Comprehensive Report and Modeling Study Focusing on SARS CoV-2

Abdalla E. M. Salih [1], Bathini Thissera [1], Mohammed Yaseen [1], Ahmed S. I. Hassane [1,2], Hesham R. El-Seedi [3,4,5], Ahmed M. Sayed [6,*] and Mostafa E. Rateb [1,*]

[1] School of Computing, Engineering & Physical Sciences, University of the West of Scotland, Paisley PA1 2BE, UK; b00382049@studentmail.uws.ac.uk (A.E.M.S.); Bathini.Thissera@uws.ac.uk (B.T.); Mohammed.Yaseen@uws.ac.uk (M.Y.); ahmedsayed.hassane@nhs.scot (A.S.I.H.)
[2] Aberdeen Royal Infirmary, Foresterhill Health Campus, Aberdeen AB25 2ZN, UK
[3] Pharmacognosy Group, Department of Pharmaceutical Biosciences, BMC, Uppsala University, Uppsala, Box 591, SE 751 24 Uppsala, Sweden; hesham.el-seedi@farmbio.uu.se
[4] International Research Center for Food Nutrition and Safety, Jiangsu University, Zhenjiang 212013, China
[5] Department of Chemistry, Faculty of Science, Menoufia University, Shebin El-Kom 32512, Egypt
[6] Department of Pharmacognosy, Faculty of Pharmacy, Nahda University, Beni-Suef 62513, Egypt
* Correspondence: Ahmed.mohamed.sayed@nub.edu.eg (A.M.S.); Mostafa.Rateb@uws.ac.uk (M.E.R.)

Citation: Salih, A.E.M.; Thissera, B.; Yaseen, M.; Hassane, A.S.I.; El-Seedi, H.R.; Sayed, A.M.; Rateb, M.E. Marine Sulfated Polysaccharides as Promising Antiviral Agents: A Comprehensive Report and Modeling Study Focusing on SARS CoV-2. *Mar. Drugs* **2021**, *19*, 406. https://doi.org/10.3390/md19080406

Academic Editors: Khaled A. Shaaban and Vassilios Roussis

Received: 31 May 2021
Accepted: 20 July 2021
Published: 22 July 2021

Publisher's Note: MDPI stays neutral with regard to jurisdictional claims in published maps and institutional affiliations.

Abstract: SARS-CoV-2 (severe acute respiratory syndrome coronavirus-2) is a novel coronavirus strain that emerged at the end of 2019, causing millions of deaths so far. Despite enormous efforts being made through various drug discovery campaigns, there is still a desperate need for treatments with high efficacy and selectivity. Recently, marine sulfated polysaccharides (MSPs) have earned significant attention and are widely examined against many viral infections. This article attempted to produce a comprehensive report about MSPs from different marine sources alongside their antiviral effects against various viral species covering the last 25 years of research articles. Additionally, these reported MSPs were subjected to molecular docking and dynamic simulation experiments to ascertain potential interactions with both the receptor-binding domain (RBD) of SARS CoV-2's spike protein (S-protein) and human angiotensin-converting enzyme-2 (ACE2). The possible binding sites on both S-protein's RBD and ACE2 were determined based on how they bind to heparin, which has been reported to exhibit significant antiviral activity against SARS CoV-2 through binding to RBD, preventing the virus from affecting ACE2. Moreover, our modeling results illustrate that heparin can also bind to and block ACE2, acting as a competitor and protective agent against SARS CoV-2 infection. Nine of the investigated MSPs candidates exhibited promising results, taking into consideration the newly emerged SARS CoV-2 variants, of which five were not previously reported to exert antiviral activity against SARS CoV-2, including sulfated galactofucan (**1**), sulfated polymannuroguluronate (SPMG) (**2**), sulfated mannan (**3**), sulfated heterorhamnan (**8**), and chondroitin sulfate E (CS-E) (**9**). These results shed light on the importance of sulfated polysaccharides as potential SARS-CoV-2 inhibitors.

Keywords: sulfated polysaccharides; antiviral; SARS-CoV-2; docking; molecular dynamic simulations

1. Introduction

Mother Nature remains an outstanding hub for valuable natural compounds that have been used for multiple purposes since ancient times by our ancestors, including pharmaceutical, biomedical, nutritional, and cosmetics applications. Among the dozens of natural product groups, polysaccharides are macromolecular polymeric carbohydrate molecules comprised of large chains of monosaccharide units. They are widely found in animals, plants, and microorganisms, and their functions are mainly either structure- or storage-related. Many studies have revealed that natural polysaccharides and their chemically modified derivatives have significant inhibitory activities against viral diseases such as human immunodeficiency virus (HIV) and herpes simplex virus (HSV) [1,2].

Sulfated polysaccharides (SPs) are a class of negatively charged polysaccharides comprising natural or modified sulfate moieties in their structural carbohydrate backbone. They possess significant biological activities such as antioxidant, anti-allergic, antiviral, anticancer, and anticoagulant abilities; hence, the study of SPs has significant importance for drug discovery campaigns [1,3]. SPs are mainly found in the cell walls of marine algae or seaweeds; they are less common in some mammals, such as fish skins, and rare in mangrove plants. The seaweed cell wall comprises about 40% of sulfated polysaccharides, which is relatively higher than the average content in other sources. The most interesting marine algal SPs are sourced from ulvans from green macroalgae, carrageenans and agar from red macroalgae, and fucoidans and laminarians from brown macroalgae. These SPs have shown antiviral activity against herpes simplex virus (HSV), human immunodeficiency virus type-1 (HIV-1), chikungunya virus, cytomegalovirus (CMV), influenza virus, and hepatitis virus, in addition to other enveloped and non-enveloped viruses. Moreover, the antiviral activity of SPs against the current COVID-19 pandemic has also been reported [1,4,5].

Marine sulfated polysaccharides (MSPs) have recently received increasing attention due to their antiviral activity. In particular, carrageenans have exhibited promising inhibitory effects on many viral strains, effectively preventing the internalization of virus particles by interfering with the interactions between the virus and host cell receptors. Carrageenan nasal spray (Boots Dual Defence® in the UK market) has been proven effective in patients with common cold infected by human coronaviruses beta and alpha. In contrast, the impact was attributed to the increased viral clearance and the reduced relapses of symptoms in children and adults [6]. Another nasal spray formulation (xylometazoline HCl) containing iota-carrageenan (**4**) can effectively relieve nasal congestion of the upper respiratory tract and protect the respiratory mucosa against viral infection [7]. Coldamaris® lozenges, comprising iota-carrageenan (**4**) as the active pharmaceutical ingredient, cause denaturation of glycoproteins on the coronavirus surface, inhibiting the virucidal effects of coronavirus [8].

It remains challenging to develop novel drugs within a limited timeframe, as drug discovery is time- and resource-consuming; however, the process of drug discovery is immensely enhanced by the modern era of computational technology, which has accelerated drug discovery and drug repurposing. Computer-assisted or in silico design employs computational methods in drug discovery and is being applied to streamline and speed up hit-to-lead optimization and hit identification. Several computational techniques have been introduced to predict and select therapeutic targets, study the interactions between drug and receptor, characterize and determine ligand binding sites on the targets, and determine hit compounds using ligand- and structure-based virtual screening [9,10].

Molecular docking and dynamic simulations are the most intriguing among the computational tools employed for structure-based drug discovery, an approach that evaluates the binding affinities between two candidates: small molecules and macromolecular targets (protein). Many drugs currently in the market were developed based on in silico strategies, such as zanamivir (used to treat influenza), nelfinavir, and saquinavir (used in the treatment of HIV); therefore, computational methods have been receiving popularity in the pharmaceutical industry as being crucial in the drug discovery process as reliable and effective techniques [11–13].

In the present investigation, we aimed to shed light on the antiviral potential of MSPs, illustrating their modes of interaction against different targets with the aid of several in silico tools, highlighting the most potential candidates for further investigation against SARS CoV-2.

2. Results and Discussion

Herein, we have summarized the findings of approximately 80 research studies published in the last 25 years on MSPs that exhibited potential antiviral effects against a total of 22 different viral strains (Table 1, Figure S1). These antiviral MSPs were sourced from various marine sources, including brown algae, red algae, green algae, blue-green algae, microalgae, sea cucumber, and squid cartilage. According to the findings, 40% of the antiviral MSPs were isolated from the red algae, followed by 24% from brown algae, 14% from green algae, 10% from blue-green algae, 9% from microalgae, 2% from squid cartilage, and 1% from sea cucumber (Figure 1). The red-algae-derived MSPs exhibited greater efficacy against several viruses; however, the distribution was highly skewed towards herpes simplex viruses (HSV), which received most of the research effort (Figure S1).

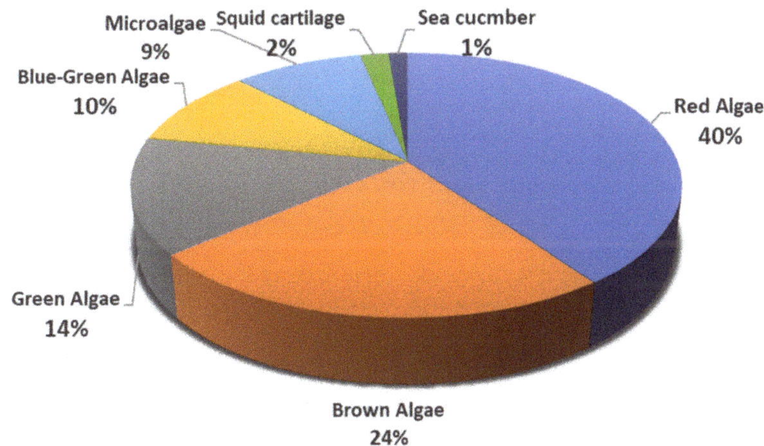

Figure 1. The systematic study of marine sources of MSPs that have shown antiviral activities in the last 25 years.

2.1. MSPs from Red Algae

Red macroalgae were commercially considered more valuable than brown and green macroalgae and are widely used in manufacturing hydrocolloids, such as carrageenan and agar, which in turn are involved many applications, such as food, pharmaceutical, and biotechnological industries. The principal polysaccharide components of the red algae are sulfated galactans that are produced extracellularly; however, they are made up of a linear backbone of alternating 3-linked β-D-galactopyranose and 4-linked α-D-galactopyranose, with a few exceptions, such as DL-hybrid sulfated galactan. The various structural types of sulfated galactan (**7**) have revealed potent antiviral activity against several types of enveloped viruses, such as HSV-1, HSV-2, DENV-2, HIV-1, and HIV-2 (Table 1).

Table 1. The antiviral activity levels and modes of action of different MSPs derived from various marine sources.

Sources	Compounds	Activity	Efficacy	Mode of Action	References
		Brown Algae			
Undaria pinnatifid	Sulfated galactofucan	HSV-1, HSV-2, and HCMV	IC_{50} = 0.2–1.1 μg/mL	Inhibit virus entry and host cell binding	[14,15]
Cystoseira indica	Sulfated Fucans	HSV-1 and HSV-2	IC_{50} = 0.50–2.8 μg/mL	Inhibit adsorption	[16]
Saccharina japonica	Fucoidan: RPI-27 and RPI-28 in a complex	SARS-CoV-2	EC_{50} = 8.3 ± 4.6 μg/mL	S-protein binding	[17]
Padina boryana	Fucoidan: α-(1→3)-linked	SARS-CoV-2	IC_{50} = 15.6 μg/mL	S-protein binding	[18,19]
Laminaria japonica	Sulfated polymannuroguluronate (SPMG)	HIV-1	IC_{50} = 30 μg/mL	Inhibit the virus entry and replication	[20]
Laminaria japonica	Sulfated polymannuroguluronate (SPMG)	HIV-1	100 μg/mL	Inhibit HIV-1 entry by suppressing rgp120 binding to sCD4	[21]
Laminaria japonica	Sulfated polymannuronate (SPMG)-derived oligosaccharides	HIV-1	ID_{50} = 5.3 nM	Inhibit HIV-1 entry by suppressing rgp120 binding to sCD4	[22]
Adenocystis utricularis	Sulfated galactofucan	HSV-1 and HSV-2	IC_{50} = 1.25–2.16 μg/mL	Unknown	[23,24]
Stoechospermum marginatum	Sulfated fucan: (1→4)- and (1→3)-linked-α-L-fucopyranosyl residues	HSV-1 and HSV-2	EC_{50} = 0.63–10.0 μg/mL	Inhibit binding of the virus to the host cell receptor	[25,26]
Sargassum horneri	Fucoidan	HSV-1, HIV-1, and HCMV	IC_{50} = 1–3.3 μg/mL	Unknown	[27]
Kjellmaniella crassifolia	Fucoidan KW	Inf A virus (H1N1, H3N2)	IC_{50} < 6.5 μg/mL	Suppress the replication	[28]
Sargassum patens	Galactofucan	HSV-1 and HSV-2	EC_{50} = 1.3–5.5 μg/mL	Inhibit adsorption	[29]
Leathesia difformis	Fucoidan	HSV-1, HSV-2, and HCMV	IC_{50} = 0.5–1.9 μg/mL	Inhibit protein synthesis and adsorption	[30]
Fucus evanescens	Fucoidan	HSV-2	10 mg/kg/day for 5 days	Unknown	[31]
Undaria pinnatifida	Fucoidan	HSV-1, HSV-2, and HCMV	IC_{50} = 1.5–2.6 μg/mL	Inhibit entry and binding	[32]
Sphacelaria indica	Xylogalactofucan and alginic acid	HSV-1	IC_{50} = 0.6–10 μg/mL	Inhibit adsorption	[33]
Sargassum mcclurei, Sargassum polycystum, and Turbinara ornata	Fucoidan	HIV-1	IC_{50} = 0.33–0.7 μg/mL	Inhibit entry	[34]
Scinaia hatei	Sulfated xylomannan	HSV-1 and HSV-2	IC_{50} = 0.5–4.6 μg/mL	Inhibit replication and virus binding	[16]

Table 1. *Cont.*

Sources	Compounds	Activity	Efficacy	Mode of Action	References
Cladosiphon okamuranus	Fucoidan	NDV	$IC_{50} = 0.75 \pm 1.6$ µg/mL	Inhibit viral-induced syncytia formation	[35]
Cladosiphon okamuranus	Fucoidan	DENV-2	Structure-based analysis: fucoidan interacts directly with envelope glycoprotein (EGP) on DEN2	Inhibit binding	[36]
Undaria pinnatifida- derived fucoidan (UPF)	Fucoidan	InfAV (H1N1, PR8)	7.0 mg/day for 7 days		[37]
Undaria pinnatifida	Fucoidan	Avian InfAV	5 mg/day for 14 days	Decrease replication	[38]
Red Algae					
Gymnogongrus griffithsiae and *Cryptonemia crenulata*	Kappa/iota/nu Carrageenan and DL-galactan hybrid	HSV-1 and HSV-2	$IC_{50} = 0.5$–5.6 µg/mL	Inhibit adsorption	[39]
Sebdenia polydactyla	Sulfated xylomannans	HSV-1	$IC_{50} = 0.35$–2.8 µg/mL	Inhibit virus attachment to the host cell	[40]
Euchema spinosum	Iota-carrageenan	SARS-CoV-2	$IC_{50} = 2.6$ µg/mL	Inhibit replication	[41,42]
Euchema spinosum	Iota-carrageenan with Xylitol® nasal spray	SARS-CoV-2	$IC_{50} < 6.0$ µg/mL	Inhibit SARS-CoV-2 in vitro	[42,43]
Chondrus crispus	lambda-carrageenan	SARS-CoV-2, and InfV A and B	$EC_{50} = 0.3$–1.4 µg/mL	Prevent the entry	[42,44]
Euchema spinosum	Iota-carrageenan containing Xylometazoline HCL	hRV1a, hRV8, and hCoV-OC43	IC_{50} for hRV1a and hRV8 = 1.56–6.2 µg/mLMIC for hCoV-OC43 = 0.024 µg/mL	Prevent the entry	[7,42]
Gigartina skottsbergii	κ/ι/µ/ν-carrageenan	HIV-1	$IC_{50} = 0.4$–3.3 µg/m	Prevent binding and replication of virions	[45]
Gymnogongrus torulosus	DL-hybrid sulfated galactan	HSV-1 and HSV-2	$IC_{50} = 1.1$ to 27.4 µg/mL	Unknown	[46,47]
Cryptonemia crenulata	DL-galactan hybrid C2S-3	DENV-2	$IC_{50} = 1$ µg/mL	Block virus multiplication	[48,49]
Stenogramme interrupta	ξ- and λ-carrageenans	HSV-1 and HSV-2	$IC_{50} = 0.65$–2.88 µg/mL	Inhibit binding and replication	[50]
Gracilaria corticata	Sulfated galactan	HSV-1 and HSV-2	$IC_{50} = 0.19$–0.24 µg/mL	Inhibit virus attachment to the host cell	[51,52]
Aghardhiella tenera	Sulfated galactan	HIV-1 and HIV-2	$IC_{50} = 0.05$–0.5 µg/mL	Inhibit cytopathic effect of HIV-1 and HIV-2 in MT-4 cells	[53]

Table 1. *Cont.*

Sources	Compounds	Activity	Efficacy	Mode of Action	References
Nemalion helminthoides	Sulfated mannan	HSV-1	IC_{50} = 5.43–9.68 µg/mL	Unknown	[54]
Nothogenia fastigiata	Sulfated xylomannan	HSV-1 and HSV-2	IC_{50} = 0.6–1.3 µg/mL	Prevent binding and replication	[55,56]
Scinaia hatei	Sulfated xylomannan	HSV	IC_{50} = 0.5 µg/mL	Inhibit replication	[57]
Nothogenia fastigiata	Sulfated xylogalactans	HSV-1	EC_{50} = 15.0–32.6 µg/mL	Inhibit virus attachment to the host cell	[58,59]
Euchema spinosum *Chondrus crispus*	Lambda/Iota-carrageenan	DENV-2 and DENV-3	EC_{50} = 0.14–4.1 µg/mL	Interference with virus adsorption	[42,60]
Euchema spinosum	Iota-carrageenan nasal spray	hRV, hCV, and InfA	3 times per day for 7 days	Inhibit virus attachment to the host cell	[6,42]
Euchema spinosum	Iota-carrageenan containing lozenges (Coldamaris®)	hRV1a, CoV OC43, and InfA H1N1n	10 mg daily	Prevent binding and inhibit replication	[8]
Euchema spinosum	Iota-carrageenan nasal spray	hRV1a, CoV OC43, and InfA H1N1n	Total of 1.0 mg daily	Prevent the virus from binding to cell surfaces or penetrating the cells	[42,61]
Euchema spinosum	Iota-carrageenan, nasal spray	H1N1 and H3N2	IC_{50} = 0.04–0.2 µg/mL	Effectively inhibit virus adsorption to host cells and reduce replication	[42,62]
Euchema spinosum	Iota-Carrageenan	hRV	10^4–10^7 $TCID_{50}$/mL; 5–50 µg/mL	Prevent binding and/or the entry into the cells, inhibit replication	[42,63]
Euchema spinosum	Iota-Carrageenan	HPV	IC_{50} = 0.005 µg/mL	Blocking the initial interaction of capsids with cells and exert a postattachment inhibitory effect.	[64]
Acanthophora specifira	Lambda-carrageenan	HSV-1 and RVFV	IC_{50} = 75.8–80.5 µg/mL	Inhibit replication	[65]
Lithothamnion muelleri	Sulfated xylogalactans	HSV-1 and HSV-2	EC_{50} = 49.64–125.79 µg/mL	Inhibit adsorption and penetration	[59,66]
Scinaia hatei	Sulfated xylan	HSV-1 and HSV-2	IC_{50} = 0.22–1.37 µg/mL	Inhibit entry	[67]
Euchema spinosum	Iota-Carrageenan	SARS-COV-2	IC_{50} ≥ 125 µg/mL	Unknown	[18,42]
Sigma (Sigma Aldrich)	Carrageenan	Japanese encephalitis virus (JEV)	EC_{50} = 15 µg/mL	Inhibit attachment and the cellular entry stages	[68]
Tichocarpus crinitus	κ/β-carrageenan	Tobacco Mosaic Virus (TMV)	1 mg/mL	Significantly inhibit virus replication	[69]
Chondrus crispus	Lambda-carrageenan	DENV-2 and DENV-3	>99% reduction in virus production at 20 µg/mL	Inhibit entry	[42,70]

Table 1. *Cont.*

Sources	Compounds	Activity	Efficacy	Mode of Action	References
Meristiella gelidium	Iota/kappa/nu-hybrid carrageenan	DENV-2	IC_{50} = 0.14–1.6 µg/mL	Unknown	[71]
	Iota/lambda/kappa-carrageenan	Hepatitis A Virus (HAV)	Ratio of CD_{50} to ED_{50} >400	Inhibit replication	[72]
Sphaerococcus coronopifolius and *Boergeseniella thuyoides*	Sulfated galactan	HIV-1 and HSV-1	EC_{50} = 4.1–17.2 µg/mL	Inhibit adsorption and replication	[73]
Green Algae					
Enteromorpha compressa	Ulvan	HSV	IC_{50} = 28.25 µg/mL	Inhibit adsorption and replication	[74,75]
Monostroma latissimum	Sulfated rhamnan	EV71	IC_{50} = 0.5 µg/mL	inhibit replication	[76]
Ulva intestinalis	Ulvan	Measles virus (MeV)	IC_{50} = 3.6 µg/mL	Reduce the formation of syncytia	[77]
Ulva clathrata	Ulvan	NDV	IC_{50} = 0.1 µg/mL	Inhibit cell-cell fusion via a direct effect on F0 protein	[78]
Ulva armoricana	Ulvan (enzymatic preparations)	HSV-1	EC_{50} = 320.9–373.0 µg/mL	Unknown	[79]
Monostroma nitidum	Rhamnan Sulfate	HSV-1, HSV-2, HCMV, MeV, MuV, HCV, HCoV and HIV	EC_{50} = 0.77–8.30 µg/mL	Inhibit viral proliferation	[80]
Monostroma nitidum	Rhamnan Sulfate	HSV-2	IC_{50} = 0.87 µg/mL	Suggested to inhibit adsorption or penetration	[81]
Gayralia oxysperma	Sulfated heterorhamnan	HSV-1 and HSV-2	IC_{50} = 0.036–0.3 µg/mL	Inhibit multiplication	[82]
Codium fragile	Sulfated galactan	HSV-2	IC_{50} = 4.7 µg/mL	Interference with the early steps such as virus adsorption to and penetration into host cells	[83]
Monostroma nitidum, C. okamurai, C. scapelliformis, Chaetomorpha crassa, C. spiralis, Codium adhaerens, C. fragile, Caulerpa brachypus and *Caulerpa latum*	Sulfated arabinoxylogalactan	HSV-1	IC_{50} = 0.38–8.5 µg/mL	Inhibit binding, penetration, and the late stages of replication	[84]
Enteromorpha compressa	Sulfated heteroglycuronan (chemically modified)	HSV-1	IC_{50} = 28.25 µg/mL	Inhibit replication	[74]
Ulva lactuca	Ulva sulfated polysaccharide extracts	JEV	0.75 mg/day for 7 days	The survival rate significantly increased	[85]

Table 1. *Cont.*

Sources	Compounds	Activity	Efficacy	Mode of Action	References
Ulva Pertusa	Ulvan	Avian InfAV	50 mg twice (INJ) immunizations dose	Enhance AIV-specific antibody production and improve the humoral immunity level	[86]
Caulerpa racemosa	Sulfated heteropolysaccharide	DENV-2	IC_{50} = 0.6 µg/mL	Interfere with virus multiplication	[87]
Blue-Green Algae					
Spirulina platensis	Calcium Spirulan (Ca-SP)	HSV-1, HCMV, MeV, MuV, InfA, HIV-1	EC_{50} = 0.92–23 µg/mL	Inhibit replication and penetration	[88]
Spirulina platensis	Calcium spirulan, Sodium spirulan, and potassium spirulan	HSV-1	IC_{50} = 0.46–0.88 µg/mL	Inhibit replication	[89]
Spirulina platensis	Spirularin® HS "Ocean Pharma (Topical cream containing Ca-SP in combination with SPME)	HSV-1	15 mg/g twice daily for 3 weeks(10 mg/g SPME)	Significantly prevent herpes labialis exacerbation	[90]
Aphanothece sacrum	Sulfated polysaccharides (ASWPH)	HSV-2 and Inf A (H1N1)	IC_{50} = 0.32–1.2 µg/mL	Inhibit adsorption	[91]
Spirulina maxima	Hot water extract (HWE)	HSV-1, HSV-2, HCMV, and pseudorabies virus (PRV)	ED_{50} = 0.069–0.333 mg/mL	Inhibit adsorption and penetration	[92]
Microalgae					
Red; *Porphyridium sp.*	cell-wall sulphated polysaccharide	HSV-1, HSV-2 and Varicella Zoster Virus (VZV)	IC_{50} = 1 µg/mL	Prevent adsorption and/or inhibit the production of new viral particles	[93]
Diatom; *Navicula directa*	Naviculan	HSV-1, HSV-2, HIV, and Inf A	IC_{50} = 7.4–170 µg/mL	Inhibit binding and penetration. Inhibit cell–cell fusion between HIV gp160- and CD4.	[94]
Gyrodinium impudium	P-KG03	InfA H1N1 and H3N2	EC_{50} = 0.19–0.48 µg/mL	Inhibit replication and entry	[95]
Gyrodinium impudicum	P-KG03	EMCV	EC_{50} = 26.9 µg/ml	Inhibit replication	[96]
Cochlodinium polykrikoides	Extracellular sulfated polysaccharides (A1 and A2)	HIV-1, HSV-1, Inf A, RSV-A, and RSV-B	EC_{50} = 1.1–4.52 µg/mL	Inhibit replication	[97]
Marine animal-derived sulfated polysaccharides					
Thelenota ananas	Sea cucumber Fucosylated chondroitin sulfate	HIV-1	EC_{50} = 0.73–31.86 µg/mL	inhibit the entry and replication	[98]

Table 1. *Cont.*

Sources	Compounds	Activity	Efficacy	Mode of Action	References
Stichopus japonicus	Sea cucumber sulfated polysaccharide (SCSP)	SARS-COV-2	$EC_{50} = 9.10$ µg/mL	Interact with the S glycoprotein	[18]
Squid cartilage	Chondroitin sulphate E (CS-E)	DENV	$EC_{50} = 0.52–11.8$ µg/mL	Inhibit the entry via targeting E protein	[99]
Squid cartilage	Chondroitin sulfate E (CS-E)	HSV-1 and HSV-2	$IC_{50} = 0.06$ to 0.2 µg/mL	Inhibit binding	[100]

Carrageenans, which are isolated from the cell walls, represent about 30–70% of the red algal dry weight, and their structure consists of linear chains of alternating galactopyranose units with linkages between 1 and 3 monomeric positions and other galactopyranose units with linkages between 1 and 4 monomeric positions or 3,6-galactopyranose units. The most important and extensively studied carrageenans are kappa (κ), iota, and lambda (λ), which vary in the number and position of sulfate ester groups (S), in accordance with the presence of 3,6-anhydrous-D-galactopyranose units [4]. Carrageenans are broadspectrum antivirals and have recently shown significant inhibitory effects by preventing the physical binding and entry of viral particles. Carrageenans were found to be effective against 12 different viruses (HSV, SARS-CoV-2, InfV, hRV, hCoV-OC43, HIV, DENV, hCV, HPV, RVFV, JEV, and TMV). Additionally, carrageenans are the most investigated class in human clinical trials against various virus diseases, such as sexually transmitted HIV, HPV, and HSV, in addition to rhinoviruses [101]. The most successful antiviral preparations of carrageenans were the nasal spray dosage forms that have recently been developed against rhinoviruses and SARS-CoV-2 [6,7,18,43]. Alongside carrageenan and sulfated galactans, other MSPs extracted from red algae and showed considerable activities against HSV, such as sulfated xylomannans, sulfated mannan, sulfated xylogalac tans, and sulfated xylan (Table 1, Figure 2).

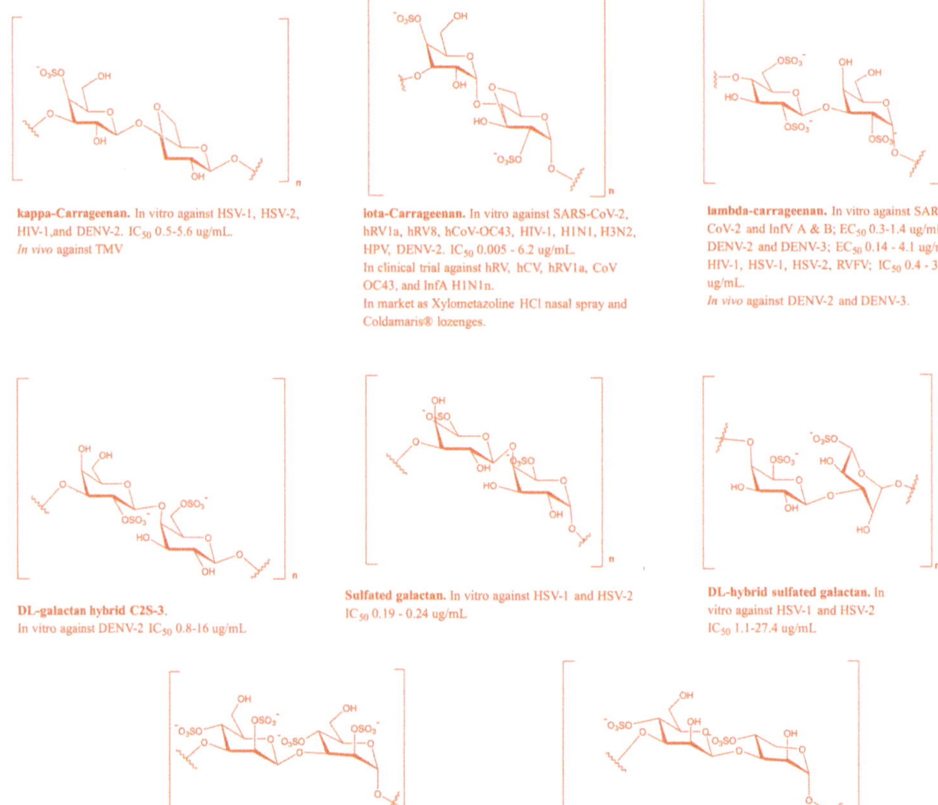

Figure 2. Representative MSPs from red algae.

2.2. MSPs from Brown Algae

Brown algae come in second place after red algae in terms of antiviral activities and research attention (Figure 1). The main reported MSPs from this genus were fucoidans, a group of heterogeneous sulfated polysaccharides that represent about 25–30% of the algae dry weight and are made up of a backbone of α-(1→3)-L-fucopyranose residues or alternating α-(1→3) and α-(1→4)-linked L-fucopyranosyls with a sulfate group mainly substituted on C-2 or C-4 fucopyranose residues [4]. Fucoidans were found to be effective against 7 different viruses (HSV, SARS-CoV-2, InfAV, HIV-1, DENV-2, NDV, and HCMV) (Table 1, Figure 3 and Figure S1).

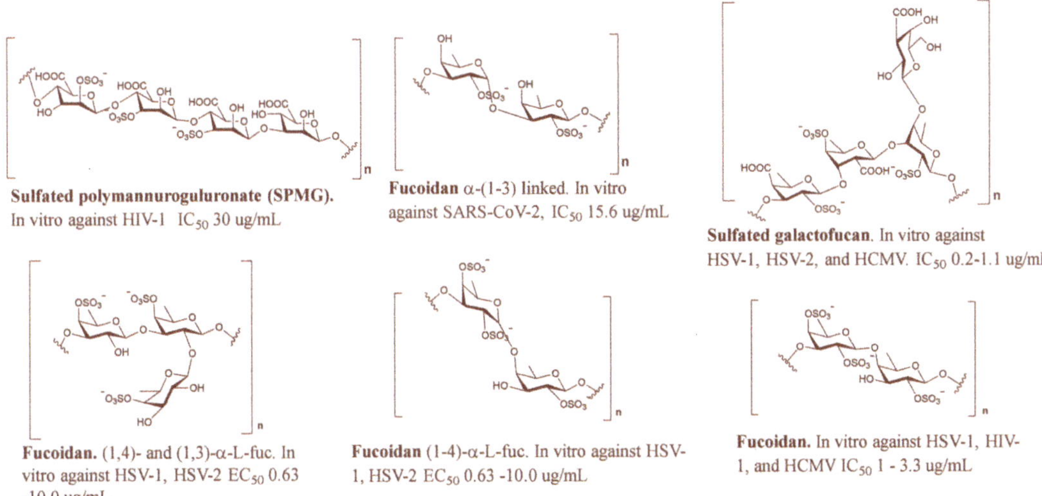

Figure 3. Representative MSPs from brown algae.

Generally, fucoidans block viral infection by preventing viral entry through competing for the positive charge attachment site of the envelope glycoproteins. The extent of antiviral activity is related to the number of sulfate groups present in the fucoidan structure [102]. As illustrated in Table 1, fucoidans have been tested in vivo using mice or mouse models for their activity against InfAV and HSV-2 and showed promising results. A recent in vitro study of two different fucoidans revealed that these compounds could be potent inhibitors of SARS-CoV-2 [17,18].

2.3. MSPs from Green Algae

Green-algae-derived MSPs come in third place as the most-investigated algae. Most of the examined compounds for antiviral activities belong to Ulvanes (Monostroma, Ulva, Enteromorpha), *Codium*, and *Caulerpa*. Different green-algae-derived MSPs were studied in vitro and in vivo against 11 viruses, including avian InfAV, HSV, HCMV, DENV-2, EV71, MeV, NDV, MuV, hCV, HIV, and JEV (Table 1, Figure 4). Song and his colleagues examined the antiviral activity of *Ulva pertusa* against avian InfAV and found that it exhibited a mild antiviral effect (40% viral inhibition). When combined with a vaccine against the same virus, it generated a synergistic effect, which significantly enhanced the production of antibodies by more than two-fold (~100%) [86]. Although the antiviral activities of green algae have received less attention than the red and brown, they possess unique antiviral properties against diverse viruses, such as NDV [78] and JEV [85].

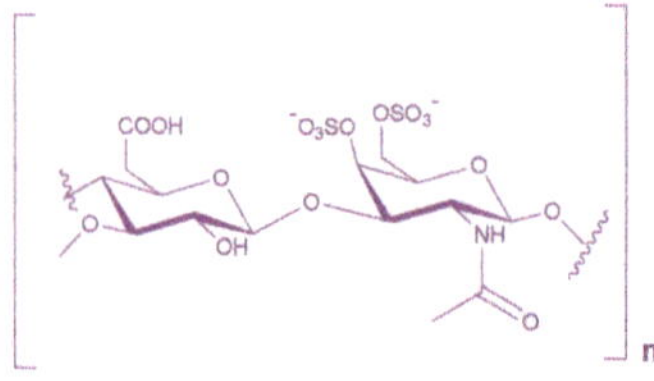

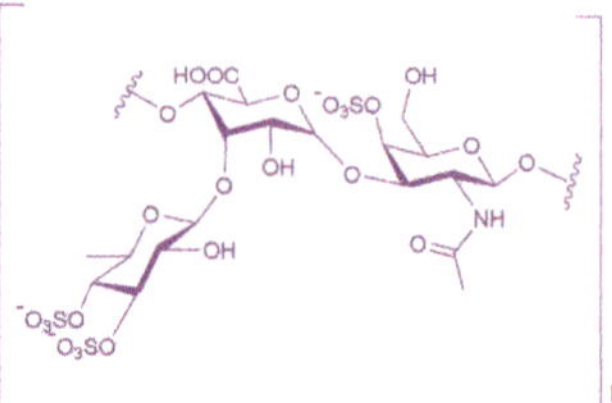

Figure 4. Representative MSPs from green algae (green structures), blue-green algae, and marine-animal-derived SPs (violet structures).

2.4. MSPs from Miscellaneous Marine Sources

Blue-green algae, microalgae, sea cucumbers, and squid cartilage (Table 1, Figure 4) are studied much less than the former algae for their MSPs as potential antiviral agents. Among them, the novel sulfated polysaccharide calcium spirulan (Ca-SP) was isolated from the blue-green alga *Spirulina platensis* and found to be an inhibitor for several viruses, including HSV-1, HCMV, InfAV, MeV, HIV-1, and Muv [88]. Recently, a clinical trial confirmed the potential of Ca-SP (Spirularin® HS) against herpes viruses through inhibiting the attachment and penetration of HSV-1 into mammalian epithelial cells and blocking the entry of Kaposi sarcoma-associated herpesvirus HSV-8 [90]. The highly sulfated polysaccharide p-KG03 isolated from the microalgae *Gyrodinium impudium* showed potent inhibitory activity against InfAV by blocking the early stage of replication and entry [95]. Sea cucumber sulfated polysaccharide (SCSP) is one of the most recent MSPs extracted from the sea cucumber *Stichopus japonicus*. Song and colleagues confirmed its activity against SARS-CoV-2 in an in vitro study through binding to the S-glycoprotein, preventing SARS-CoV-2 host cell entry [18].

2.5. In Silico Investigation of MSPs against SARS CoV-2

Molecular modeling of polysaccharides is not an easy task due to their high diversity, complexity, and flexibility; however, the recent advances with in silico tools can relieve the complexity of the process to a greater extent. Recently, heparin (**10**), an example of sulfated polysaccharide, was reported to exhibit a promising antiviral activity against SARS CoV-2 (EC_{50} = 36 µg/mL) through S-protein binding; hence, it suppresses the viral attachment to ACE2 (SARS CoV-2 S-protein receptor) and subsequently its entry inside the host cell (Figure 5) [17,103]. Similarly, most of the reported MSPs were found to mediate their antiviral activity via the exact mechanism, particularly those reported as anti-SARS CoV-2.

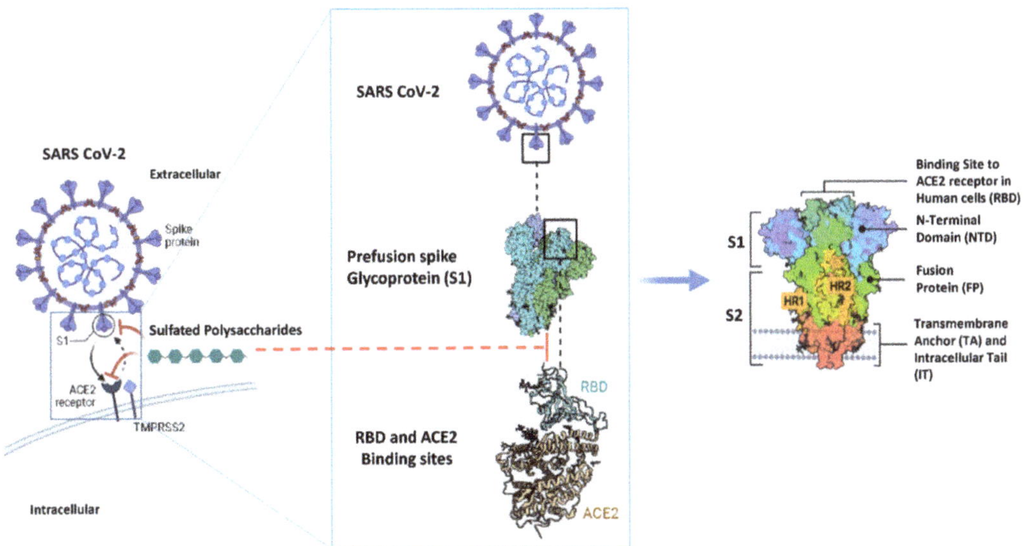

Figure 5. Schematic representation of the structure of SARS CoV-2 S-protein and how it can bind to the human ACE2. (PDB code: 6VXX) [104].

Accordingly, in this study, we shed light on MSPs as potential SARS CoV-2 antiviral agents by speculating on their plausible mode of action at the molecular level using a series of molecular docking and dynamic simulation experiments. Firstly, we determined the possible binding sites on both the S-protein receptor-binding domain (S-RBD) and ACE2 for MSPs. To do so, we utilized ClusPro [105], a software specialized in the prediction of heparin (**10**) (i.e., as an example of sulfated polysaccharide) binding sites in any given protein via molecular docking. The predicted heparin-S-RBD and heparin-ACE2 complexes were then subjected to 50 ns molecular dynamic simulation (MDS) experiments to select the most stable binding modes with each protein (S-RBD and ACE2). As shown in Figure 6, heparin was predicted to achieve stable binding with S-RBD at two sites (sites 1 and 2). Site 1 is located in a region that can interact with ACE2 directly and has a moderate positive charge. In contrast, site 2 is shallower than site 1 and is located in a wide positively charged region. Heparin-S-RBD complexes in these two sites were significantly stable over the course of MDS with low deviations from the original poses (RMSD ~ 2.5 Å) and minimal fluctuations (site 1 and site 2, Figure 6). This apparent stability resulted from the networks of H-bonds and ionic interactions formed between heparin and each binding site (Tables 2 and 3).

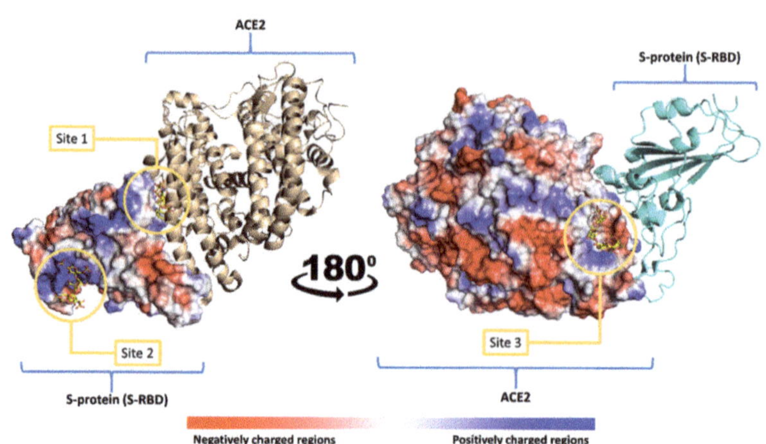

Figure 6. SARS CoV-2′s S-RBD–ACE2 complex showing heparin binding sites (sites 1, 2 and 3).

Table 2. Docking scores and binding free energies of the top-scoring MSPS against S-RBD of the original and mutated SARS CoV-2 S-protein (site 1).

No.	Compound	Vina Score (kcal/mol) [#]		ΔG [*]		Average RMSD (Å)		Reported Activity against SARS CoV-2
		Original Strain	Mutated Strain [**]	Original Strain	Mutated Strain [**]	Original Strain	Mutated Strain [**]	
1	Sulfated galactofucan: α- (1,3)- and (1,4)-α-L- (alternating)	−6.3	−6.4	−6.0	−6.3	5.1	4.5	No
2	Sulfated polyman-nuroguluronate (SPMG)	−7.5	−6.4	−7.3	−5.9	2.6	4.1	Yes
3	Sulfated mannan	−7.6	−7.7	−7.2	−7.4	2.3	1.9	No
4	iota-carrageenan	−6.0	−6.0	−0.8	−0.9	>15 [***]	>15 [***]	Yes
5	lambda-carrageenan	−7.0	−7.0	−5.4	−5.8	4.6	4.3	Yes
6	kappa-carrageenan	−6.6	−6.4	−1.3	−1.9	>15 [***]	>15 [***]	Yes
7	Sulfated galactan	−6.2	−6.3	−0.4	−0.1	>15 [***]	>15 [***]	No
8	Sulfated heterorhamnan	−6.1	−6.2	−0.1	−0.7	>15 [***]	>15 [***]	No
9	Chondroitin sulphate E (CS-E)	−7.6	−7.5	−7.1	−6.1	1.1	2.8	No
10	Heparin	−6.5	−6.4	−6.1	−6.1	3.3	3.3	Yes

Note: [*] ΔG was calculated using FEP method (see Materials and Methods for further information). [**] N501Y-mutated strain of SARS CoV-2. [***] This compound dissociated early at the beginning of MDS experiments (at ~25 ns). [#] The reported scores are the averages of three independent docking experiments (standard errors were between 0.1 and 0.3).

On the other hand, heparin was predicted to interact with a small pocket (site 3) on ACE2 located near the S-RBD binding region (Figure 6). Similar to sites 1 and 2, heparin achieved stable binding with site 3 during the MDS (RMSD ~ 1.9 Å) through extensive H-bonds and ionic interactions (site 3, Table 4 and Figure 6).

Table 3. Docking scores and binding free energies of top-scoring MSPS against RBD of SARS CoV-2 S-protein (site 2).

No.	Compound	Vina Score (kcal/mol)	ΔG *	Average RMSD (Å)	Reported Activity against SARS CoV-2
1	Sulfated galactofucan: α- (1,3)- and (1,4)-α-L- (alternating)	−5.3	−5.7	2.4	No
2	Sulfated polymannuroguluronate (SPMG)	−5.5	−5.9	2.1	No
3	Sulfated mannan	−5.4	−5.1	2.6	No
4	iota-carrageenan	−5.0	−1.7	>15 **	Yes
5	lambda-carrageenan	−5.5	−5.4	2.5	Yes
10	Heparin	−5.0	−5.8	1.9	Yes

Note: * ΔG was calculated using the FEP method (see Materials and Methods for further information). ** This compound dissociated early at the beginning of MDS (at ~25 ns).

Table 4. Docking scores and binding free energies of the top-scoring MSPS against ACE2 (site 3).

No.	Compound	Vina Score (kcal/mol)	ΔG *	Average RMSD (Å)	Reported Activity against SARS CoV-2
2	Sulfated polymannuroguluronate (SPMG)	−5.9	−5.4	2.8	No
5	lambda -carrageenan	−5.5	−5.0	3.1	Yes
10	Heparin	−5.3	−5.0	3.0	Yes

Note: * ΔG was calculated using FEP method (see Materials and Methods for further information).

The complex stability of RDB–ACE2 was also studied upon heparin binding to each site. As depicted in Figure 4, the distance between S-RBD and ACE2 (calculated as the distance between GLN-493 and GLU-35, respectively) remained constant (~5.1 Å) over 30 ns of MDS, while binding of heparin to S-RBD or ACE2 via sites 1, 2, or 3 led to significant instability of the complex and gradual dissociation (i.e., increased distance between GLN-493 and GLU-35). Heparin binding to site 1 on S-RBD or site 3 on ACE2 showed the greatest effects as S-RBD dissociated almost completely from ACE2 after 20 ns of simulation (Figure 7). Accordingly, it can be concluded that stable binding of SPs to sites 1, 2, or 3 destabilizes the S-RBD–ACE2 complex.

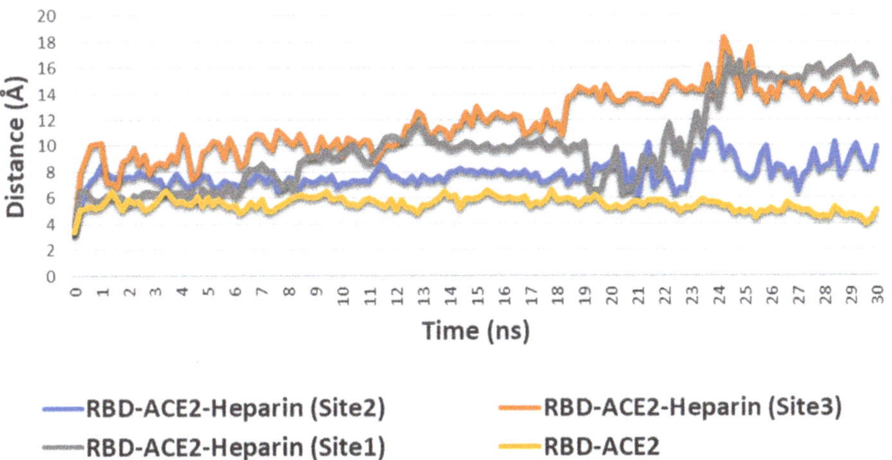

Figure 7. The calculated distance between S-RBD and ACE2 (i.e., between GLN-493 and GLU-35, respectively) during 30 ns of MDS in the absence and presence of heparin in site 1, 2, or 3.

It was recently reported that heparin can bind to S-RBD, preventing SARS CoV-2 from reaching ACE2 and entering the host cell; however, according to our modeling results, heparin can also bind to and block ACE2, acting as a protective agent against SARS CoV-2 infection. To study the mode of action of all reported MSPs against SARS CoV-2 to suggest previously unreported candidates, we subjected all collected compounds (Table 1) to dock against the proposed binding sites (sites 1, 2, and 3). Top-scoring hits (Figure 8) were selected according to the following criteria: (i) Docking score < −5 kcal/mol. Scores >−5 (i.e., −4.9 to −1.2 kcal/mol) showed unstable binding with the corresponding protein (RMSD > 20 Å at the first 20 ns); (ii) ΔG value < −5 kcal/mol, (iii) the docking pose remains stable over 50 ns of MDS. Some compounds achieved docking scores <−5 kcal mol (e.g., −6.4 kcal mol), although they were significantly unstable during the MDS experiments (RMSD > 15 Å at the first 20 ns) and presented significantly higher ΔG values (~−1.3 kcal/mol). As such, we made the binding stability at least 50 ns alongside ΔG value < −5 kcal/mol another selection criterion to discriminate between binders from non-binders, ensuring that all selected hits can achieve stable binding with the corresponding protein.

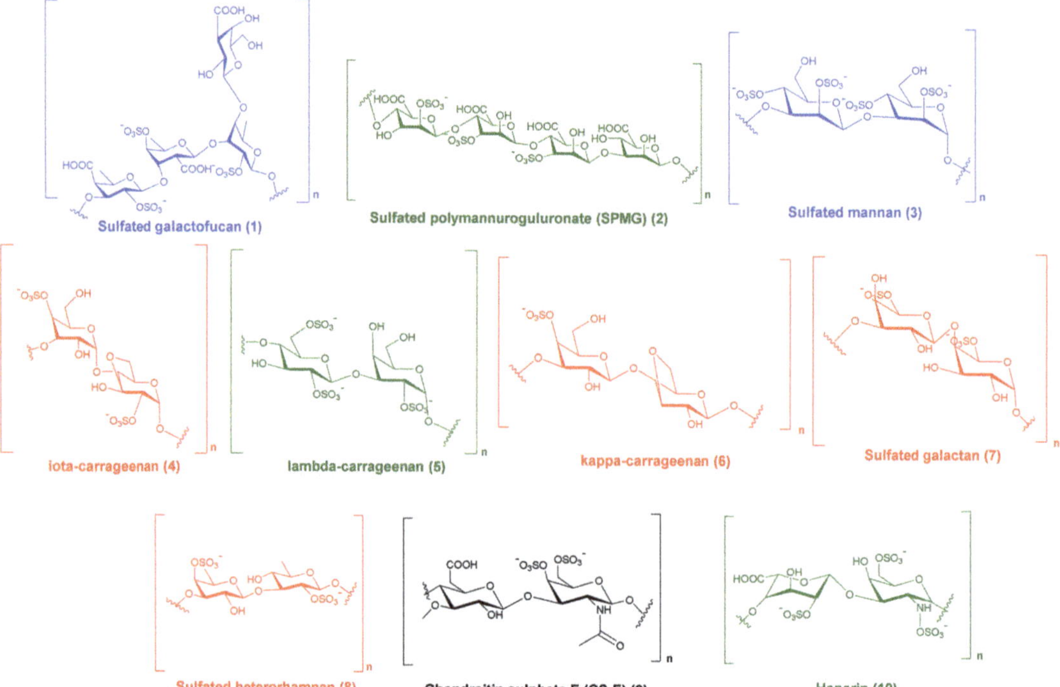

Figure 8. Sulphated polysaccharides with docking scores <−5 kcal/mol. Red compounds were significantly unstable during the course of MDS experiments. Black compound (compound 9) achieved stable binding with site 1 only. Blue compounds achieved stable bindings with sites 1 and 2. Green compounds achieved stable bindings with sites 1, 2, and 3.

Additionally, we took into consideration the newly emerged SARS CoV-2 variants during our docking experiments. Upon reviewing the recent mutations of the viral S-protein, we found that the UK variant (i.e., B.1.1.7) has a mutation in its S-RBD (i.e., N501Y) that was associated with its higher affinity to the ACE2 receptor [106]. This mutation was the replacement of the amino acid ASP-501 with TYR. We applied this mutation to the S-RBD structure to study its effect on the binding with MSPs upon docking and MDS.

As shown in Table 2, docking against site 1 (with and without the N501Y mutation) resulted in nine SPs with scores <−5 kcal/mol (Figure 8). Four of these hits were previously reported to exert antiviral activity against SARS CoV-2. Further MDS and ΔG experiments revealed that the best binding modes for iota-carrageenan (**4**), kappa-carrageenan (**6**), sulfated galactan (**7**), and sulfated heterorhamnan (**8**) (Figure 8) with site 1 were significantly unstable (their average RMSD values were higher than 15 Å and their ΔG values were higher than −5 kcal/mol) and they could easily dissociate from site 1 (with and without the N501Y mutation). Sulfated galactofucan (**1**), sulfated mannan (**3**), and chondroitin sulphate E (CS-E) (**9**) were among the best-scoring hits that were significantly stable during the MDS experiments, either with the mutated or non-mutated RBD. Additionally, they had not been previously reported for their antiviral efficacy against SARS CoV-2; hence, they were considered good candidates for future evaluation.

As shown in Figures 9–14, the prevalent interactions of these MSPs and RBD site 1 in its mutated and non-mutated forms were H-bonds and water bridges, where the sulfate esters were the main contributors in these interactions with site 1 key amino acid residues (Table 2). Moreover, their fluctuations and deviations from the starting docking poses were almost identical to both the mutated and non-mutated form of site 1 during the MDS, except for chondroitin sulphate E, which achieved slightly more stability and less deviation from the starting binding pose upon binding to the non-mutated form of site 1 (Figure 13).

Regarding site 2, only five sulfated polysaccharides presented docking scores <−5 kcal/mol, of which iota-carrageenan (**4**) was unstable on MDS (got an average RMSD >15 Å) and presented a ΔG value higher than −5 kcal/mol (i.e., 1.7 kcal/mol). Sulfated galactofucan (**1**), sulfated polymannuroguluronate (SPMG) (**2**), and sulfated mannan (**3**) were among the best MSPs that achieved stable binding with S-RBD site 2 and were not previously reported against inhibitory effects of SARS CoV-2 (Table 3). In this site (i.e., site 2), ionic interactions between the negatively charged sulfate moieties and the positively charged key amino acid residues of this site were crucial, along with the other polar interactions (e.g., H-bonding and water bridging), in maintaining stable binding with such shallow binding site (Figures 15–17).

On the other hand, docking against the ACE2 binding site (site 3) revealed that both (SPMG) and lambda-carrageenan were the best-performing MSPs. Additionally, they were also stable over the course of MDS, achieving low deviations from their starting binding poses (RMSD ~ 2.9 Å) and ΔG values <−5 kcal/mol (~−5.2 kcal/mol). Lambda-carrageenan was previously identified to bind to SARS CoV-2 S-protein [44]; however, there have been no reports on the ACE2 binding potential of MSPs so far. As such, focusing on this human protein target will be of great interest in future investigations of SPs as anti-COVID-19 therapeutics. Similarly to site 1 and 2, the dominant interactions between these MSPs and the key amino acid residues inside site 3 were also of the polar type (e.g., H-bonds, water bridges, and ionic interactions) (Figures 18 and 19).

2.6. Structure–Activity Relationship

The structural complexity of SPs has made the study of their structure–activity relationship quite challenging, and until now it has not been entirely understood; however, certain key bioactivity-related structural features could be concluded from the previously reported studies and the present modeling study.

The first structural determinant that may affect the antiviral potential of this class of compounds is the number of negatively charged groups (e.g., sulfates or carboxylates). Polysaccharides with higher numbers of sulfate or carboxylate groups per monosaccharide were more bioactive. Highly sulfated glucans (either natural or chemically synthesized derivatives) were found to be far more bioactive as antiviral agents than those with lower degrees of sulfation, while non-sulfated glucans were practically inactive [107,108]. Moreover, in our modeling study, MSPs with a single sulfate or carboxylate group per monosaccharide did not achieve stable binding with either S-RBD or ACE2 (ΔG values > −5 kcal/mol); hence, we could conclude that electrostatic and other polar interactions mediated by these negatively charged groups play a significant role in stabilizing the binding of this class of compounds to S-RBD and ACE2.

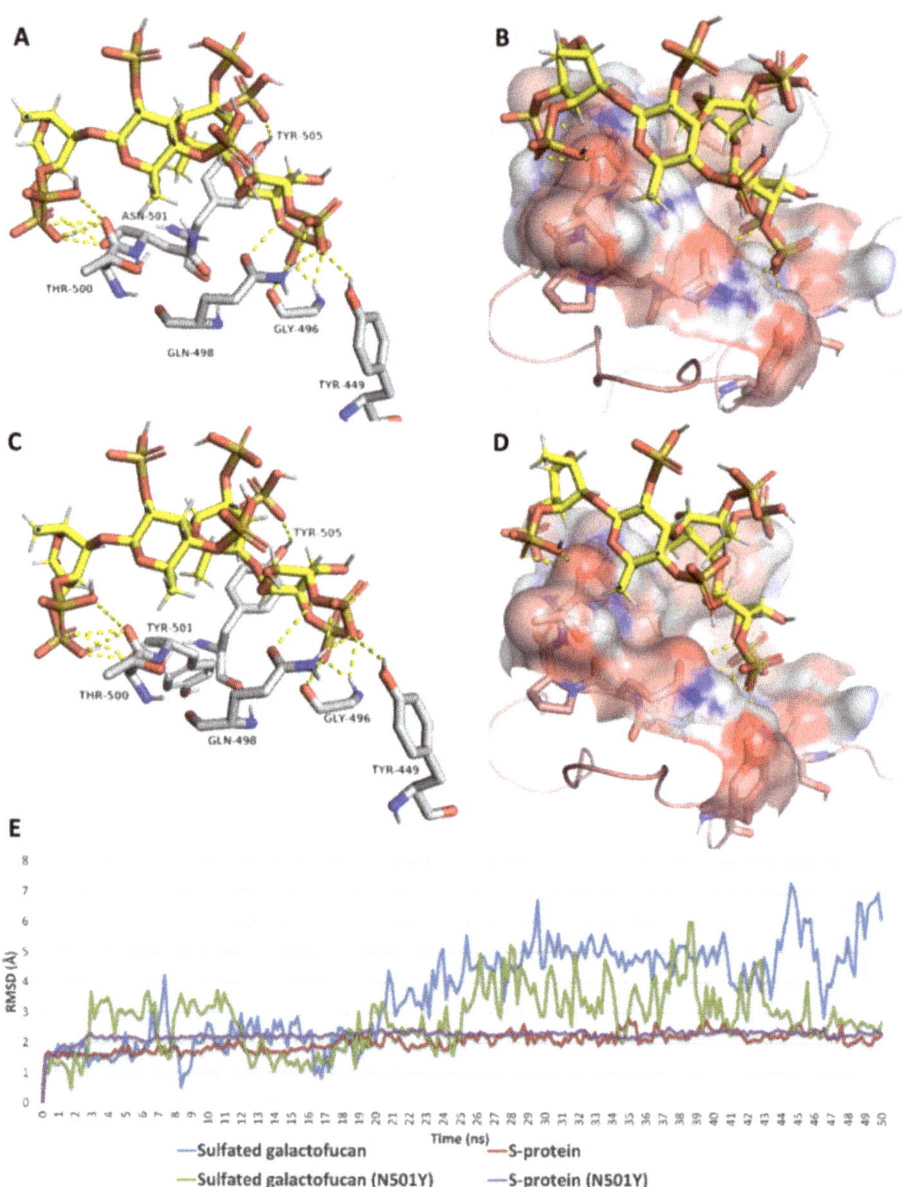

Figure 9. Binding mode of sulphated galactofucan (compound **1**) inside site 1 (**A**,**B**) and its mutated form (**C**,**D**) together with its RMSDs during 50 ns of MDSs (**E**).

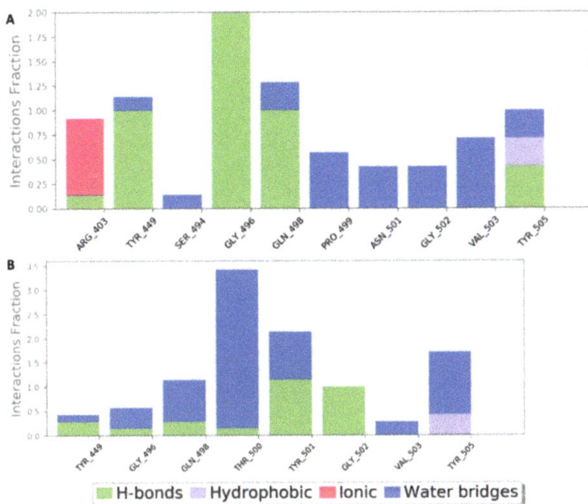

Figure 10. Protein–ligand contacts of sulphated galactofucan (compound 1) inside site 1 (**A**) and its mutated form (**B**) during the MDS.

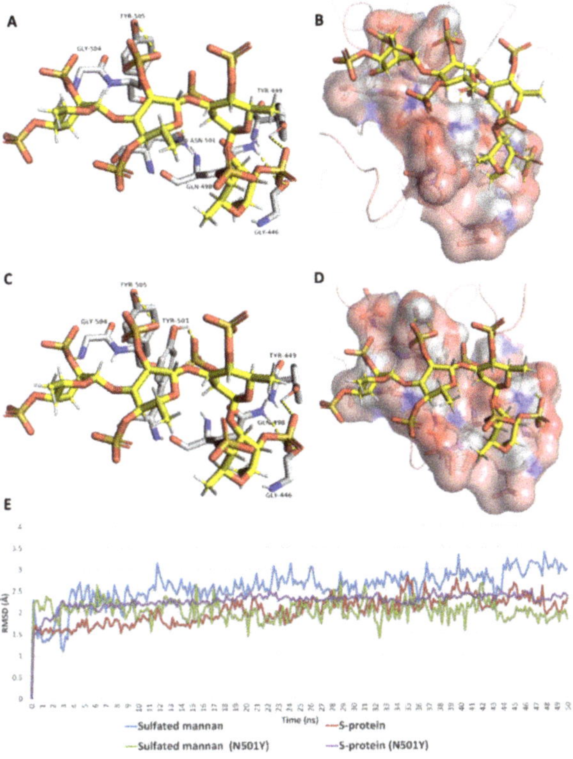

Figure 11. Binding mode of sulfated mannan (**3**) inside site 1 (**A,B**) and its mutated form (**C,D**) together with its RMSDs during 50 ns of MDSs (**E**).

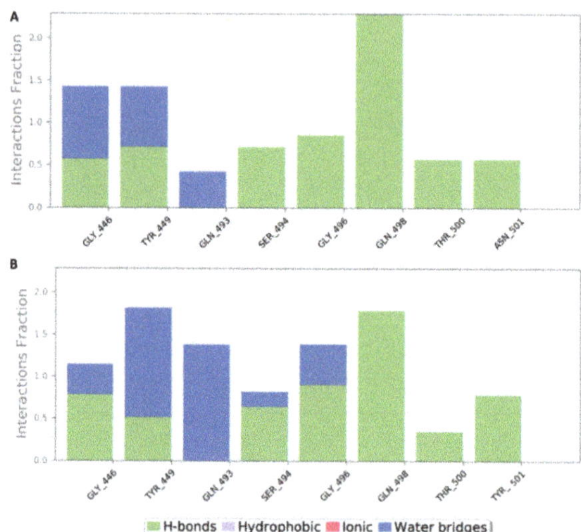

Figure 12. Protein–ligand contacts of sulfated mannan (**3**) inside site 1 (**A**) and its mutated form (**B**) during the MDS.

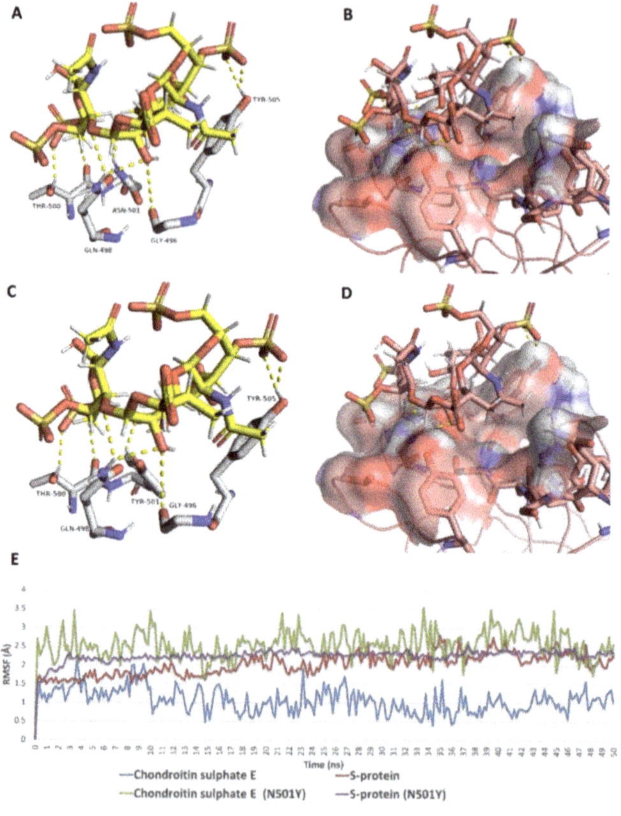

Figure 13. Binding mode of chondroitin sulphate E (compound **9**) inside site 1 (**A,B**) and its mutated form (**C,D**) together with its RMSDs during 50 ns of MDSs (**E**).

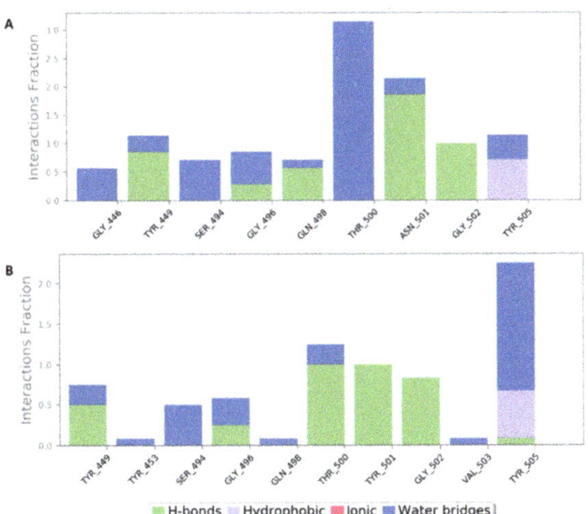

Figure 14. Protein–ligand contacts of chondroitin sulphate E (compound 9) inside site 1 (**A**) and its mutated form (**B**) during the MDS.

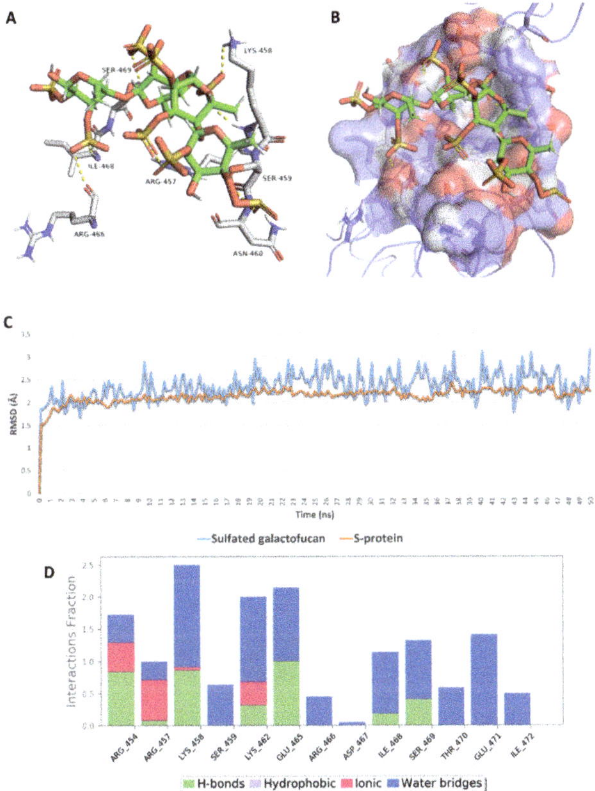

Figure 15. Binding mode of sulfated galactofucan (**1**) inside site 2 (**A,B**) together with its RMSDs and protein–ligand contacts during 50 ns of MDSs (**C,D**).

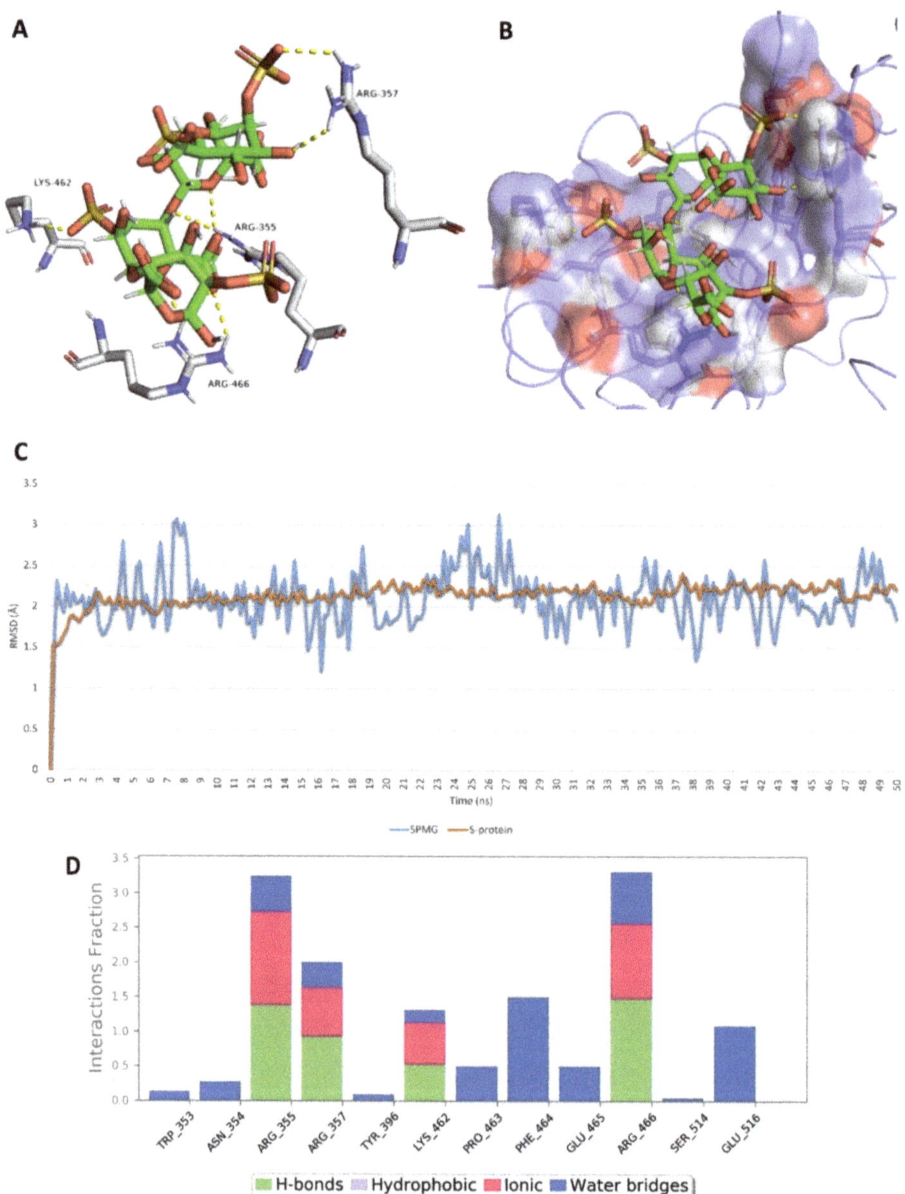

Figure 16. Binding mode of sulfated polymannuroguluronate (SPMG) (compound 2) inside site 2 (**A**,**B**) together with its RMSDs and protein–ligand contacts during 50 ns of MDSs (**C**,**D**).

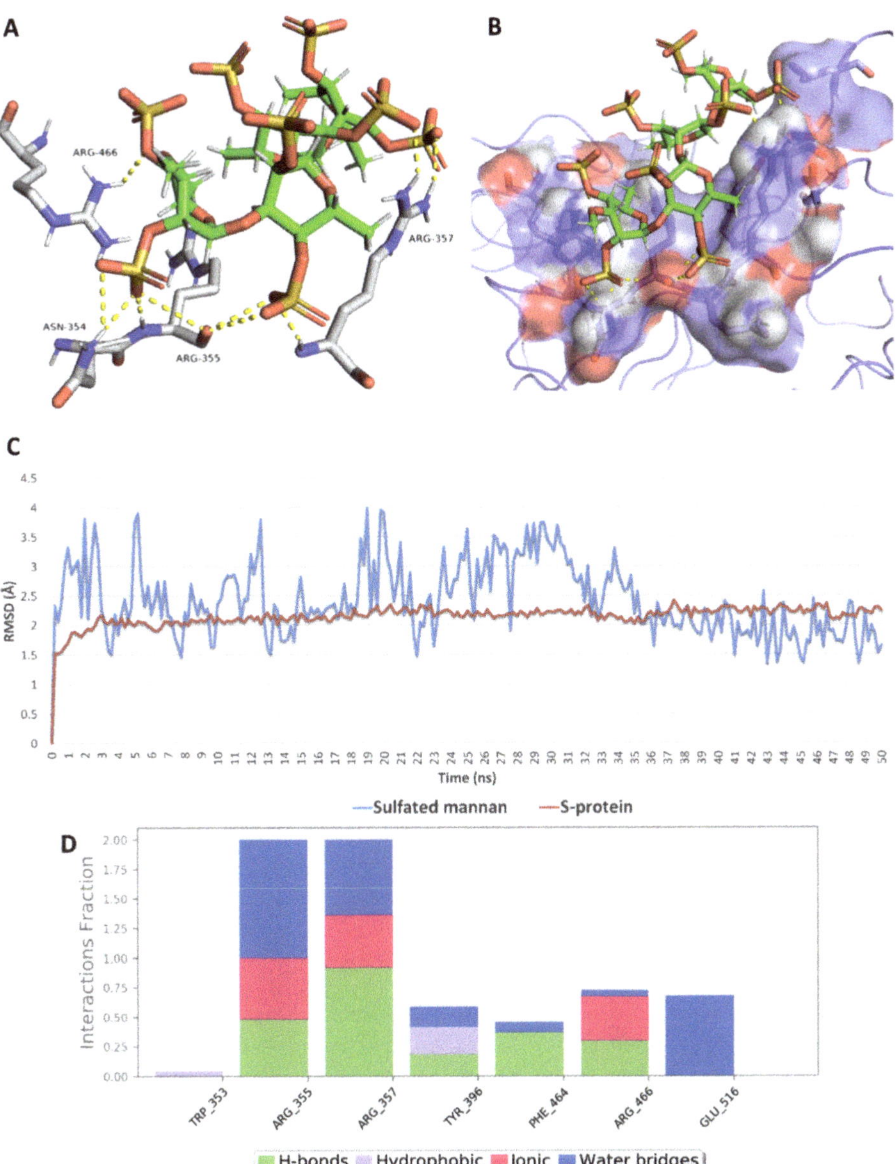

Figure 17. Binding mode of sulfated mannan (compound 3) inside site 2 (**A**,**B**) together with its RMSDs and protein–ligand contacts during 50 ns of MDSs (**C**,**D**).

Secondly, the distribution of these negatively charged moieties played an essential role in their antiviral activity. In previous studies, carrageenans with similar degrees of sulfation (50 mol%) but isolated from different sources showed varying antiviral activities [39,109,110]. Different charge densities were proposed to explain these findings, and accordingly to explain the antiviral activity [71]. Our modeling study found that neither iota- nor kappa-carrageenan achieved stable binding with any proposed binding sites.

This could be attributed to their relatively low degree of sulfation (one sulfate group per onosaccharide); however, both were reported to be bioactive against SARS CoV-2 and other pathogenic viruses. This could be explained by the unique distribution of these limited sulfate groups, which might cause them to be bioactive via different modes of action. In contrast, sulfated galactan (**7**) has a high degree of sulfation (Figure 8); however, it did not achieve stable binding with the proposed binding sites (ΔG value >-5 kcal/mol). Accordingly, the presence of many negatively charged moieties does not guarantee good antiviral activity without the proper orientations and distribution over the polysaccharide backbone.

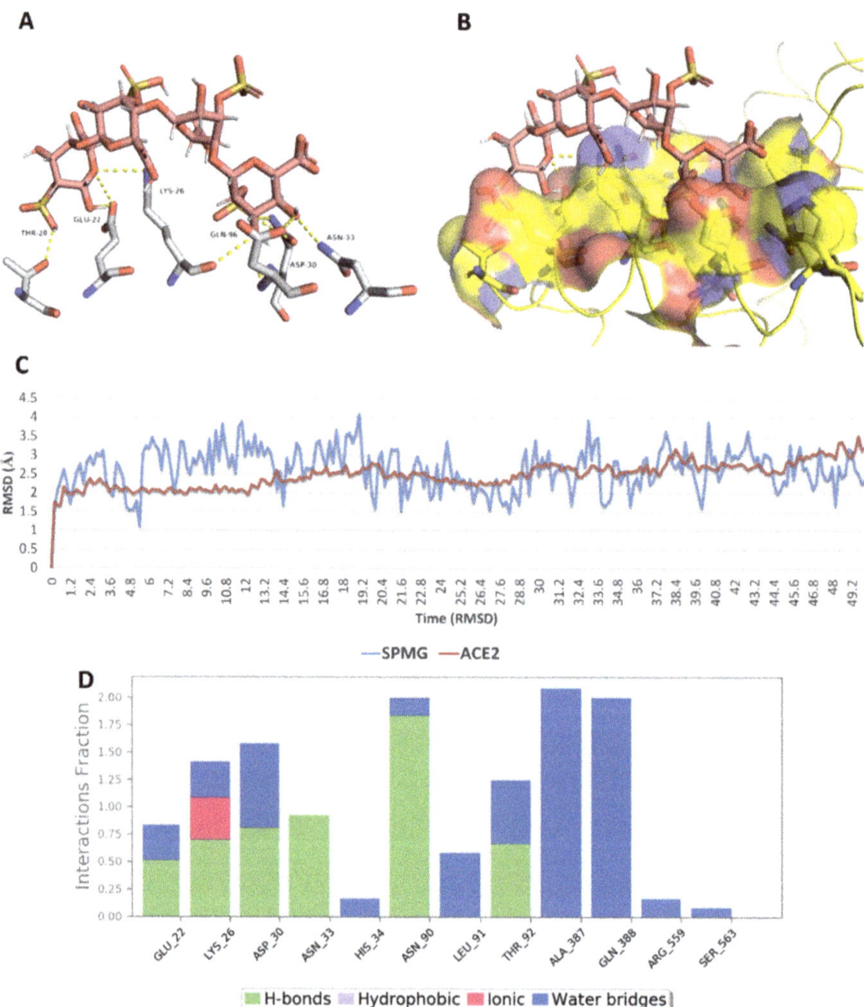

Figure 18. Binding mode of sulfated polymannuroguluronate (SPMG) (compound 2) inside site 3 (**A,B**) together with its RMSDs and protein–ligand contacts during 50 ns of MDSs (**C,D**).

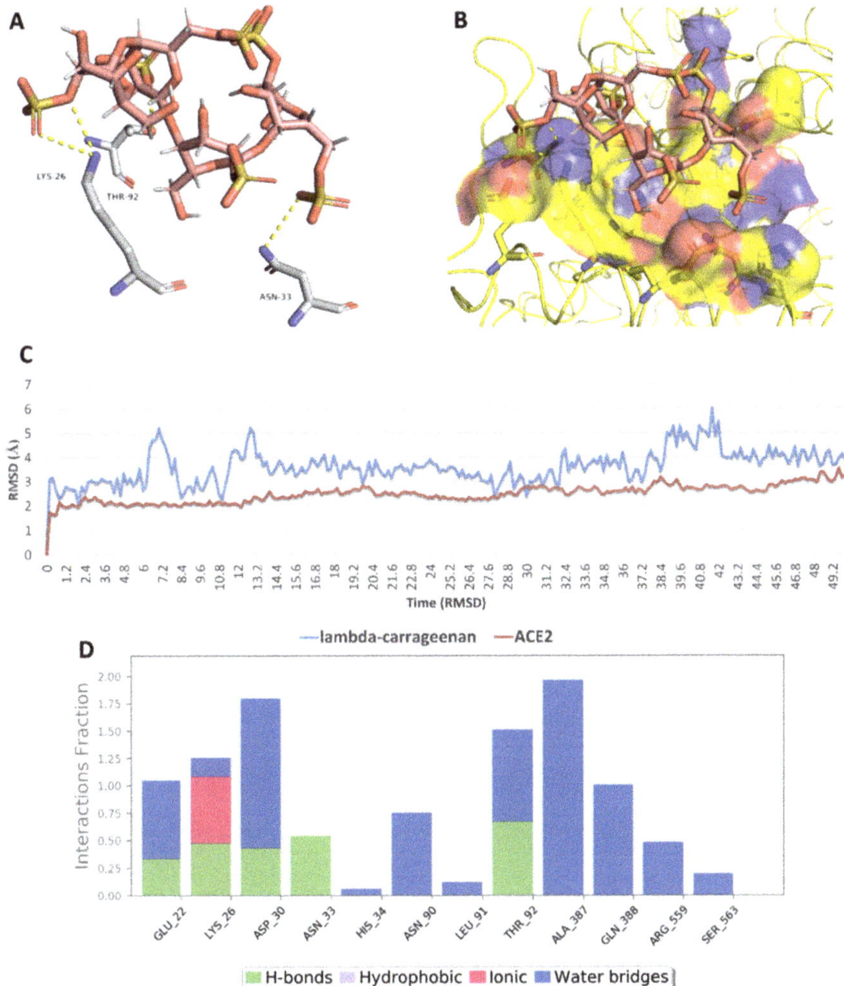

Figure 19. Binding mode of lambda-carrageenan (compound 5) inside site 3 (**A**,**B**) together with its RMSDs and protein–ligand contacts during 50 ns of MDSs (**C**,**D**).

Finally, the general chemical structure of the polysaccharides (e.g., the stereochemistry of the glycosidic linkage and that of monosaccharides and the presence of branching points in the main backbone) has a general impact on the degree of their antiviral activity [111].

3. Methodology

3.1. Databases Used

The search for MSPs in this study was conducted using SciFinder (https://www.cas.org/solutions/cas-scifinder-discovery-platform/cas-scifinder, accessed on 15 May 2021), Scopus (https://www.scopus.com), Web of Science (https://login.webofknowledge.com), Google Scholar, and DNP (Dictionary of Natural Products) databases. The keywords "antiviral", "SARS CoV-2", marine-derived", and "COVID-19" were paired with "sulphated polysaccharide" to obtain published records up until 2021.

3.2. Molecular Modeling

We utilized the carbohydrate modeling protocol used by Sapay and his co-workers in our in silico study of MSPs [112].

3.2.1. Molecular Docking

The docking study was carried out against both the receptor-binding domain (RBD) of SARS CoV-2's spike protein (S-protein) and human angiotensin-converting enzyme-2 (ACE2) (PDB codes: 6lZG and 1R42, respectively) [113,114]. An Autodock Vina docking machine [105] was used for docking experiments. To determine the best binding site for docking, we docked heparin (i.e., sulphated poly saccharide analogue) against both S-protein and ACE2 using the ClusPro server [14] by uploading each crystal structure to the website and choosing heparin as a ligand. The retrieved docking poses were then arranged from the highest to the lowest scores. We selected the docking pose of the highest score and subjected it to a short molecular dynamic simulation (25 ns) experiment to study the stability of heparin in this pose. Heparin showed acceptable stability in this orientation, where it deviated from the initial pose by an average RMSD of 2.1 Å (Figure 6); hence, the binding site of heparin in this pose was used for docking in the subsequent docking experiments. Heparin docking was carried out using its tetrasaccharide form (i.e., the default form in the ClusPro software). Accordingly, we also used the tetrasaccharide form of each MSP for docking experiments to reduce the computational cost required for docking and molecular dynamic simulations. All results were visualized using PyMol software [115].

3.2.2. Molecular Dynamic Simulation

MD simulation experiments were carried out using the MDS machine in Maestro software, Desmond v. 2. [116–118], using its default force field (i.e., OPLS-AA). Protein–ligand systems were constructed using the System Builder function. Thereafter, these systems were embedded in an orthorhombic box consisting of TIP3P water and 0.15 M Na+ and Cl- ions (the default dimensions were used). Subsequently, the prepared systems were energy-minimized and equilibrated for 10 ns. Ligand parameterization was carried out during the system building step according to the OPLS force field. For MD simulations carried out using NAMD software, the parameters and topologies of the ligands were computed using the Charmm36 force field. The online software Ligand Reader and Modeler (http://www.charmm-gui.org/?doc=input/ligandrm, accessed on 15 May 2021) [119] and the VMD plugin Force Field Toolkit (ffTK) [120] were utilized for this regard. The generated parameters and topology files were then loaded into VMD to read the protein–ligand complexes without errors, then the simulation step was performed. MD simulations were run for 50 ns at 310 K in the NPT ensemble with the Nose–Hoover thermostat and Martyna-Tobias-Klein barostat using anisotropic coupling. We selected the best binding poses for each compound as starting co-ordinates to investigate their binding stability and mode of interaction.

3.2.3. Binding Free Energy Calculations

Binding free energy calculations (ΔG) were performed using the free energy perturbation (FEP) method [121]. Briefly, this method estimates the binding free energy (i.e., $\Delta G_{binding}$) according to the following equation: $\Delta G_{binding} = \Delta G_{Complex} - \Delta G_{Ligand}$. These estimations were derived from separate simulations (NAMD software was used for these experiments). All input files required for simulation by NAMD were papered using the online website Charmm-GUI (https://charmm-gui.org/?doc=input/afes.abinding, accessed on 30 April 2021). Subsequently, we loaded these files into NAMD in order to produce the required simulations. The FEP function in NAMD was used to accomplish this experiment. The equilibration step was achieved in the NPT ensemble at 300 K and 1 atm (1.01325 bar) with Langevin piston pressure (for "complex" and "ligand") in the presence of the TIP3P water model. Then, 10 ns FEP simulations were carried out for each ligand and the last

5 ns of the free energy values was measured for the final free energy estimation [121]. All resulting trajectories were visualized and analyzed by VMD software.

4. Conclusions

To challenge highly virulent SARS CoV-2 and its emerging mutations, which have cost millions of lives, there is an urgent need to ramp up drug discovery pipelines within a short timeframe. In order to shorten the drug development cycle, screening of existing chemical libraries holds greater potential than starting from scratch. The levels of structural and functional diversity of MSPs have been widely discussed for their antiviral properties; however, the depth of these activities in terms of possible mechanisms against potential drug targets has been poorly studied. A large number of members belong to the class of SPs being overlooked as potential candidates to suppress the virulence of SARS CoV-2.

In the present investigation, we aimed to point out the most probable binding sites on S-protein's RBD and ACE2 using a well-known sulfated polysaccharide (i.e., heparin) as a molecular probe. Interestingly, the binding of heparin to these sites led to dissociation of the RBD–ACE2 complex. These MDS-derived results indicated and validated these proposed binding sites as potential targets for further ligands from the same chemical class. Accordingly, these identified binding sites were used in a series of docking experiments using most of the collected MSPs in this report. A number of MSPs were then found to be potential binders to these sites. Most of the identified hits have been reported to exert antiviral activity against SARS CoV-2; however, some of them have not, meaning they are considered very promising candidates for experimental testing. Overall, the present investigation shed light on the huge potential of MSPs as antiviral chemical entities and provided the scientific community hints about their potential against SARS CoV-2, with some important structural information to be utilized in further experimental research.

Supplementary Materials: The following are available online at https://www.mdpi.com/article/10.3390/md19080406/s1, Figure S1–S8 indicating details about the reported SMPs in addition to the reported SMPs that were reported in the manuscripts.

Author Contributions: Conceptualization, A.M.S., M.E.R.; methodology, A.E.M.S., B.T, A.S.I.H., A.M.S.; software, A.E.M.S., A.M.S.; validation, A.E.M.S., B.T., H.R.E.-S., A.M.S., M.Y., M.E.R.; formal analysis, A.E.M.S., A.M.S.; investigation, A.E.M.S., A.M.S., M.E.R.; data curation, B.T., M.Y., A.S.I.H., H.R.E.-S.., A.M.S., M.E.R.; writing—original draft preparation, A.E.M.S., A.M.S.; writing—review and editing, B.T., M.Y., H.R.E.-S., M.E.R.; supervision, M.E.R. All authors have read and agreed to the published version of the manuscript.

Funding: This research received no external funding.

Data Availability Statement: Not applicable.

Conflicts of Interest: The authors declare no conflict of interest.

References

1. Chen, L.; Huang, G. The antiviral activity of polysaccharides and their derivatives. *Int. J. Biol. Macromol.* **2018**, *115*, 77–82. [CrossRef] [PubMed]
2. Lee, S. Chapter 4–Strategic Design of Delivery Systems for Nutraceuticals. In *Nanotechnology Applications in Food*; Oprea, A.E., Grumezescu, A.M., Eds.; Academic Press: Cambridge, MA, USA, 2017; pp. 65–86. [CrossRef]
3. Ngo, D.-H.; Kim, S.-K. Sulfated polysaccharides as bioactive agents from marine algae. *Int. J. Biol. Macromol.* **2013**, *62*, 70–75. [CrossRef]
4. Hans, N.; Malik, A.; Naik, S. Antiviral activity of sulfated polysaccharides from marine algae and its application in combating COVID-19: Mini review. *Bioresour. Technol. Rep.* **2021**, *13*, 100623. [CrossRef] [PubMed]
5. Muthukumar, J.; Chidambaram, R.; Sukumaran, S. Sulfated polysaccharides and its commercial applications in food industries—A review. *J. Food Sci. Technol.* **2020**. [CrossRef]
6. Koenighofer, M.; Lion, T.; Bodenteich, A.; Prieschl-Grassauer, E.; Grassauer, A.; Unger, H.; Mueller, C.A.; Fazekas, T. Carrageenan nasal spray in virus confirmed common cold: Individual patient data analysis of two randomized controlled trials. *Multidiscip. Respir. Med.* **2014**, *9*, 57. [CrossRef] [PubMed]

7. Graf, C.; Bernkop-Schnürch, A.; Egyed, A.; Koller, C.; Prieschl-Grassauer, E.; Morokutti-Kurz, M. Development of a nasal spray containing xylometazoline hydrochloride and iota-carrageenan for the symptomatic relief of nasal congestion caused by rhinitis and sinusitis. *Int. J. Gen. Med.* **2018**, *11*, 275. [CrossRef]
8. Morokutti-Kurz, M.; Graf, C.; Prieschl-Grassauer, E. Amylmetacresol/2, 4-dichlorobenzyl alcohol, hexylresorcinol, or carrageenan lozenges as active treatments for sore throat. *Int. J. Gen. Med.* **2017**, *10*, 53. [CrossRef]
9. Ekins, S.; Mestres, J.; Testa, B. In silico pharmacology for drug discovery: Methods for virtual ligand screening and profiling. *Br. J. Pharmacol.* **2007**, *152*, 9–20. [CrossRef]
10. Sliwoski, G.; Kothiwale, S.; Meiler, J.; Lowe, E.W. Computational methods in drug discovery. *Pharmacol. Rev.* **2014**, *66*, 334–395. [CrossRef]
11. Makhouri, F.R.; Ghasemi, J.B. In Silico Studies in Drug Research Against Neurodegenerative Diseases. *Curr. Neuropharmacol.* **2018**, *16*, 664–725. [CrossRef] [PubMed]
12. Phillips, M.A. *Has Molecular Docking Ever Brought Us a Medicine?* IntechOpen: London, UK, 2018. [CrossRef]
13. Fadda, E.; Woods, R.J. Molecular simulations of carbohydrates and protein–carbohydrate interactions: Motivation, issues and prospects. *Drug Discov. Today* **2010**, *15*, 596–609. [CrossRef] [PubMed]
14. Hemmingson, J.A.; Falshaw, R.; Furneaux, R.H.; Thompson, K. Structure and Antiviral Activity of the Galactofucan Sulfates Extracted from UndariaPinnatifida (Phaeophyta). *J. Appl. Phycol.* **2006**, *18*, 185. [CrossRef]
15. Vishchuk, O.S.; Ermakova, S.P.; Zvyagintseva, T.N. Sulfated polysaccharides from brown seaweeds Saccharina japonica and Undaria pinnatifida: Isolation, structural characteristics, and antitumor activity. *Carbohydr. Res.* **2011**, *346*, 2769–2776. [CrossRef]
16. Mandal, P.; Mateu, C.G.; Chattopadhyay, K.; Pujol, C.A.; Damonte, E.B.; Ray, B. Structural Features and Antiviral Activity of Sulphated Fucans from the Brown Seaweed Cystoseira Indica. *Antivir. Chem. Chemother.* **2007**, *18*, 153–162. [CrossRef]
17. Kwon, P.S.; Oh, H.; Kwon, S.-J.; Jin, W.; Zhang, F.; Fraser, K.; Hong, J.J.; Linhardt, R.J.; Dordick, J.S. Sulfated polysaccharides effectively inhibit SARS-CoV-2 in vitro. *Cell Discov.* **2020**, *6*, 50. [CrossRef]
18. Song, S.; Peng, H.; Wang, Q.; Liu, Z.; Dong, X.; Wen, C.; Ai, C.; Zhang, Y.; Wang, Z.; Zhu, B. Inhibitory activities of marine sulfated polysaccharides against SARS-CoV-2. *Food Funct.* **2020**, *11*, 7415–7420. [CrossRef]
19. Usoltseva, R.V.; Anastyuk, S.D.; Ishina, I.A.; Isakov, V.V.; Zvyagintseva, T.N.; Thinh, P.D.; Zadorozhny, P.A.; Dmitrenok, P.S.; Ermakova, S.P. Structural characteristics and anticancer activity in vitro of fucoidan from brown alga Padina boryana. *Carbohydr. Polym.* **2018**, *184*, 260–268. [CrossRef] [PubMed]
20. Miao, B.; Geng, M.; Li, J.; Li, F.; Chen, H.; Guan, H.; Ding, J. Sulfated polymannuroguluronate, a novel anti-acquired immune deficiency syndrome (AIDS) drug candidate, targeting CD4 in lymphocytes. *Biochem. Pharmacol.* **2004**, *68*, 641–649. [CrossRef]
21. Meiyu, G.; Fuchuan, L.; Xianliang, X.; Jing, L.; Zuowei, Y.; Huashi, G. The potential molecular targets of marine sulfated polymannuroguluronate interfering with HIV-1 entry: Interaction between SPMG and HIV-1 rgp120 and CD4 molecule. *Antivir. Res.* **2003**, *59*, 127–135. [CrossRef]
22. Liu, H.; Geng, M.; Xin, X.; Li, F.; Zhang, Z.; Li, J.; Ding, J. Multiple and multivalent interactions of novel anti-AIDS drug candidates, sulfated polymannuronate (SPMG)-derived oligosaccharides, with gp120 and their anti-HIV activities. *Glycobiology* **2004**, *15*, 501–510. [CrossRef]
23. Ponce, N.M.A.; Pujol, C.A.; Damonte, E.B.; Flores, M.L.; Stortz, C.A. Fucoidans from the brown seaweed Adenocystis utricularis: Extraction methods, antiviral activity and structural studies. *Carbohydr. Res.* **2003**, *338*, 153–165. [CrossRef]
24. Ale, M.T.; Mikkelsen, J.D.; Meyer, A.S. Important Determinants for Fucoidan Bioactivity: A Critical Review of Structure-Function Relations and Extraction Methods for Fucose-Containing Sulfated Polysaccharides from Brown Seaweeds. *Mar. Drugs* **2011**, *9*, 2106–2130. [CrossRef] [PubMed]
25. Adhikari, U.; Mateu, C.G.; Chattopadhyay, K.; Pujol, C.A.; Damonte, E.B.; Ray, B. Structure and antiviral activity of sulfated fucans from Stoechospermum marginatum. *Phytochemistry* **2006**, *67*, 2474–2482. [CrossRef] [PubMed]
26. Shen, P.; Yin, Z.; Qu, G.; Wang, C. 11-fucoidan and its health benefits. In *Bioactive Seaweeds for Food Applications*; Qin, Y., Ed.; Academic Press: Cambridge, MA, USA, 2018; pp. 223–238.
27. Hoshino, T.; Hayashi, T.; Hayashi, K.; Hamada, J.; Lee, J.-B.; Sankawa, U. An antivirally active sulfated polysaccharide from Sargassum horneri (TURNER) C. AGARDH. *Biol. Pharm. Bull.* **1998**, *21*, 730–734. [CrossRef]
28. Wang, W.; Wu, J.; Zhang, X.; Hao, C.; Zhao, X.; Jiao, G.; Shan, X.; Tai, W.; Yu, G. Inhibition of Influenza A Virus Infection by Fucoidan Targeting Viral Neuraminidase and Cellular EGFR Pathway. *Sci. Rep.* **2017**, *7*, 40760. [CrossRef] [PubMed]
29. Zhu, W.; Ooi, V.E.C.; Chan, P.K.S.; Ang, P.O., Jr. Isolation and characterization of a sulfated polysaccharide from the brown alga Sargassum patens and determination of its anti-herpes activity. *Biochem. Cell Biol.* **2003**, *81*, 25–33. [CrossRef] [PubMed]
30. Feldman, S.C.; Reynaldi, S.; Stortz, C.A.; Cerezo, A.S.; Damonte, E.B. Antiviral properties of fucoidan fractions from Leathesia difformis. *Phytomedicine* **1999**, *6*, 335–340. [CrossRef]
31. Krylova, N.V.; Ermakova, S.P.; Lavrov, V.F.; Leneva, I.A.; Kompanets, G.G.; Iunikhina, O.V.; Nosik, M.N.; Ebralidze, L.K.; Falynskova, I.N.; Silchenko, A.S.; et al. The Comparative Analysis of Antiviral Activity of Native and Modified Fucoidans from Brown Algae Fucus evanescens In Vitro and In Vivo. *Mar. Drugs* **2020**, *18*, 224. [CrossRef]
32. Lee, J.-B.; Hayashi, K.; Hashimoto, M.; Nakano, T.; Hayashi, T. Novel Antiviral Fucoidan from Sporophyll of Undaria pinnatifida (Mekabu). *Chem. Pharm. Bull.* **2004**, *52*, 1091–1094. [CrossRef] [PubMed]
33. Bandyopadhyay, S.S.; Navid, M.H.; Ghosh, T.; Schnitzler, P.; Ray, B. Structural features and in vitro antiviral activities of sulfated polysaccharides from Sphacelaria indica. *Phytochemistry* **2011**, *72*, 276–283. [CrossRef]

34. Thuy, T.T.T.; Ly, B.M.; Van, T.T.T.; Van Quang, N.; Tu, H.C.; Zheng, Y.; Seguin-Devaux, C.; Mi, B.; Ai, U. Anti-HIV activity of fucoidans from three brown seaweed species. *Carbohydr. Polym.* **2015**, *115*, 122–128. [CrossRef]

35. Elizondo-Gonzalez, R.; Cruz-Suarez, L.E.; Ricque-Marie, D.; Mendoza-Gamboa, E.; Rodriguez-Padilla, C.; Trejo-Avila, L.M. In vitro characterization of the antiviral activity of fucoidan from Cladosiphon okamuranus against Newcastle Disease Virus. *Virol. J.* **2012**, *9*, 307. [CrossRef] [PubMed]

36. Hidari, K.I.P.J.; Takahashi, N.; Arihara, M.; Nagaoka, M.; Morita, K.; Suzuki, T. Structure and anti-dengue virus activity of sulfated polysaccharide from a marine alga. *Biochem. Biophys. Res. Commun.* **2008**, *376*, 91–95. [CrossRef] [PubMed]

37. Richards, C.; Williams, N.A.; Fitton, J.H.; Stringer, D.N.; Karpiniec, S.S.; Park, A.Y. Oral Fucoidan Attenuates Lung Pathology and Clinical Signs in a Severe Influenza a Mouse Model. *Mar. Drugs* **2020**, *18*, 246. [CrossRef]

38. Synytsya, A.; Bleha, R.; Synytsya, A.; Pohl, R.; Hayashi, K.; Yoshinaga, K.; Nakano, T.; Hayashi, T. Mekabu fucoidan: Structural complexity and defensive effects against avian influenza A viruses. *Carbohydr. Polym.* **2014**, *111*, 633–644. [CrossRef]

39. Talarico, L.B.; Zibetti, R.G.M.; Faria, P.C.S.; Scolaro, L.A.; Duarte, M.E.R.; Noseda, M.D.; Pujol, C.A.; Damonte, E.B. Anti-herpes simplex virus activity of sulfated galactans from the red seaweeds Gymnogongrus griffithsiae and Cryptonemia crenulata. *Int. J. Biol. Macromol.* **2004**, *34*, 63–71. [CrossRef] [PubMed]

40. Ghosh, T.; Pujol, C.A.; Damonte, E.B.; Sinha, S.; Ray, B. Sulfated Xylomannans from the Red Seaweed Sebdenia Polydactyla: Structural Features, Chemical Modification and Antiviral Activity. *Antivir. Chem. Chemother.* **2009**, *19*, 235–242. [CrossRef]

41. Morokutti-Kurz, M.; Fröba, M.; Graf, P.; Große, M.; Grassauer, A.; Auth, J.; Schubert, U.; Prieschl-Grassauer, E. Iota-carrageenan neutralizes SARS-CoV-2 and inhibits viral replication in vitro. *PLoS ONE* **2021**, *16*, e0237480. [CrossRef] [PubMed]

42. Hongu, T.; Phillips, G.O.; Takigami, M. 7-Polymer fibers for health and nutrition. In *New Millennium Fibers*; Hongu, T., Phillips, G.O., Takigami, M., Eds.; Woodhead Publishing: Sawston, UK, 2005; pp. 218–246.

43. Bansal, S.; Jonsson, C.B.; Taylor, S.L.; Figueroa, J.M.; Dugour, A.V.; Palacios, C.; César Vega, J. Iota-carrageenan and Xylitol inhibit SARS-CoV-2 in cell culture. *bioRxiv* **2020**. [CrossRef]

44. Jang, Y.; Shin, H.; Lee, M.K.; Kwon, O.S.; Shin, J.S.; Kim, Y.; Kim, M. Antiviral activity of lambda-carrageenan against influenza viruses in mice and severe acute respiratory syndrome coronavirus 2 in vitro. *bioRxiv* **2020**. [CrossRef]

45. Carlucci, M.J.; Ciancia, M.; Matulewicz, M.C.; Cerezo, A.S.; Damonte, E.B. Antiherpetic activity and mode of action of natural carrageenans of diverse structural types. *Antivir. Res.* **1999**, *43*, 93–102. [CrossRef]

46. Chattopadhyay, K.; Ghosh, T.; Pujol, C.A.; Carlucci, M.J.; Damonte, E.B.; Ray, B. Polysaccharides from Gracilaria corticata: Sulfation, chemical characterization and anti-HSV activities. *Int. J. Biol. Macromol.* **2008**, *43*, 346–351. [CrossRef] [PubMed]

47. Pujol, C.A.; Estevez, J.M.; Carlucci, M.J.; Ciancia, M.; Cerezo, A.S.; Damonte, E.B. Novel DL-Galactan Hybrids from the Red Seaweed Gymnogongrus Torulosus are Potent Inhibitors of Herpes Simplex Virus and Dengue Virus. *Antivir. Chem. Chemother.* **2002**, *13*, 83–89. [CrossRef]

48. Kang, H.-K.; Seo, C.; Park, Y. The Effects of Marine Carbohydrates and Glycosylated Compounds on Human Health. *Int. J. Mol. Sci.* **2015**, *16*, 6018–6056. [CrossRef]

49. Talarico, L.B.; Pujol, C.A.; Zibetti, R.G.M.; Faría, P.C.S.; Noseda, M.D.; Duarte, M.E.R.; Damonte, E.B. The antiviral activity of sulfated polysaccharides against dengue virus is dependent on virus serotype and host cell. *Antivir. Res.* **2005**, *66*, 103–110. [CrossRef] [PubMed]

50. Cáceres, P.J.; Carlucci, M.a.J.; Damonte, E.B.; Matsuhiro, B.; Zúñiga, E.A. Carrageenans from chilean samples of Stenogramme interrupta (Phyllophoraceae): Structural analysis and biological activity. *Phytochemistry* **2000**, *53*, 81–86. [CrossRef]

51. Mazumder, S.; Ghosal, P.K.; Pujol, C.A.; Carlucci, M.a.J.; Damonte, E.B.; Ray, B. Isolation, chemical investigation and antiviral activity of polysaccharides from Gracilaria corticata (Gracilariaceae, Rhodophyta). *Int. J. Biol. Macromol.* **2002**, *31*, 87–95. [CrossRef]

52. Pereira, M.G.; Benevides, N.M.B.; Melo, M.R.S.; Valente, A.P.; Melo, F.R.; Mourão, P.A.S. Structure and anticoagulant activity of a sulfated galactan from the red alga, Gelidium crinale. Is there a specific structural requirement for the anticoagulant action? *Carbohydr. Res.* **2005**, *340*, 2015–2023. [CrossRef]

53. Witvrouw, M.; Este, J.A.; Mateu, M.Q.; Reymen, D.; Andrei, G.; Snoeck, R.; Ikeda, S.; Pauwels, R.; Bianchini, N.V.; Desmyter, J.; et al. Activity of a Sulfated Polysaccharide Extracted from the Red Seaweed Aghardhiella Tenera against Human Immunodeficiency Virus and Other Enveloped Viruses. *Antivir. Chem. Chemother.* **1994**, *5*, 297–303. [CrossRef]

54. Pérez Recalde, M.; Noseda, M.D.; Pujol, C.A.; Carlucci, M.J.; Matulewicz, M.C. Sulfated mannans from the red seaweed Nemalion helminthoides of the South Atlantic. *Phytochemistry* **2009**, *70*, 1062–1068. [CrossRef]

55. Kolender, A.A.; Pujol, C.A.; Damonte, E.B.; Matulewicz, M.C.; Cerezo, A.S. The system of sulfated α-(1→ 3)-linked D-mannans from the red seaweed Nothogenia fastigiata: Structures, antiherpetic and anticoagulant properties. *Carbohydr. Res.* **1997**, *304*, 53–60. [CrossRef]

56. Erra-Balsells, R.; Kolender, A.A.; Matulewicz, M.a.C.; Nonami, H.; Cerezo, A.S. Matrix-assisted ultraviolet laser-desorption ionization time-of-flight mass spectrometry of sulfated mannans from the red seaweed Nothogenia fastigiata. *Carbohydr. Res.* **2000**, *329*, 157–167. [CrossRef]

57. Mandal, P.; Pujol, C.A.; Carlucci, M.J.; Chattopadhyay, K.; Damonte, E.B.; Ray, B. Anti-herpetic activity of a sulfated xylomannan from Scinaia hatei. *Phytochemistry* **2008**, *69*, 2193–2199. [CrossRef]

58. Damonte, E.B.; Matulewicz, M.C.; Cerezob, A.S.; Coto, C.E. Herpes simplex Virus-Inhibitory Sulfated Xylogalactans from the Red Seaweed Nothogenia fastigiata. *Chemotherapy* **1996**, *42*, 57–64. [CrossRef]

59. Navarro, D.A.; Stortz, C.A. The system of xylogalactans from the red seaweed Jania rubens (Corallinales, Rhodophyta). *Carbohydr. Res.* **2008**, *343*, 2613–2622. [CrossRef]

60. Talarico, L.B.; Damonte, E.B. Interference in dengue virus adsorption and uncoating by carrageenans. *Virology* **2007**, *363*, 473–485. [CrossRef]

61. Hemilä, H.; Chalker, E. Carrageenan nasal spray may double the rate of recovery from coronavirus and influenza virus infections: Re-analysis of randomized trial data. *Res. Sq.* **2020**. [CrossRef]

62. Leibbrandt, A.; Meier, C.; König-Schuster, M.; Weinmüllner, R.; Kalthoff, D.; Pflugfelder, B.; Graf, P.; Frank-Gehrke, B.; Beer, M.; Fazekas, T. Iota-carrageenan is a potent inhibitor of influenza A virus infection. *PLoS ONE* **2010**, *5*, e14320. [CrossRef] [PubMed]

63. Grassauer, A.; Weinmuellner, R.; Meier, C.; Pretsch, A.; Prieschl-Grassauer, E.; Unger, H. Iota-Carrageenan is a potent inhibitor of rhinovirus infection. *Virol. J.* **2008**, *5*, 107. [CrossRef]

64. Buck, C.B.; Thompson, C.D.; Roberts, J.N.; Müller, M.; Lowy, D.R.; Schiller, J.T. Carrageenan is a potent inhibitor of papillomavirus infection. *PLoS Pathog* **2006**, *2*, e69. [CrossRef] [PubMed]

65. Gomaa, H.H.; Elshoubaky, G.A. Antiviral activity of sulfated polysaccharides carrageenan from some marine seaweeds. *Int. J. Curr. Pharm. Rev. Res.* **2016**, *7*, 34–42.

66. Malagoli, B.G.; Cardozo, F.T.G.S.; Gomes, J.H.S.; Ferraz, V.P.; Simões, C.M.O.; Braga, F.C. Chemical characterization and antiherpes activity of sulfated polysaccharides from Lithothamnion muelleri. *Int. J. Biol. Macromol.* **2014**, *66*, 332–337. [CrossRef] [PubMed]

67. Mandal, P.; Pujol, C.A.; Damonte, E.B.; Ghosh, T.; Ray, B. Xylans from Scinaia hatei: Structural features, sulfation and anti-HSV activity. *Int. J. Biol. Macromol.* **2010**, *46*, 173–178. [CrossRef] [PubMed]

68. Nor Rashid, N.; Yusof, R.; Rothan, H.A. Antiviral and virucidal activities of sulphated polysaccharides against Japanese encephalitis virus. *Trop. Biomed.* **2020**, *37*, 713–721. [CrossRef] [PubMed]

69. Reunov, A.; Nagorskaya, V.; Lapshina, L.; Yermak, I.; Barabanova, A. Effect of κ/ß-Carrageenan from red alga Tichocarpus crinitus (Tichocarpaceae) on infection of detached tobacco leaves with tobacco mosaic virus. *J. Plant Dis. Prot.* **2004**, *111*, 165–172. [CrossRef]

70. Piccini, L.E.; Carro, A.C.; Quintana, V.M.; Damonte, E.B. Antibody-independent and dependent infection of human myeloid cells with dengue virus is inhibited by carrageenan. *Virus Res.* **2020**, *290*, 198150. [CrossRef] [PubMed]

71. De S.F-Ticher, P.C.; Talarico, L.B.; Noseda, M.D.; Pita, B.; Guimarães, S.M.; Damonte, E.B.; Duarte, M.E.R. Chemical structure and antiviral activity of carrageenans from Meristiella gelidium against herpes simplex and dengue virus. *Carbohydr. Polym.* **2006**, *63*, 459–465. [CrossRef]

72. Girond, S.; Crance, J.M.; Van Cuyck-Gandre, H.; Renaudet, J.; Deloince, R. Antiviral activity of carrageenan on hepatitis A virus replication in cell culture. *Res. Virol.* **1991**, *142*, 261–270. [CrossRef]

73. Bouhlal, R.; Haslin, C.; Chermann, J.-C.; Colliec-Jouault, S.; Sinquin, C.; Simon, G.; Cerantola, S.; Riadi, H.; Bourgougnon, N. Antiviral Activities of Sulfated Polysaccharides Isolated from Sphaerococcus coronopifolius (Rhodophyta, Gigartinales) and Boergeseniella thuyoides (Rhodophyta, Ceramiales). *Mar. Drugs* **2011**, *9*, 1187–1209. [CrossRef]

74. Lopes, N.; Ray, S.; Espada, S.F.; Bomfim, W.A.; Ray, B.; Faccin-Galhardi, L.C.; Linhares, R.E.C.; Nozawa, C. Green seaweed Enteromorpha compressa (Chlorophyta, Ulvaceae) derived sulphated polysaccharides inhibit herpes simplex virus. *Int. J. Biol. Macromol.* **2017**, *102*, 605–612. [CrossRef]

75. Robic, A.; Rondeau-Mouro, C.; Sassi, J.F.; Lerat, Y.; Lahaye, M. Structure and interactions of ulvan in the cell wall of the marine green algae Ulva rotundata (Ulvales, Chlorophyceae). *Carbohydr. Polym.* **2009**, *77*, 206–216. [CrossRef]

76. Wang, S.; Wang, W.; Hao, C.; Yunjia, Y.; Qin, L.; He, M.; Mao, W. Antiviral activity against enterovirus 71 of sulfated rhamnan isolated from the green alga Monostroma latissimum. *Carbohydr. Polym.* **2018**, *200*, 43–53. [CrossRef] [PubMed]

77. Morán-Santibañez, K.; Cruz-Suárez, L.E.; Ricque-Marie, D.; Robledo, D.; Freile-Pelegrín, Y.; Peña-Hernández, M.A.; Rodríguez-Padilla, C.; Trejo-Avila, L.M. Synergistic Effects of Sulfated Polysaccharides from Mexican Seaweeds against Measles Virus. *BioMed Res. Int.* **2016**, *2016*, 8502123. [CrossRef]

78. Aguilar-Briseño, J.A.; Cruz-Suarez, L.E.; Sassi, J.-F.; Ricque-Marie, D.; Zapata-Benavides, P.; Mendoza-Gamboa, E.; Rodríguez-Padilla, C.; Trejo-Avila, L.M. Sulphated Polysaccharides from Ulva clathrata and Cladosiphon okamuranus Seaweeds both Inhibit Viral Attachment/Entry and Cell-Cell Fusion, in NDV Infection. *Mar. Drugs* **2015**, *13*, 697–712. [CrossRef] [PubMed]

79. Hardouin, K.; Bedoux, G.; Burlot, A.-S.; Donnay-Moreno, C.; Bergé, J.-P.; Nyvall-Collén, P.; Bourgougnon, N. Enzyme-assisted extraction (EAE) for the production of antiviral and antioxidant extracts from the green seaweed Ulva armoricana (Ulvales, Ulvophyceae). *Algal Res.* **2016**, *16*, 233–239. [CrossRef]

80. Terasawa, M.; Hayashi, K.; Lee, J.-B.; Nishiura, K.; Matsuda, K.; Hayashi, T.; Kawahara, T. Anti-Influenza A Virus Activity of Rhamnan Sulfate from Green Algae Monostroma nitidum in Mice with Normal and Compromised Immunity. *Mar. Drugs* **2020**, *18*, 254. [CrossRef] [PubMed]

81. Lee, J.-B.; Koizumi, S.; Hayashi, K.; Hayashi, T. Structure of rhamnan sulfate from the green alga Monostroma nitidum and its anti-herpetic effect. *Carbohydr. Polym.* **2010**, *81*, 572–577. [CrossRef]

82. Cassolato, J.E.F.; Noseda, M.D.; Pujol, C.A.; Pellizzari, F.M.; Damonte, E.B.; Duarte, M.E.R. Chemical structure and antiviral activity of the sulfated heterorhamnan isolated from the green seaweed Gayralia oxysperma. *Carbohydr. Res.* **2008**, *343*, 3085–3095. [CrossRef] [PubMed]

83. Ohta, Y.; Lee, J.-B.; Hayashi, K.; Hayashi, T. Isolation of Sulfated Galactan from Codium fragile and Its Antiviral Effect. *Biol. Pharm. Bull.* **2009**, *32*, 892–898. [CrossRef]

84. Lee, J.B.; Hayashi, K.; Maeda, M.; Hayashi, T. Antiherpetic activities of sulfated polysaccharides from green algae. *Planta Med* **2004**, *70*, 813–817. [CrossRef]

85. Chiu, Y.-H.; Chan, Y.-L.; Li, T.-L.; Wu, C.-J. Inhibition of Japanese Encephalitis Virus Infection by the Sulfated Polysaccharide Extracts from Ulva lactuca. *Mar. Biotechnol.* **2012**, *14*, 468–478. [CrossRef]

86. Song, L.; Chen, X.; Liu, X.; Zhang, F.; Hu, L.; Yue, Y.; Li, K.; Li, P. Characterization and Comparison of the Structural Features, Immune-Modulatory and Anti-Avian Influenza Virus Activities Conferred by Three Algal Sulfated Polysaccharides. *Mar. Drugs* **2015**, *14*, 4. [CrossRef]

87. Pujol, C.A.; Ray, S.; Ray, B.; Damonte, E.B. Antiviral activity against dengue virus of diverse classes of algal sulfated polysaccharides. *Int. J. Biol. Macromol.* **2012**, *51*, 412–416. [CrossRef] [PubMed]

88. Hayashi, T.; Hayashi, K.; Maeda, M.; Kojima, I. Calcium Spirulan, an Inhibitor of Enveloped Virus Replication, from a Blue-Green Alga Spirulina platensis. *J. Nat. Prod.* **1996**, *59*, 83–87. [CrossRef] [PubMed]

89. Lee, J.-B.; Srisomporn, P.; Hayashi, K.; Tanaka, T.; Sankawa, U.; Hayashi, T. Effects of Structural Modification of Calcium Spirulan, a Sulfated Polysaccharide from Spirulina Platensis, on Antiviral Activity. *Chem. Pharm. Bull.* **2001**, *49*, 108–110. [CrossRef]

90. Mader, J.; Gallo, A.; Schommartz, T.; Handke, W.; Nagel, C.-H.; Günther, P.; Brune, W.; Reich, K. Calcium spirulan derived from Spirulina platensis inhibits herpes simplex virus 1 attachment to human keratinocytes and protects against herpes labialis. *J. Allergy Clin. Immunol.* **2016**, *137*, 197–203. [CrossRef] [PubMed]

91. Ogura, F.; Hayashi, K.; Lee, J.-B.; Kanekiyo, K.; Hayashi, T. Evaluation of an Edible Blue-Green Alga, Aphanothece sacrum, for Its Inhibitory Effect on Replication of Herpes Simplex Virus Type 2 and Influenza Virus Type A. *Biosci. Biotechnol. Biochem.* **2010**, *74*, 1687–1690. [CrossRef]

92. Hernández-Corona, A.; Nieves, I.; Meckes, M.; Chamorro, G.; Barron, B.L. Antiviral activity of Spirulina maxima against herpes simplex virus type 2. *Antivir. Res.* **2002**, *56*, 279–285. [CrossRef]

93. Huleihel, M.; Ishanu, V.; Tal, J.; Arad, S. Antiviral effect of red microalgal polysaccharides on Herpes simplex and Varicella zoster viruses. *J. Appl. Phycol.* **2001**, *13*, 127–134. [CrossRef]

94. Lee, J.-B.; Hayashi, K.; Hirata, M.; Kuroda, E.; Suzuki, E.; Kubo, Y.; Hayashi, T. Antiviral Sulfated Polysaccharide from Navicula directa, a Diatom Collected from Deep-Sea Water in Toyama Bay. *Biol. Pharm. Bull.* **2006**, *29*, 2135–2139. [CrossRef]

95. Kim, M.; Yim, J.H.; Kim, S.-Y.; Kim, H.S.; Lee, W.G.; Kim, S.J.; Kang, P.-S.; Lee, C.-K. In vitro inhibition of influenza A virus infection by marine microalga-derived sulfated polysaccharide p-KG03. *Antivir. Res.* **2012**, *93*, 253–259. [CrossRef]

96. Yim, J.H.; Kim, S.J.; Ahn, S.H.; Lee, C.K.; Rhie, K.T.; Lee, H.K. Antiviral Effects of Sulfated Exopolysaccharide from the Marine Microalga Gyrodinium impudicum Strain KG03. *Mar. Biotechnol.* **2004**, *6*, 17–25. [CrossRef]

97. Hasui, M.; Matsuda, M.; Okutani, K.; Shigeta, S. In vitro antiviral activities of sulfated polysaccharides from a marine microalga (Cochlodinium polykrikoides) against human immunodeficiency virus and other enveloped viruses. *Int. J. Biol. Macromol.* **1995**, *17*, 293–297. [CrossRef]

98. Huang, N.; Wu, M.-Y.; Zheng, C.-B.; Zhu, L.; Zhao, J.-H.; Zheng, Y.-T. The depolymerized fucosylated chondroitin sulfate from sea cucumber potently inhibits HIV replication via interfering with virus entry. *Carbohydr. Res.* **2013**, *380*, 64–69. [CrossRef] [PubMed]

99. Kato, D.; Era, S.; Watanabe, I.; Arihara, M.; Sugiura, N.; Kimata, K.; Suzuki, Y.; Morita, K.; Hidari, K.I.P.J.; Suzuki, T. Antiviral activity of chondroitin sulphate E targeting dengue virus envelope protein. *Antivir. Res.* **2010**, *88*, 236–243. [CrossRef] [PubMed]

100. Bergefall, K.; Trybala, E.; Johansson, M.; Uyama, T.; Naito, S.; Yamada, S.; Kitagawa, H.; Sugahara, K.; Bergström, T. Chondroitin sulfate characterized by the E-disaccharide unit is a potent inhibitor of herpes simplex virus infectivity and provides the virus binding sites on gro2C cells. *J. Biol. Chem.* **2005**, *280*, 32193–32199. [CrossRef]

101. Perino, A.; Consiglio, P.; Maranto, M.; De Franciscis, P.; Marci, R.; Restivo, V.; Manzone, M.; Capra, G.; Cucinella, G.; Calagna, G. Impact of a new carrageenan-based vaginal microbicide in a female population with genital HPV-infection: First experimental results. *Eur. Rev. Med. Pharm. Sci.* **2019**, *23*, 6744–6752. [CrossRef]

102. Damonte, E.B.; Matulewicz, M.C.; Cerezo, A.S. Sulfated seaweed polysaccharides as antiviral agents. *Curr. Med. Chem.* **2004**, *11*, 2399–2419. [CrossRef] [PubMed]

103. Clausen, T.M.; Sandoval, D.R.; Spliid, C.B.; Pihl, J.; Perrett, H.R.; Painter, C.D.; Narayanan, A.; Majowicz, S.A.; Kwong, E.M.; McVicar, R.N.; et al. SARS-CoV-2 Infection Depends on Cellular Heparan Sulfate and ACE2. *Cell* **2020**, *183*, 1043–1057. [CrossRef]

104. Walls, A.C.; Park, Y.J.; Tortorici, M.A.; Wall, A.; McGuire, A.T.; Veesler, D. Structure, Function, and Antigenicity of the SARS-CoV-2 Spike Glycoprotein. *Cell* **2020**, *181*, 281–292. [CrossRef]

105. Mottarella, S.E.; Beglov, D.; Beglova, N.; Nugent, M.A.; Kozakov, D.; Vajda, S. Docking server for the identification of heparin binding sites on proteins. *J. Chem. Inf. Model* **2014**, *54*, 2068–2078. [CrossRef]

106. Singh, J.; Samal, J.; Kumar, V.; Sharma, J.; Agrawal, U.; Ehtesham, N.Z.; Sundar, D.; Rahman, S.A.; Hira, S.; Hasnain, S.E. Structure-Function Analyses of New SARS-CoV-2 Variants B.1.1.7, B.1.351 and B.1.1.28.1: Clinical, Diagnostic, Therapeutic and Public Health Implications. *Viruses* **2021**, *13*, 439. [CrossRef]

107. Witvrouw, M.; De Clercq, E. Sulfated polysaccharides extracted from sea algae as potential antiviral drugs. *Gen. Pharm.* **1997**, *29*, 497–511. [CrossRef]

108. Mukherjee, S.; Ghosh, K.; Hahn, F.; Wangen, C.; Strojan, H.; Müller, R.; Anand, N.; Ali, I.; Bera, K.; Ray, B.; et al. Chemically sulfated polysaccharides from natural sources: Assessment of extraction-sulfation efficiencies, structural features and antiviral activities. *Int. J. Biol. Macromol.* **2019**, *136*, 521–530. [CrossRef] [PubMed]

109. Carlucci, M.J.; Scolaro, L.A.; Errea, M.I.; Matulewicz, M.C.; Damonte, E.B. Antiviral activity of natural sulphated galactans on herpes virus multiplication in cell culture. *Planta Med.* **1997**, *63*, 429–432. [CrossRef]
110. Rodríguez, M.C.; Merino, E.R.; Pujol, C.A.; Damonte, E.B.; Cerezo, A.S.; Matulewicz, M.C. Galactans from cystocarpic plants of the red seaweed Callophyllis variegata (Kallymeniaceae, Gigartinales). *Carbohydr. Res.* **2005**, *340*, 2742–2751. [CrossRef] [PubMed]
111. Jiao, G.; Yu, G.; Zhang, J.; Ewart, H.S. Chemical structures and bioactivities of sulfated polysaccharides from marine algae. *Mar. Drugs* **2011**, *9*, 196–223. [CrossRef]
112. Sapay, N.; Nurisso, A.; Imberty, A. Simulation of carbohydrates, from molecular docking to dynamics in water. *Methods Mol. Biol.* **2013**, *924*, 469–483. [CrossRef] [PubMed]
113. Wang, Q.; Zhang, Y.; Wu, L.; Niu, S.; Song, C.; Zhang, Z.; Lu, G.; Qiao, C.; Hu, Y.; Yuen, K.-Y.; et al. Structural and Functional Basis of SARS-CoV-2 Entry by Using Human ACE2. *Cell* **2020**, *181*, 894–904. [CrossRef] [PubMed]
114. Towler, P.; Staker, B.; Prasad, S.G.; Menon, S.; Tang, J.; Parsons, T.; Ryan, D.; Fisher, M.; Williams, D.; Dales, N.A.; et al. ACE2 X-ray structures reveal a large hinge-bending motion important for inhibitor binding and catalysis. *J. Biol. Chem.* **2004**, *279*, 17996–18007. [CrossRef]
115. Seeliger, D.; De Groot, B.L. Ligand docking and binding site analysis with PyMOL and Autodock/Vina. *J. Comput. Aided Mol. Des.* **2010**, *24*, 417–422. [CrossRef]
116. Bowers, K.J.; Chow, D.E.; Xu, H.; Dror, R.O.; Eastwood, M.P.; Gregersen, B.A.; Klepeis, J.L.; Kolossvary, I.; Moraes, M.A.; Sacerdoti, F.D. Scalable algorithms for molecular dynamics simulations on commodity clusters. In Proceedings of the SC'06: 2006 ACM/IEEE Conference on Supercomputing, Tampa, FL, USA, 11–17 November 2006; p. 43.
117. Horner-Miller, B. CellSs: A programming model for the cell BE architecture. In Proceedings of the 2006 ACM/IEEE Conference on Supercomputing, Tampa, FL, USA, 11–17 November 2006; Association for Computing Machinery: New York, NY, USA, 2006.
118. *Schrodinger Maestro*; LLC: New York, NY, USA, 2009. Available online: https://www.schrodinger.com/products/maestro (accessed on 18 July 2021).
119. Jo, S.; Kim, T.; Iyer, V.G.; Im, W. CHARMM-GUI: A web-based graphical user interface for CHARMM. *J. Comput. Chem.* **2008**, *29*, 1859–1865. [CrossRef] [PubMed]
120. Humphrey, W.; Dalke, A.; Schulten, K. VMD: Visual molecular dynamics. *J. Mol. Graph.* **1996**, *14*, 33–38. [CrossRef]
121. Kim, S.; Oshima, H.; Zhang, H.; Kern, N.R.; Re, S.; Lee, J.; Roux, B.; Sugita, Y.; Jiang, W.; Im, W. CHARMM-GUI Free Energy Calculator for Absolute and Relative Ligand Solvation and Binding Free Energy Simulations. *J. Chem. Theory Comput.* **2020**, *16*, 7207–7218. [CrossRef] [PubMed]

marine drugs

Review

Culturable Microorganisms Associated with Sea Cucumbers and Microbial Natural Products

Lei Chen *, Xiao-Yu Wang, Run-Ze Liu and Guang-Yu Wang *

Department of Bioengineering, School of Marine Science and Technology, Harbin Institute of Technology at Weihai, Weihai 264209, China; wangxiaoyu2020@stu.hit.edu.cn (X.-Y.W.); liurunze2017@stu.hit.edu.cn (R.-Z.L.)
* Correspondence: chenlei18@hit.edu.cn or chenleihit@163.com (L.C.); wanggy@hit.edu.cn or wanggy18_2007@163.com (G.-Y.W.); Tel.: +86-631-5687076 (L.C.); +86-631-5682925 (G.-Y.W.)

Abstract: Sea cucumbers are a class of marine invertebrates and a source of food and drug. Numerous microorganisms are associated with sea cucumbers. Seventy-eight genera of bacteria belonging to 47 families in four phyla, and 29 genera of fungi belonging to 24 families in the phylum Ascomycota have been cultured from sea cucumbers. Sea-cucumber-associated microorganisms produce diverse secondary metabolites with various biological activities, including cytotoxic, antimicrobial, enzyme-inhibiting, and antiangiogenic activities. In this review, we present the current list of the 145 natural products from microorganisms associated with sea cucumbers, which include primarily polyketides, as well as alkaloids and terpenoids. These results indicate the potential of the microorganisms associated with sea cucumbers as sources of bioactive natural products.

Keywords: sea cucumber; bioactivity; diversity; microorganism; polyketides; alkaloids

Citation: Chen, L.; Wang, X.-Y.; Liu, R.-Z.; Wang, G.-Y. Culturable Microorganisms Associated with Sea Cucumbers and Microbial Natural Products. *Mar. Drugs* **2021**, *19*, 461. https://doi.org/10.3390/md19080461

Academic Editor: Khaled A. Shaaban

Received: 20 July 2021
Accepted: 13 August 2021
Published: 16 August 2021

Publisher's Note: MDPI stays neutral with regard to jurisdictional claims in published maps and institutional affiliations.

1. Introduction

Sea cucumbers are marine invertebrates that belong to the class Holothuroidea of the phylum Echinodermata. Globally, there are about 1500 species of sea cucumbers [1], which are divided into three subclasses: Aspidochirotacea, Apodacea, and Dendrochirotacea, and can be further divided into six orders: Aspidochirotida, Elasipodida, Apodida, Molpadida, Dendrochirotida, and Dactylochirotida [2].

Sea cucumbers are found in benthic areas and the deep sea worldwide [3]. They play an important role in marine ecosystems and occupy a similar niche to earthworms in terrestrial ecosystems [4]. Sea cucumbers obtain food by ingesting marine sediments or filtering seawater [5] and provide a unique, fertile habitat for a variety of microorganisms, including bacteria and fungi [6]. However, since most microorganisms are unculturable under conventional laboratory conditions [7], this review primarily focuses on culturable sea-cucumber-associated microorganisms.

Sea cucumbers have been used in medicine in Asia for a long time [8]. For example, an ointment derived from the sea cucumber *Stichopus* sp. 1 is used to treat back and joint pain in Malaysia [9]. Compounds isolated from sea cucumbers have a variety of biological and pharmacological activities, such as anticancer, antiangiogenic, anticoagulant/antithrombotic, antioxidant, antiinflammatory, antimicrobial, antihypertension, and radioprotective properties [10,11]. A phase II clinical trial of a sea cucumber extract, called TBL-12, has been conducted in patients with untreated asymptomatic myeloma [12]. Many studies have shown that the microorganisms associated with marine animals, such as sponges and ascidians, are the true producers of marine natural products [13–16]. Therefore, investigating sea-cucumber-associated microorganisms is essential for discovering new compounds with potential as novel active drugs. For the past 20 years, there has been an increasing effort made by researchers on diversity and bioactive compounds of microorganisms associated with sea cucumber. However, previously, no comprehensive review article as such has ever been published about this field.

This review discusses the biodiversity of the culturable microorganisms associated with sea cucumbers and the chemical structure and bioactive properties of the secondary metabolites produced by these microorganisms.

2. Microorganisms Associated with Sea Cucumbers

2.1. Geographical Distribution of Microorganisms Associated with Sea Cucumbers

Although sea cucumbers are distributed in oceans worldwide [3], most studies on the biological and chemical diversity of sea-cucumber-associated microorganisms have focused on species in the northern temperate areas and tropical areas of the eastern hemisphere [17–26]. More than 80% of the sampling sites are located on the west coast of the Pacific Ocean. However, a small number of sampling sites are also located in the Atlantic, Indian, and Antarctic Oceans [17–26] (Figure 1 and Table S1). Sea cucumber samples are typically collected from the coast at a depth of less than 20 m [17–21].

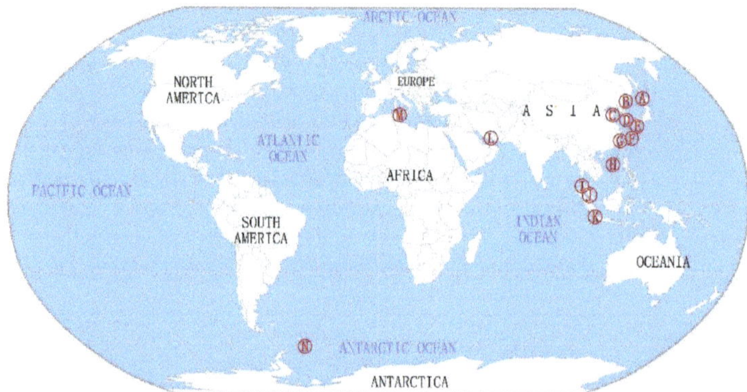

Figure 1. Geographical distribution of sea cucumber samples used for studies of culturable microorganisms. The red circles represent sampling sites: (A) Funka Bay and Ainuma fishing port, Hokkaido, Japan; (B) Sea of Japan, Russia; (C) Yellow Sea, China; (D) Geomun-do, Yeosu, Korea; (E) Kushima, Omura; Koecho; Nagasaki; Japan; (F) Coast of Aka Island, Okinawa prefecture, Japan; (G) Ningde, Fujian, China; (H) South China Sea, China; (I) Dayang Bunting Island, Yan, Kedah Darul Aman, Malaysia; (J) Tioman Island, Pahang Darul Makmur; Peninsular Malaysia; Pangkor Island, Perak; Malaysia; (K) Sari Ringgung, Lampung, Indonesia; (L) Larak Island, Persian Gulf, Iran; (M) Tabarka, Tunisia; and (N) the Antarctic.

2.2. Culturable Microorganisms Associated with Sea Cucumbers

The sea cucumbers used for the isolation of culturable microorganisms belong to five genera (*Holothuria, Cucumaria, Stichopus, Apostichopus,* and *Eupentacta*) in four families (Holothuriidae, Stichopodidae, Cucumariidae, and Sclerodactylidae) (Table 1). The dominant species is *Apostichopus japonicus*, which accounts for about 41% of the total sea cucumber population. In second place, *Holothuria leucospilota* accounts for about 27% of the total sea cucumber population (Table S1).

In studies on microorganisms associated with sea cucumbers, samples are primarily obtained from the following body parts: the body wall [22,23], body surface [18,21,24–29], inner body tissue [30], coelomic fluid [24,31], stomach [30], intestines [4,6,17,19,25,32–35], brown gastrointestinal tissue [30], and feces [20,22].

Sea cucumbers harbor a rich and diverse assortment of microorganisms. A variety of microorganisms, including bacteria and fungi, have been isolated from sea cucumbers. Most of the isolation conditions (medium, temperature, and aeration) are common. There are some papers on the diversity of culturable bacteria associated with sea cucumbers, which plays a very important role in understanding the digestion and diseases of sea cucumbers [4,6,17,25,33]. Because marine-derived fungi had shown potential to synthesize pharmaceutical compounds with bioactivities, researchers usually directly isolate fungi associated with sea cucumbers for the separation of active natural products [21,28,29], except one paper about the diversity and bioactivity of fungi associated with sea cucumbers [22].

2.2.1. Bacteria

To date, 78 genera belonging to 47 families in four phyla have been cultured from sea cucumbers (Table 2) [4,6,17–19,23–26,30–34]. The phylum Proteobacteria was represented by 34 genera, 23 genera belong to the phylum Actinobacteria, 13 genera belong to the phylum Firmicutes, and only eight genera were from the phylum Bacteroidetes. The bacteria isolated from sea cucumbers are mainly the genus *Bacillus*, followed by *Vibrio*, and *Pseudoalteromona* (Table S1).

Bacteria have been isolated from seven species in three genera of sea cucumbers: *Apostichopus japonicus, Holothuria atra,* Holothuria edulis, *Holothuria leucospilota, Stichopus badionotus, Stichopus chloronotus,* and *Stichopus vastus* [4,6,17–19,23–26,30–34]. A. japonicus displayed a high bacterial diversity, and 54 bacterial genera were isolated from this species. Thirty-six genera were isolated from *H. leucospilota,* and fifteen genera were isolated from *S. vastus.* Two, one, six, and three genera of bacteria were isolated from *H. atra, H. edulis, S. badionotus,* and *S. chloronotus,* respectively (Table 1 and Table S1).

2.2.2. Fungi

Sea-cucumber-associated fungi belong to 29 genera in 24 families (Table 2). All of them are in the phylum Ascomycota [20–22,27–29,35–45]. The dominant genus was *Aspergillus,* followed by *Penicillium* (Table S1).

Fungi were isolated from six species in five genera of sea cucumbers: *A. japonicus, Cucumaria japonica, Eupentacta fraudatrix, Holothuria nobilis, Holothuria poli,* and *Stichopus japonicus* [20,22,29,35–42]. Among them, the greatest number of fungal species was isolated from *H. poli,* with 16 genera. Thirteen genera were isolated from *E. fraudatrix,* and twelve genera were isolated from *A. japonicus.* Two, three, and one genera of fungi were isolated from the sea cucumbers *C. japonica, H. nobilis,* and *S. japonicus,* respectively (Table 1 and Table S1).

Table 1. Sea cucumbers used for the isolation of culturable microorganisms.

Sea Cucumbers			Microorganism Genera		References
Family	**Genus**	**Species**	**Bacteria**	**Fungi**	
Cucumariidae	*Cucumaria*	*japonica*	0	2	[20,36]
Holothuriidae	*Holothuria*	*atra*	2	0	[30]
		edulis	1	0	[18]
		leucospilota	36	0	[4,19,25,31]
		nobilis	0	3	[35,37,38]
		poli	0	16	[22]
Sclerodactylidae	*Eupentacta*	*fraudatrix*	0	13	[20,21]
Stichopodidae	*Apostichopus*	*japonicus*	54	12	[6,17,20,23,24,29,32–34,39,40]
	Stichopus	*badionotus*	6	0	[26]
		chloronotus	3	0	[31]
		japonicus	0	1	[41,42]
		vastus	15	0	[25]

Table 2. Culturable microorganisms associated with sea cucumbers.

Kingdom	Phylum	Class	Family	Genus	References
Bacteria	Actinobacteria	Acidimicrobiia	Iamiaceae	*Iamia*	[18]
		Actinomycetia	Brevibacteriaceae	*Brevibacterium*	[23,25]
			Corynebacteriaceae	*Corynebacterium*	[25]
			Dermabacteraceae	*Brachybacterium*	[6]
			Dermacoccaceae	*Dermacoccus*	[25]
			Dietziaceae	*Dietzia*	[25]
			Gordoniaceae	*Williamsia*	[24]
			Intrasporangiaceae	*Janibacter*	[25]
			Kytococcaceae	*Kytococcus*	[25,31]
			Microbacteriaceae	*Microbacterium*	[6,32]
			Micrococcaceae	*Glutamicibacter*	[6,25]
				Kocuria	[25]
				Micrococcus	[4,6,24,25,31,33]
				Rothia	[24,25,31]
			Nocardioidaceae	*Nocardioides*	[25]
			Nocardiopsaceae	*Nocardiopsis*	[4,6,17]
			Oerskoviaceae	*Paraoerskovia*	[4]
			Ornithinimicrobiaceae	*Ornithinimicrobium*	[25]
				Serinicoccus	[25]
			Promicromonosporaceae	*Cellulosimicrobium*	[6,25]
				Isoptericola	[25]
			Propionibacteriaceae	*Pseudopropionibacterium*	[25]
			Streptomycetaceae	*Streptomyces*	[6,17,19,25]
	Bacteroidetes	Cytophagia	Cytophagaceae	*Cytophaga*	[24]
		Flavobacteriia	Flavobacteriaceae	*Flavobacterium*	[33]
				Lacinutrix	[24]
				Maribacter	[24]
				Psychroserpens	[24]
				Ulvibacter	[24]
				Winogradskyella	[24]
				Zobellia	[24]
	Firmicutes	Bacilli	Bacillaceae	*Bacillus*	[4,6,17,24,25,30–33]
				Geomicrobium	[4,17]
				Gracilibacillus	[4,17]
				Halobacillus	[4,6,17]
				Halolactibacillus	[17]
				Oceanobacillus	[4,17]
				Salsuginibacillus	[17]
				Virgibacillus	[4,6,17]
			Planococcaceae	*Lysinibacillus*	[17]
				Planococcus	[26]

Table 2. *Cont.*

Kingdom	Phylum	Class	Family	Genus	References
				Sporosarcina	[4,17]
			Staphylococcaceae	*Staphylococcus*	[4,25]
	Proteobacteria	Alphaproteobacteria	Unidentified Ahrensiaceae	*Exiguobacterium* *Ahrensia*	[26,31] [24]
			Erythrobacteraceae	*Erythrobacter*	[25]
			Rhizobiaceae	*Agrobacterium*	[24]
			Rhodobacteraceae	*Epibacterium*	[25]
				Marinosulfonomonas	[24]
				Octadecabacter	[24]
				Paracoccus	[25]
				Roseobacter	[24]
				Ruegeria	[4]
			Sphingomonadaceae	*Sphingomonas*	[24,26]
			Stappiaceae	*Pseudovibrio*	[17]
		Betaproteobacteria	Comamonadaceae	*Acidovorax*	[24]
		Gammaproteobacteria	Aeromonadaceae	*Aeromonas*	[33]
				Oceanisphaera	[32]
			Alteromonadaceae	*Alteromonas*	[24]
			Colwelliaceae	*Colwellia*	[24]
			Enterobacteriaceae	*Enterobacter*	[33]
				Klebsiella	[30]
			Erwiniaceae	*Pantoea*	[25]
			Ferrimonadaceae	*Ferrimonas*	[17]
			Halomonadaceae	*Halomonas*	[4,33]
			Idiomarinaceae	*Pseudidiomarina*	[32]
			Lysobacteraceae	*Stenotrophomonas*	[31]
			Moraxellaceae	*Acinetobacter*	[25,32]
				Psychrobacter	[24–26]
			Oceanospirillaceae	*Marinobacterium*	[32]
				Marinomonas	[24,32]
			Pseudoalteromonadaceae	*Pseudoalteromonas*	[4,17,24,26,32–34]
			Pseudomonadaceae	*Pseudomonas*	[6,17,24,25,31–33]
			Psychromonadaceae	*Psychromonas*	[24]
			Shewanellaceae	*Shewanella*	[4,6,24,32]
			Vibrionaceae	*Aliivibrio*	[24]
				Photobacterium	[4]
				Vibrio	[4,6,24–26,31–33]
Fungi	Ascomycota	Dothideomycetes	Cladosporiaceae	*Cladosporium*	[20,22]
			Didymellaceae	*Epicoccum*	[20,40,43]
			Pleosporaceae	*Alternaria*	[20,22,27,28]
				Ulocladium	[20]

Table 2. *Cont.*

Kingdom	Phylum	Class	Family	Genus	References
			Saccotheciaceae	*Aureobasidium*	[22]
			Torulaceae	*Dendryphiella*	[20]
		Eurotiomycetes	Aspergillaceae	*Aspergillus*	[20,22,35,36,39,41,42]
				Emericella	[22]
				Paecilomyces	[22]
				Penicillium	[20,22]
			Onygenaceae	*Auxarthron*	[22]
		Leotiomycetes	Myxotrichaceae	*Oidiodendron*	[20]
			Ploettnerulaceae	*Cadophora*	[22]
			Sclerotiniaceae	*Botryophialophora*	[20]
		Sordariomycetes	Bionectriaceae	*Dendrodochium*	[37]
			Cephalothecaceae	*Phialemonium*	[38]
			Chaetomiaceae	*Chaetomium*	[20,22,29]
			Cordycipitaceae	*Beauveria*	[20]
			Hypocreaceae	*Acrostalagmus*	[22]
				Trichoderma	[20,22,44]
			Nectriaceae	*Fusarium*	[45]
			Plectosphaerellaceae	*Verticillium*	[20]
			Stachybotryaceae	*Stachybotrys*	[22]
			Tilachlidiaceae	*Tilachlidium*	[20]
			Unidentified	*Acremonium*	[20–22]
			Unidentified	*Myrothecium*	[22]
			Unidentified	*Stilbella*	[20]
		Unidentified	Unidentified	*Myriodontium*	[22]
		Unidentified	Unidentified	*Phialophorophoma*	[20]

3. Structures and Bioactivities of Natural Products

To date, 145 natural products have been isolated from sea-cucumber-associated microorganisms (Figure 2). These compounds include polyketides, alkaloids, and terpenoids, among others. These natural products have diverse properties, such as cytotoxic [37,39,45], antimicrobial [44], enzyme-inhibiting [46], and antiangiogenic activities [47].

3.1. Polyketides

Polyketides are a class of secondary metabolites that are produced by bacteria, fungi, actinobacteria, and plants [48,49]. They include polyphenols, macrolides, polyenes, anthraquinones, enediynes, and other compounds [50,51]. Polyketides have diverse bioactive properties, including antibiotic, antifungal, immunosuppressant, antiparasitic, cholesterol-lowering, and antitumoral activities [50,52].

The polyketones territrem A (**1**), territrem B (**2**), dihydrogeodin (**3**), emodin (**4**), questin (**5**), and 1-(2,4-dihydroxyphenyl)-ethanone (**6**) were isolated from the marine fungus *Aspergillus terreus*, associated with the sea cucumber *A. japonicus*, collected from Zhifu Island in Yantai, China [39]. Compounds **4** and **5** are common quinone compounds, and compound **4** has cytotoxic effects on human oral epithelial cancer cells (KB) and multidrug-resistant cells (KBv200), with IC_{50} values of 32.97 and 16.15 µg/mL, respectively [39]. Compound **4** was also isolated from sea-cucumber-derived fungus *Trichoderma* sp., and it showed weak inhibitory effects against *Pseudomonas putida*, with a minimum inhibitory concentration (MIC) of 25 μM [44]. Compound **5** has weak cytotoxicity in KB and KBv200 cells, with IC_{50} values > 50 μg/mL [39].

Three additional compounds, 1-hydroxyl-3-methylanthracene-9,10-dione (**7**), chrysophanol (**8**), and sterigmatocystin (**9**), are secondary metabolites of the fungus *Alternaria* sp., isolated from sea cucumber in the sea surrounding Zhifu Island in Yantai, China [28]. Compound **8** was also isolated from a sea-cucumber-associated fungus *Trichoderma* sp. and showed weak inhibitory effects against *Vibrio parahaemolyticus*, with an MIC value of 25 μM [44].

The anthraquinone compounds coniothyrinone A (**10**) and lentisone (**11**) were isolated from the fungus *Trichoderma* sp. associated with a sea cucumber that was collected from Chengshantou Island in the Yellow Sea in Weihai City, China [44]. Compounds **10** and **11** were isolated for the first time from fungi of the genus *Trichoderma*, and they had weak antiangiogenicc activity. Compound **10** showed pronounced antibacterial activity against three common marine pathogens, *Vibrio parahaemolyticus*, *Vibrio anguillarum*, and *Pseudomonas putida*, and the MIC values were 6.25, 1.56, and 3.13 μM, respectively. Compound **11** showed inhibitory effect against *V. parahaemolyticus*, *V. anguillarum*, and *P. putida*, with MIC values of 12.5, 1.56, and 6.25 μM, respectively [44].

Six compounds, javanicin (**12**), norjavanicin (**13**), fusarubin (**14**), terrain (**15**), sclerin (**16**), and 5-hydroxy-7-methoxy-3-methyl-2-(2-oxopropyl) naphthalene-1,4-dione (**17**), were isolated from the sea-cucumber-associated fungus *Fusarium* sp. from the Yantai Sea, China [45]. Compounds **12–14** showed moderate cytotoxicity in KB cells, with IC_{50} values of 2.90, 10.6, and 9.61 g/mL, respectively, and they also showed moderate cytotoxic effects in KBv200 cells, with IC_{50} values of 5.91, 12.12, and 6.74 g/mL, respectively [45].

Four new polyhydroxy cyclohexanol analogues, named dendrodochol A–D (**18–21**), were isolated from the fungus *Dendrodochium* sp. associated with the sea cucumber *H. nobilis*, which was collected from the South China Sea [53]. Compounds **18** and **20** showed modest antifungal activity against *Candida* strains, *Cryptococcus neoformans*, and *Trichophyton rubrum* (MIC_{80} = 8–16 μg/mL) in an in vitro bioassay [53]. Additionally, thirteen new 12-membered macrolides, dendrodolides A–M (**22–34**), were isolated from the fungus *Dendrodochium* sp. associated with the sea cucumber *H. nobilis* [37]. Compounds **22–25, 29, 30**, and **32** showed cytotoxic effects on SMMC-7721 tumor cells, with IC_{50} values of 19.2, 24.8, 18.0, 15.5, 21.8, 14.7, and 21.1 μg/mL, respectively [37]. Compounds **24, 26, 28, 30, 32**, and **33** had cytotoxic effects on HCT116 tumor cells, with IC_{50} values of 13.8, 5.7, 9.8, 11.4, 15.9, and 26.5 μg/mL, respectively [37].

Aspergillolide (**35**), a newly discovered 12-membered macrolide, was isolated from the fungus *Aspergillus* sp. S-3-75, associated with the sea cucumber *H. nobilis* that was collected from the Antarctic [35].

Azaphilone compounds are fungal polyketide pigments produced by a variety of ascomycetes and basidiomycetes [54]. Four previously known azaphilones, chaetoviridin A (**36**), chaetoviridin E (**37**), chaetoviridin B (**38**), and chaetomugilin A (**39**), and a known cochliodinol (**40**), were produced by the fungus *Chaetomium globosum*, associated with the sea cucumber *A. japonicus*, which was collected from Chengshantou Island, Weihai, China [29].

3.2. Alkaloids

Alkaloids have been identified as a class of nitrogenous organic compounds derived from plants [55,56]; although they are most commonly found in plants, alkaloids can also be isolated from marine organisms and marine microorganisms [57,58].

Chaetoglobosins, which are a large class of secondary metabolites that are cytochalasin alkaloids, have been isolated mainly from the fungus *Chaetomium globosum* [59]. Three previously known chaetoglobosins, chaetoglobosin Fex (**41**), G (**42**), and B (**43**), and one new chaetoglobosin, cytoglobosin X (**44**), were isolated from the fungus *Chaetomium globosum*, associated with the sea cucumber *A. japonicus*, on Chengshantou Island, China [29]. Compound **43** has some inhibitory effects against *Staphylococcus aureus* and methicillin-resistant *Staphylococcus aureus* (MRSA), with MIC values of 47.3 and 94.6 µM, respectively, and weak activity against *Candida albicans* SC5314, *Candida albicans* 17#, *Pseudomonas aeruginosa*, and *Bacillus Calmette–Guérin* (BCG), with MIC values >100 µg/mL for all organisms [29].

Nineteen compounds were isolated from the fungus *Aspergillus fumigatus*, associated with the sea cucumber *S. japonicus*, collected near Lingshan Island, Qingdao, China [33]. Among these 19 compounds are seven new prenylated indole diketopiperazine alkaloids, including compound **45**, three spirotryprostatins (C–E) (**46–48**), two derivatives of fumitremorgin B (**49 and 50**), and 13-oxoverruculogen (**51**), along with 12 known compounds, including spirotryprostatin A (**52**), 13-oxofumitremorgin B (**53**), fumitremorgin B (**54**), verruculogen (**55**), 3-β hydroxy cyclo-L-tryptophyl-L-proline (**56**), cyclo-L-tryptophyl-L-proline (**57**), tryprostatin B (**58**), tryprostatin A (**59**), N-prenyl-cyclo-L-tryptophyl-L-proline (**60**), fumitremorgin C (**61**), 12,13-dihydroxyfumitremorgin C (**62**), and cyclotryprostatin A (**63**) [41]. Compound **45** showed weak cytotoxicity in HL–60 cells, with an IC_{50} value of 125.3 µM. Compounds **46–51** exhibited some cytotoxicity in MOLT–4 cells, HL–60 cells, A–549 cells, and BEL-7402 cells. Compound **48** showed higher activity in MOLT–4 and A–549 cells than the others, with an IC_{50} value of 3.1 µM for both cell types. Compound **49** showed higher activity in BEL–7402 cells than the others, with an IC_{50} value of 7.0 µM. Compound **51** showed higher activity in HL–60 cells than the others, with an IC_{50} value of 1.9 µM [41]. Compound **53** was also isolated from the fungus *Aspergillus* sp., associated with the sea cucumber *S. japonicus*, collected from Lingshan Island, Qingdao, China [42]. Two new compounds, pseurotin A_1 (**64**) and A_2 (**65**), as well as pseurotin A (**66**) were also isolated from the fungus *Aspergillus fumigatus*, associated with the sea cucumber *S. japonicus*. Compound **65** exhibited slight cytotoxicity in A549 and HL-60 cells, with IC_{50} values of 48.0 and 70.8 µmol/L, respectively, and compound **66** showed slight cytotoxicity in HL-60 cells, with an IC_{50} value of 67.0 µmol/L [60].

3.3. Terpenoids

Terpenoids, which are widely found in nature and in numerous species, have various structures and are divided into monoterpenes (C_{10}), sesquiterpenes (C_{15}), diterpenes (C_{20}), and sesterterpenes (C_{25}) [61]. Although most known terpenoids have been isolated from plants [62], they are also produced by marine microorganisms [63].

Three new pimarane diterpenes, aspergilone A (**67**) and compounds **68** and **69**, one new isopimarane diterpene (**70**), and four known compounds, diaporthin B (**71**), diaporthein B (**72**), 11-deoxydiaporthein A (**73**), and isopimara-8(14),15-diene (**74**), were obtained from the fungus *Epicoccum* sp., associated with the sea cucumber *A. japonicus*, which was collected from Yantai, Shandong Province, China [40,64,65]. Compounds **67**, **68**, and **71** exhibited cytotoxicity in KB cells, with IC_{50} values of 3.51, 20.74, and 3.86 µg/mL, respectively, and in KBv200 cells, with IC_{50} values of 2.34, 14.47, and 6.52 µg/mL, respectively [40]. Compounds **70** and **73** exhibited effective inhibitory activities against α-glucosidase, with IC_{50} values of 4.6 and 11.9 µM, respectively [65].

The fungus *Aspergillus* sp. H30, derived from the sea cucumber *Cucumaria japonica*, which was collected from the South China Sea, produced a meroterpenoid called chevalone B (**75**) that exhibited weak antibacterial activity [36].

Terpene glycosides are a group of natural products with a triterpene or sterol core, and marine diterpene glycosides (MDGs) are a subset of terpene glycosides [66]. Thirty-one new diterpene glycosides, including virescenosides M–R (**76–81**), R_1–R_3 (**82–84**), S–X (**85–90**), Z (**91**), and Z_4–Z_{18} (**92–106**), and three known diterpenic glycosides, virescenosides A (**107**), B (**108**), and C (**109**), together with three known analogues, virescenoside F (**110**), G (**111**), a lactone of virescenoside G (**112**), and the aglycon of virescenoside A (**113**), were isolated from the fungus *Acremonium striatisporum* KMM 4401, associated with the sea cucumber *Eupentacta fraudatrix*, which was collected from Kitovoe Rebro Bay in the Sea of Japan [21,46,67–71]. Compounds **76**, **77**, **79**, and **107–109** showed cytotoxic effects on developing eggs of the sea urchin *Strongylocentrotus intermedius* (MIC_{50} = 2.7–20 µM) [21,67]. Compounds **76–81**, **85–87**, and **107–109** exhibited cytotoxic activities against Ehrlich carcinoma tumor cells (IC$_{50}$ = 10–100 μM) in vitro [21,67,68]. Compounds **81** and **85–87** showed weak cytotoxic effects on developing eggs of the sea urchin *S. intermedius* (IC$_{50}$ = 100–150 μM) [68]. At a concentration of 100 mg/mL, compounds **82–84** and **91** inhibited esterase activity by 56%, 58%, 36%, and 40%, respectively [46]. The aglycon **113** inhibited urease activity, with an IC$_{50}$ value of 138.8 μM [71]. Compounds **97**, **98**, **100**, **101**, **104**, and **110–113**, at 10 μM, downregulated reactive oxygen species (ROS) production in lipopolysaccharide (LPS)-stimulated macrophages [71]. At 1 μM, compounds **98** and **101** induced moderate downregulation of NO production in LPS-stimulated macrophages [71].

3.4. Other Types of Compounds Isolated from Sea-Cucumber-Associated Microorganisms

Other secondary metabolites, including cyclo-(L-Pro-L-Phe) (**114**), cyclo-(L-Pro-L-Met) (**115**), cyclo-(L-Pro-L-Tyr) (**116**), cyclo-(L-Pro-L-Val) (**117**), cyclo-(L-Pro-L-Pro) (**118**), cyclo-(L-Val-L-Gly) (**119**), and cyclo-(L-Pro-L-Leu) (**120**), have been isolated from the actinomycete *Brevibacterium* sp., associated with the sea cucumber *A. japonicus* [23].

Four compounds, 5-methyl-6-hydroxy-8-methyoxy-3-methylisochroman (**121**), peroxyergosterol (**122**), succinic acid (**123**), and 8-hydroxy-3-methylisochroman-1-one (**124**), were isolated from the fungus *Epicoccum* spp., associated with sea cucumber collected in the Yellow Sea, China [43]. Compound **121** is a pheromone [43] that was also isolated from the fungus *Alternaria* sp., associated with the sea cucumber collected from the Yellow Sea in Weihai, China [27]. The fungus *Alternaria* sp., associated with sea cucumber, also produced a new benzofuran derivative, 4-acetyl-5-hydroxy-3,6,7-trimethylbenzofuran-2(3H)-one (**125**), and a known compound, 2-carboxy-3-(2-hydroxypropanyl) phenol (**126**) [27].

Two depsidones, emeguisin A (**127**) and aspergillusidone C (**128**), were isolated from the fungus *Phialemonium* sp., associated with the sea cucumber *H. nobilis*, collected in South China [38].

Three compounds, (+)-butyrolactone IV (**129**), butyrolactone I (**130**), and terrelactone A (**131**), were isolated from the fungus *Aspergillus terreus*, associated with the sea cucumber *A. japonicus*, collected from the Yellow Sea in China [47]. Compounds **129** and **130** showed moderate antiangiogenic activity when evaluated using a zebrafish assay. The inhibition ratio of compound **129**, at a concentration of 100 μg/mL, was 43.4% and that of compound **130**, at a concentration of 10 μg/mL, was 28.7% [47].

Nine known compounds, 2,4-dihydroxy-6-methylaceto-phenone (**132**), pannorin (**133**), 2-hydroxy-4-(3-hydroxy-5-methylphenoxy)-6-methylbenzoic acid (**134**), 3,3′-dihydroxy-5,5′-dimethyldiphenyl ether (**135**), aloesone (**136**), aloesol (**137**), acremolin (**138**), cyclo-(L-Trp-L-Phe) (**139**), and cyclo-(L-Trp-L-Leu) (**140**), were isolated from the fungus *Aspergillus* sp. S-3-75, associated with the sea cucumber *H. nobilis*, which was collected from the Antarctic [35].

Cerebroside (**141**) was isolated from the fungus *Alternaria* sp., associated with sea cucumber from the sea near Zhifu Island in Yantai, China [28].

Three known compounds, streptodepsipeptide P11B (**142**), streptodepsipeptide P11A (**143**), and valinomycin (**144**), and one novel valinomycin analogue, streptodepsipeptide SV21 (**145**), were produced by the actinobacteria *Streptomyces* sp. SV 21, isolated from the sea cucumber *S. vastus* in Lampung, Indonesia [72]. Compounds **142–145** exhibited antifungal activity against *Mucor hiemalis*, with MIC values of 16.6, 8.3, 2.1, and 16.6 μg/mL, respectively. These four compounds also exhibited antifungal activity against *Ruegeria glutinis*, with MIC values of 33.3, 8.3, 4.2, and 16.6 μg/mL, respectively. Compounds **144** and **145** showed activities against the Gram-positive bacterium *Staphylococcus aureus*, with MIC values of 4.2 and 16.6 μg/mL, respectively. Compound **145** showed activity against the Gram-positive bacterium *Bacillus subtilis*, with an MIC value of 33.3 μg/mL. Compounds **143–145** showed pronounced antiinfectivity effects against hepatitis C virus (HCV). Compound **142** showed weak antiinfectivity effects against HCV [72].

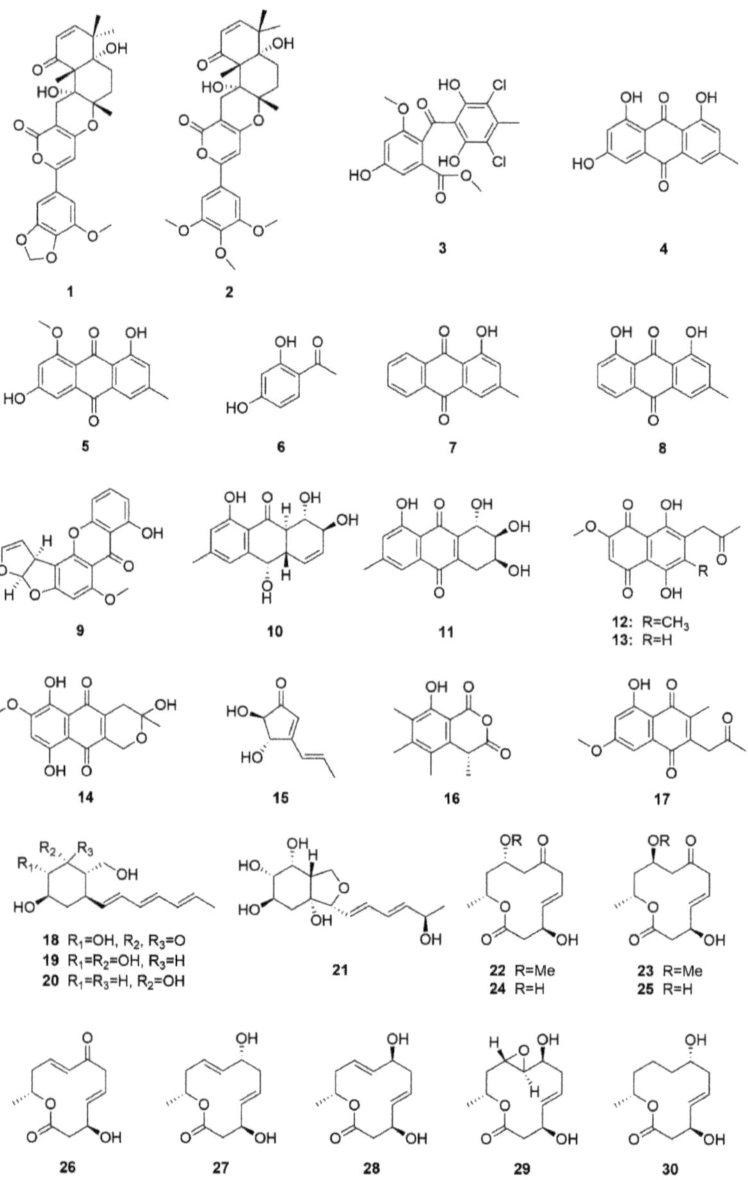

Figure 2. *Cont.*

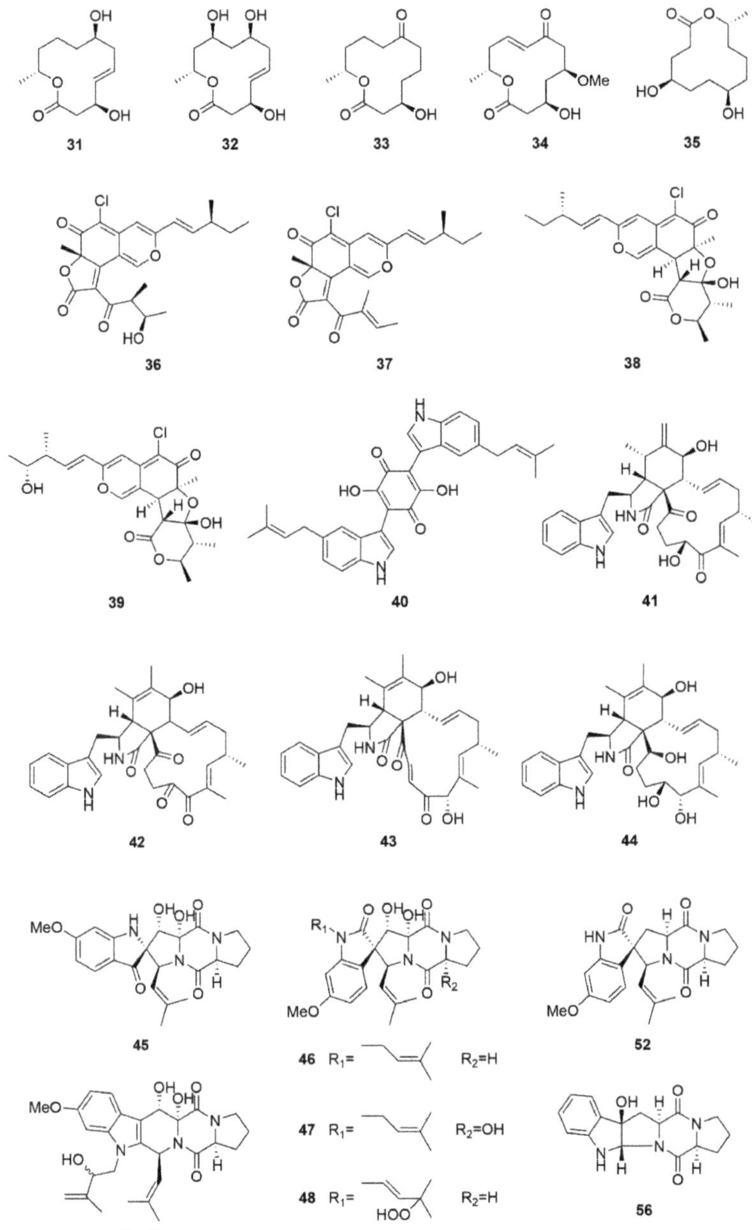

Figure 2. *Cont.*

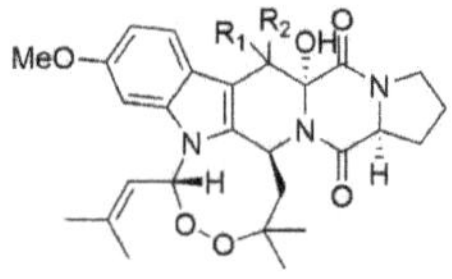

51 R_1=R_2=O
55 R_1=β-H R_2=α-OH

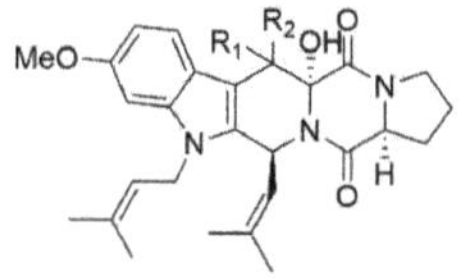

53 R_1=R_2=O
54 R_1=β-H R_2=α-OH

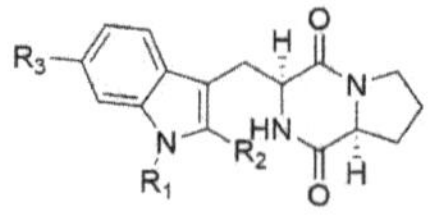

57 R_1=R_2=R_3=H
58 R_1=R_3=H R_2=
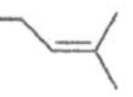
59 R_1=H R_3=OMe R_2=
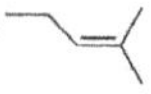
60 R_2=R_3=H R_1=
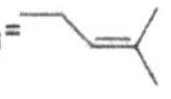

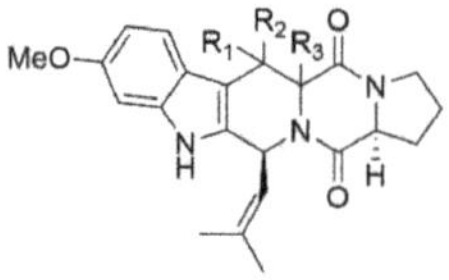

61 R_1=R_2=H R_3=β-H
62 R_1=β-H R_2=α-OH R_3=α-OH
63 R_1=β-H R_2=α-OH R_3=β-OH

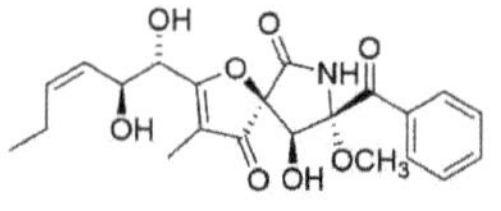

64

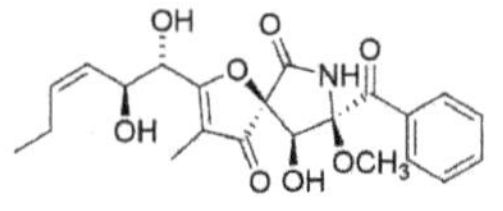

65

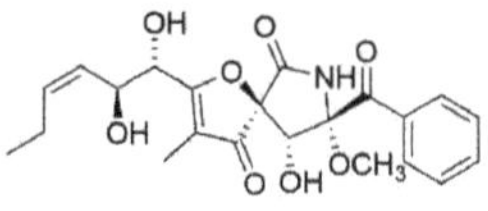

66

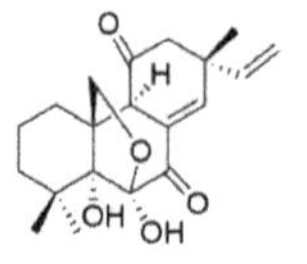

67

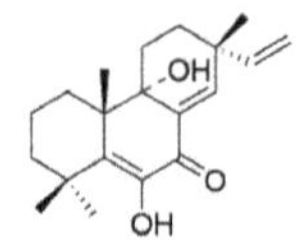

68

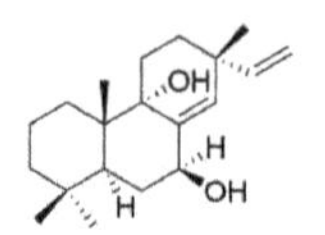

69

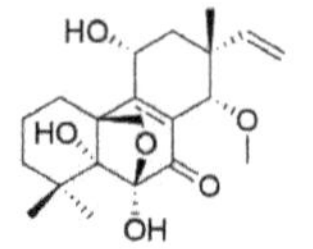

70

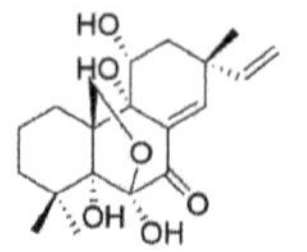

71

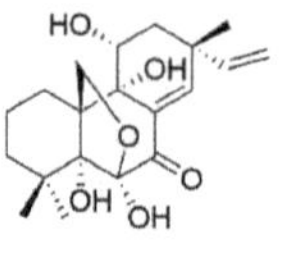

72

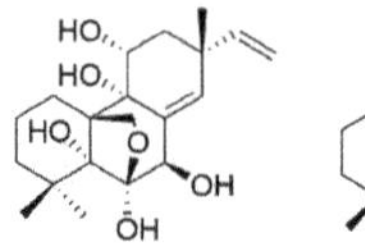

73

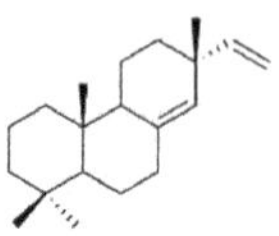

74

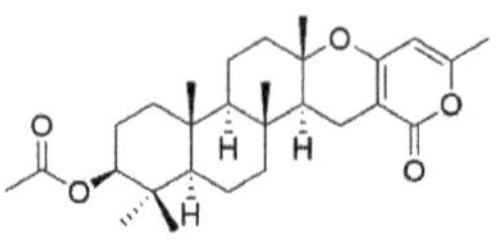

75

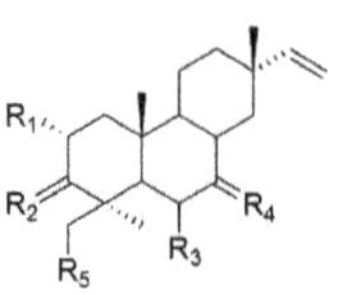

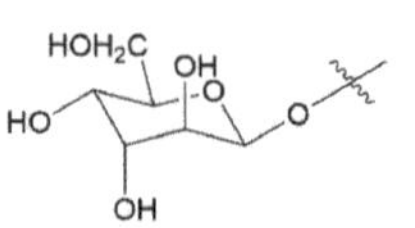

β-D-altropyranosyl-

	R_1	R_2	R_3	R_4	R_5	
76	OH	H, β-OH	H	O	β-D-altropyranosyl-	$\Delta^{8,9}$
77	OH	H, β-OH	β-OH	H, H	β-D-altropyranosyl-	$\Delta^{7,8}$
107	OH	H, β-OH	H	H, H	β-D-altropyranosyl-	$\Delta^{7,8}$
108	H	H, β-OH	H	H, H	β-D-altropyranosyl-	$\Delta^{7,8}$
109	H	O	H	H, H	β-D-altropyranosyl-	$\Delta^{7,8}$

Figure 2. *Cont.*

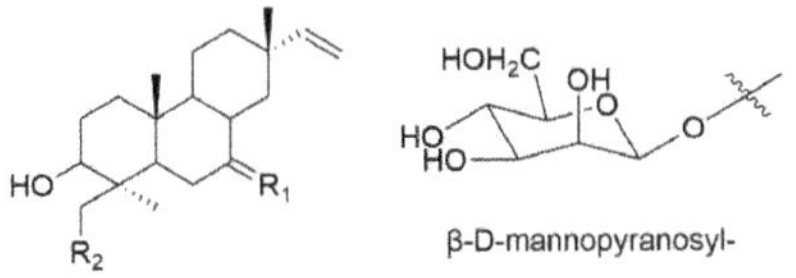

	R₁	R₂	
78	H, α-OH	β-D-altropyranosyl-	$\Delta^{8,14}$
79	O	β-D-altropyranosyl-	$\Delta^{8,9}$
80	H	β-D-mannopyranosyl-	$\Delta^{7,8}$

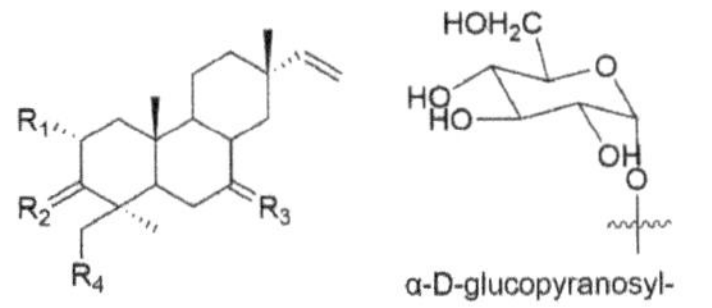

	R₁	R₂	R₃	R₄	
81	OH	H, β-OH	H	α-D-glucopyranosyl(1→6)- β-D-altropyranosyl-	$\Delta^{7,8}$
85	H	O	H, α-OH	β-D-altropyranosyl-	$\Delta^{8,14}$
86	H	O	O	β-D-altropyranosyl-	$\Delta^{8,9}$
87	H	O	O	β-D-altropyranosyl-	$\Delta^{8,14}$

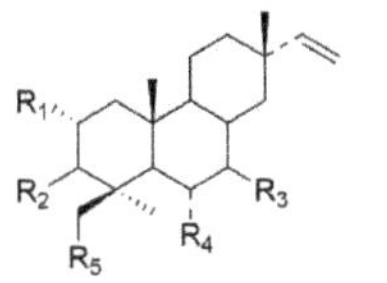

	R₁	R₂	R₃	R₄	R₅	
88	OH	H, β-OH	O	H	β-D-altropyranosyl-	$\Delta^{8,14}$
89	OH	H, β-OH	H	OH	β-D-altropyranosyl-	$\Delta^{7,8}$
90	OH	H, β-OH	H, α-OH	H	β-D-altropyranosyl-	$\Delta^{8,9}$

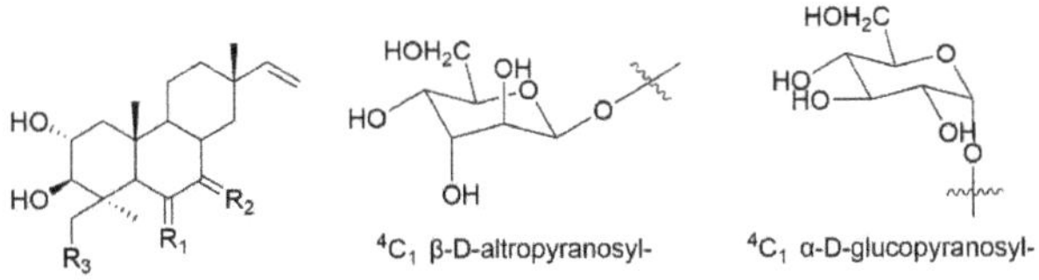

	R₁	R₂	R₃	
82	H	H	α-D-glucopyranosyl-(1→6)-β-D-altropyranosyl-	$\Delta^{6,8(14)}$
83	H, H	O	α-D-glucopyranosyl-(1→6)-β-D-altropyranosyl-	$\Delta^{8,9}$
84	H, H	H	α-D-glucopyranosyl-(1→4)-β-D-altropyranosyl-	$\Delta^{7,8}$
91	O	H	β-D-altropyranosyl-, ⁴C₁	$\Delta^{7,8}$

Figure 2. *Cont.*

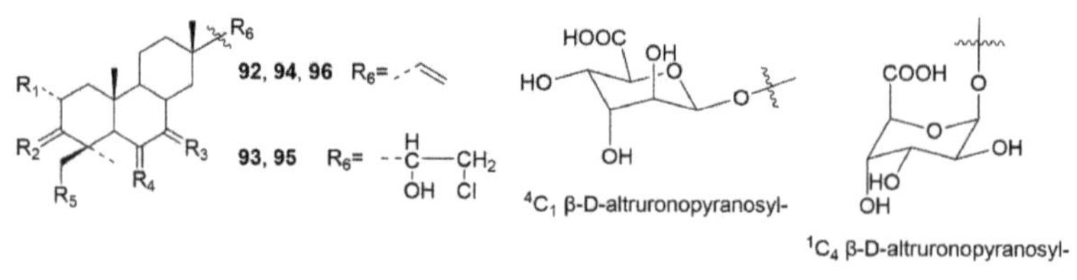

92, 94, 96 R₆=

93, 95 R₆=

$^4C_1$ β-D-altruronopyranosyl-

$^1C_4$ β-D-altruronopyranosyl-

	R₁	R₂	R₃	R₄	R₅	Δ
92	H	O	H	H, H	β-D-altruronopyranosyl-, $^4C_1$	Δ^7
93	H	O	H	H, H	β-D-altruronopyranosyl-, $^4C_1$	Δ^7
94	H	H, β-OH	H, α-OH	H, H	β-D-altruronopyranosyl-, $^4C_1$	$\Delta^{8(14)}$
95	OH	H, β-OH	H	H, H	β-D-altruronopyranosyl-, $^1C_4$	Δ^7
96	OH	H, β-OH	H	H	β-D-altruronopyranosyl-, $^1C_4$	$\Delta^{6,8(14)}$

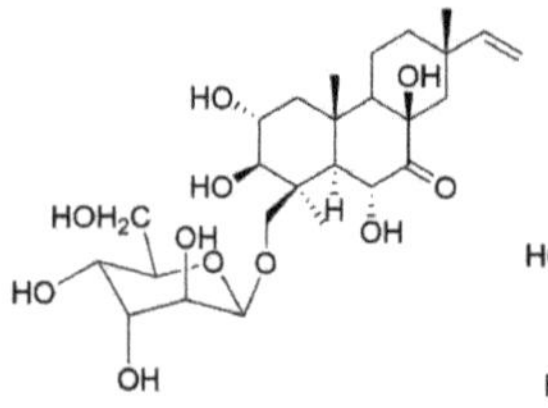

97

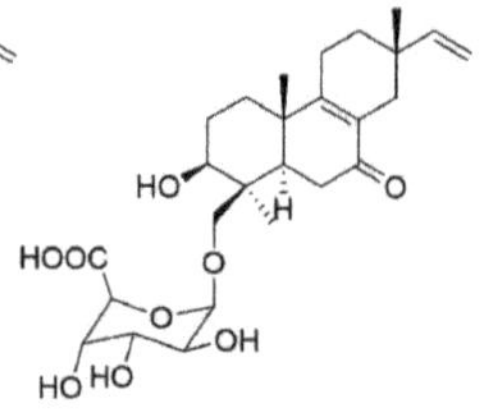

98

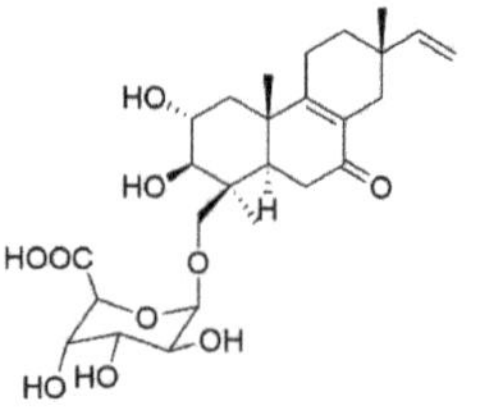

99

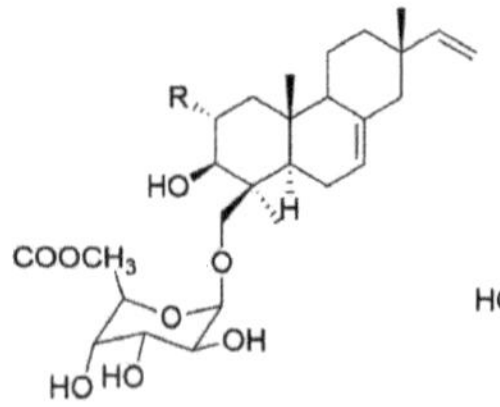

100 R=H
101 R=OH

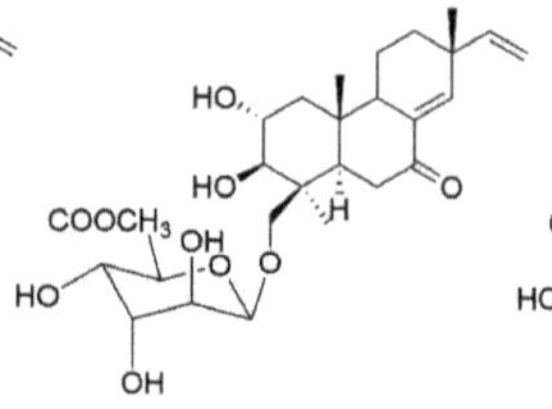

102

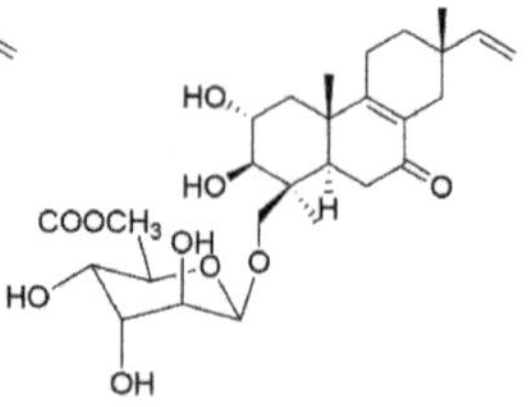

103

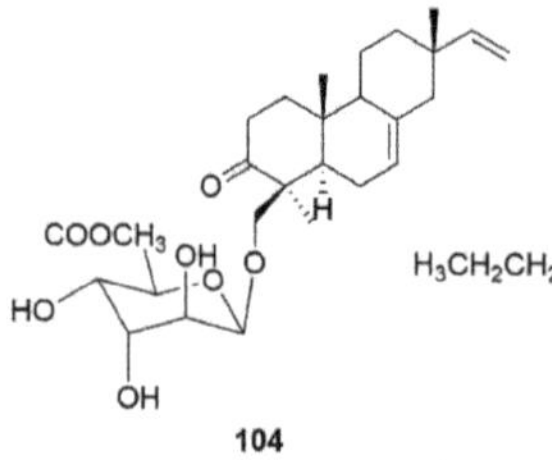

104

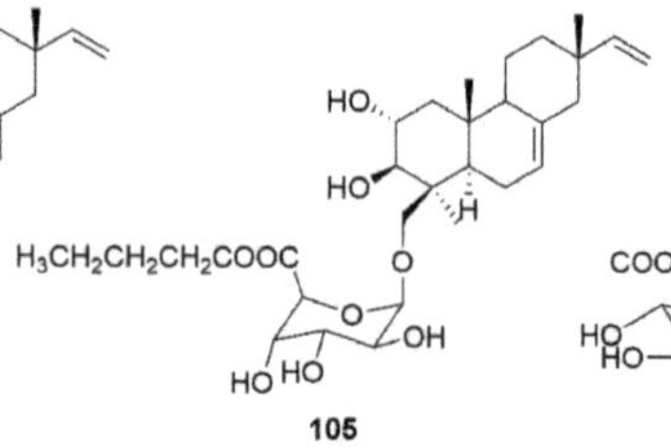

105

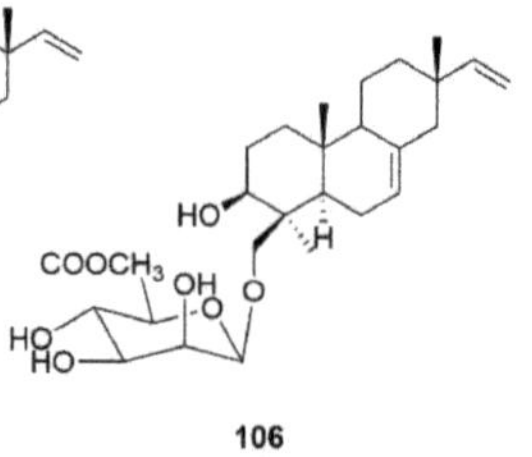

106

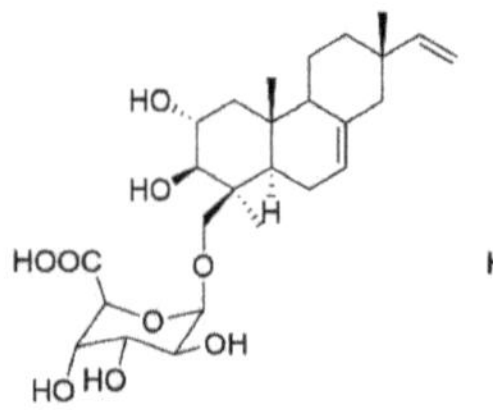

110

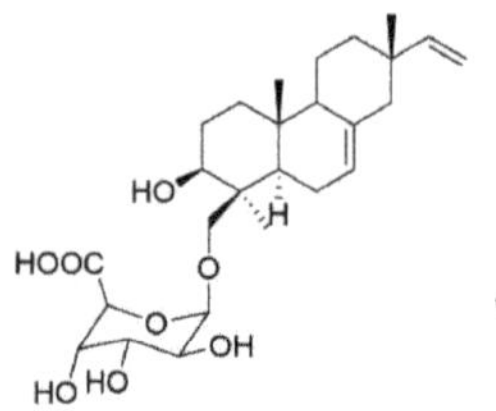

111

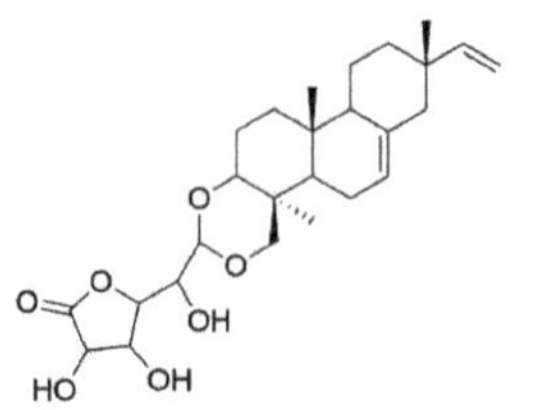

112

Figure 2. *Cont.*

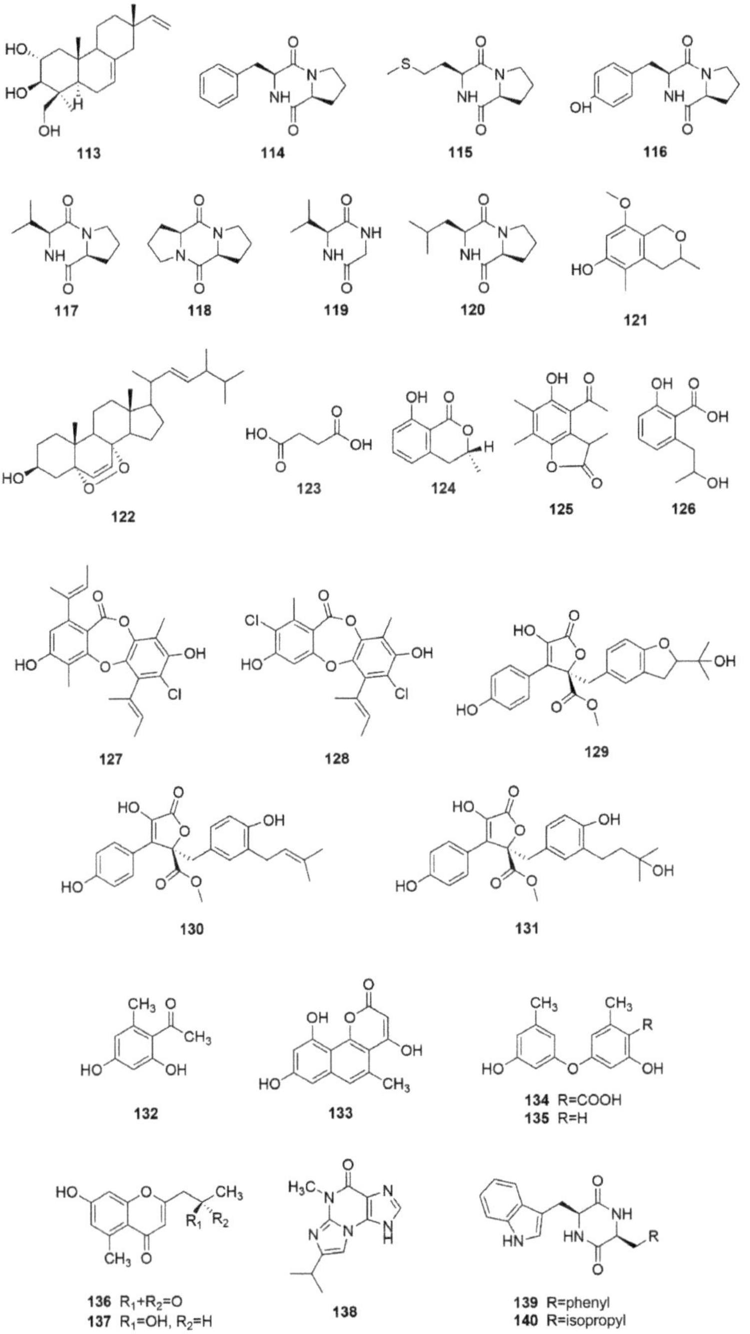

Figure 2. *Cont.*

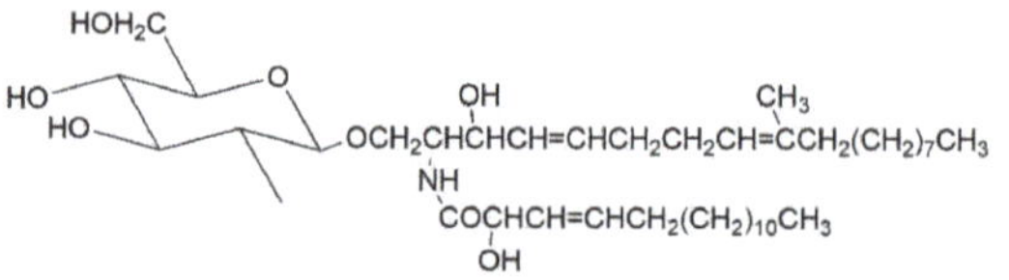

141

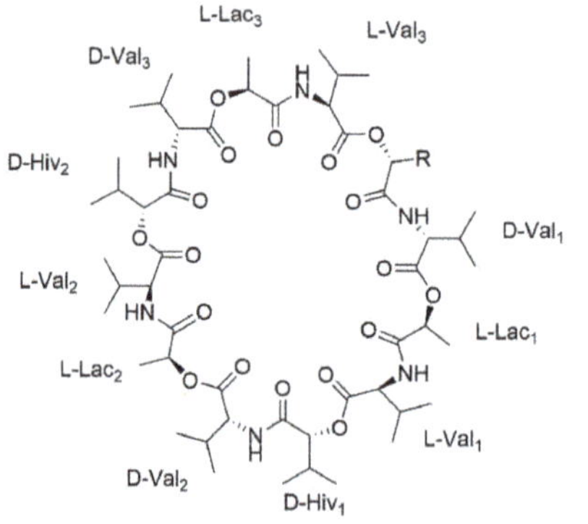

142 R=*L*-Lac
143 R=*D*-Hba
144 R=*D*-Hiv
145 R=*D*-Hmpa

Figure 2. Chemical structures of the 145 compounds isolated from sea-cucumber-associated microorganisms.

3.5. Summary of the Natural Products Isolated from Microorganisms Associated with Sea Cucumbers

From 2000 to 2021, 145 natural products were isolated from microorganisms associated with sea cucumbers. The numbers of compounds isolated in 2008, 2014, and 2020 were significantly higher than the numbers isolated in other years (Figure 3). The compounds isolated from sea-cucumber-associated microorganisms are mainly polyketides, alkaloids, and terpenoids (Figures 4 and 5), which account for 28%, 18%, and 32% of the total isolated compounds, respectively (Figure 4). Most of these compounds were isolated from sea-cucumber-associated fungi (Figure 4), and many of them have demonstrated bioactivities, including cytotoxicity, antimicrobial, enzyme-inhibiting, antiviral, and antiangiogenic activities, and the downregulation of ROS and NO production (Figure 6).

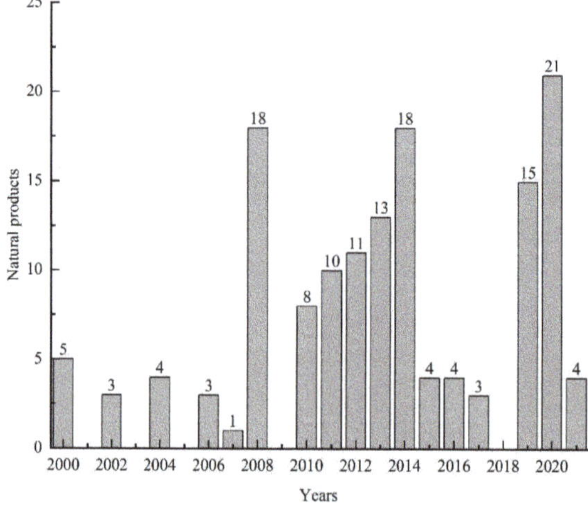

Figure 3. Natural products isolated from sea-cucumber-associated microorganisms from 2000 to 2021.

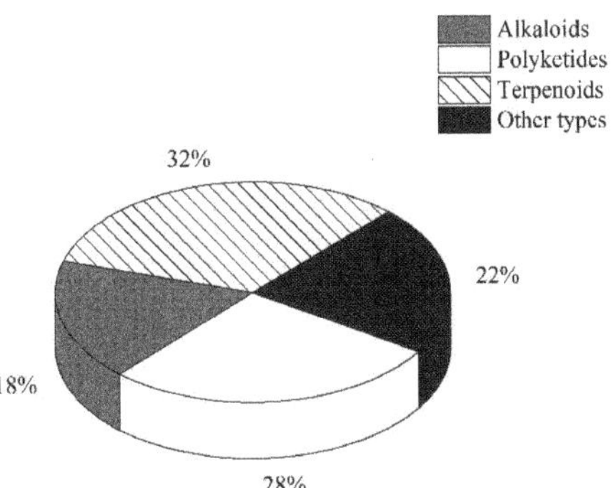

Figure 4. Percentage distribution of the natural products isolated from sea-cucumber-associated microorganisms.

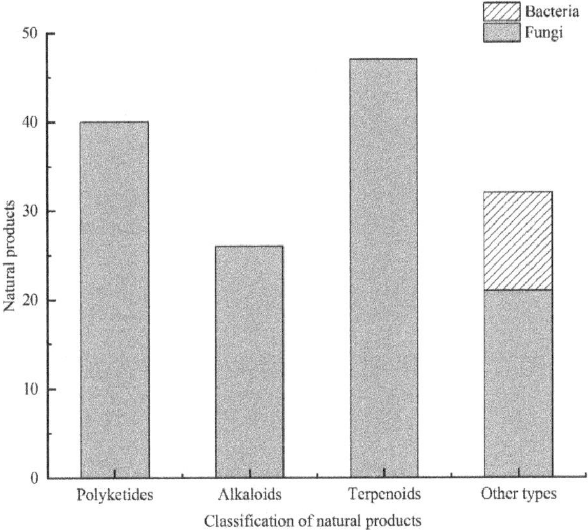

Figure 5. Natural products isolated from sea-cucumber-associated microorganisms.

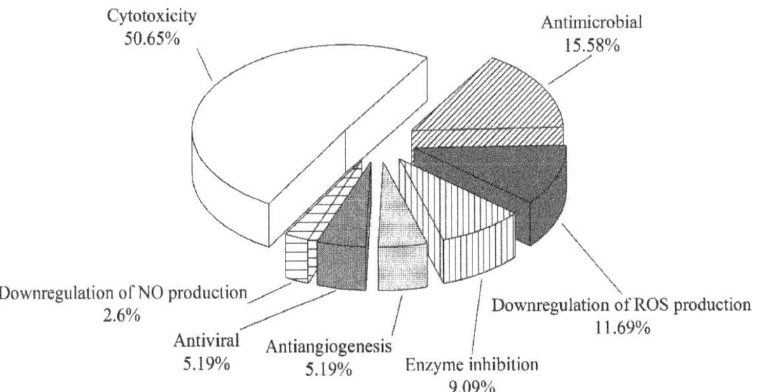

Figure 6. Percentage distribution of the bioactivities of the natural products isolated from sea-cucumber-associated microorganisms.

4. Conclusions

Sea cucumbers have been extensively utilized in medicine in Asia for a long time, and a variety of compounds with pharmacological activities have been isolated from sea cucumbers [10]. The actual producers of these marine natural products may be sea-cucumber-associated microorganisms. Sea cucumbers harbor a rich and diverse assortment of microorganisms. Over the past 20 years, seventy-eight genera of bacteria belonging to 47 families in four phyla, and 29 genera of fungi belonging to 24 families in the phylum Ascomycota have been cultured from sea cucumbers. A total of 145 natural products have been isolated from sea-cucumber-associated microorganisms. These compounds are polyketides, terpenoids, alkaloids, and others, and many have been shown to have various biological activities. Sea-cucumber-associated microorganisms have great potential for the production and isolation of high-value bioactive compounds.

Supplementary Materials: The following are available online at https://www.mdpi.com/article/10.3390/md19080461/s1, Table S1. Microorganism genera associated with sea cucumbers.

Author Contributions: L.C. and G.-Y.W. conceived and designed the format of the manuscript. L.C., X.-Y.W. and R.-Z.L. analyzed the data and drafted and edited the manuscript. X.-Y.W. drew the chemical structure of compounds. L.C. and G.-Y.W. reviewed the manuscript. All the authors contributed in terms of critical reading and discussion of the manuscript. All authors have read and agreed to the published version of the manuscript.

Funding: This study was funded by the Discipline Construction Guide Foundation in Harbin Institute of Technology at Weihai (No. WH20150204 and No. WH20160205), and the Research Innovation Foundation in Harbin Institute of Technology at Weihai (No. 2019KYCXJJYB15).

Conflicts of Interest: The authors declare no conflict of interest.

References

1. Kiew, P.L.; Don, M.M. Jewel of the seabed: Sea cucumbers as nutritional and drug candidates. *Int. J. Food Sci. Nutr.* **2012**, *63*, 616–636. [CrossRef]
2. Mondol, M.A.M.; Shin, H.J.; Rahman, M.A.; Islam, M.T. Sea cucumber glycosides: Chemical structures, producing species and important biological properties. *Mar. Drugs* **2017**, *15*, 317. [CrossRef]
3. Bordbar, S.; Anwar, F.; Saari, N. High-value components and bioactives from sea cucumbers for functional foods—A review. *Mar. Drugs* **2011**, *9*, 1761–1805. [CrossRef] [PubMed]
4. Zhang, X.; Nakahara, T.; Miyazaki, M.; Nogi, Y.; Taniyama, S.; Arakawa, O.; Inoue, T.; Kudo, T. Diversity and function of aerobic culturable bacteria in the intestine of the sea cucumber *Holothuria leucospilota*. *J. Gen. Appl. Microbiol.* **2012**, *58*, 447–456. [CrossRef]
5. Gao, F.; Li, F.; Tan, J.; Yan, J.; Sun, H. Bacterial community composition in the gut content and ambient sediment of sea cucumber *Apostichopus japonicus* revealed by 16S rRNA gene pyrosequencing. *PLoS ONE* **2014**, *9*, e100092. [CrossRef]
6. Chen, L.; Du, S.; Qu, W.Y.; Guo, F.R.; Wang, G.Y. Biosynthetic potential of culturable bacteria associated with *Apostichopus japonicus*. *J. Appl. Microbiol.* **2019**, *127*, 1686–1697. [CrossRef] [PubMed]
7. Jung, D.; Seo, E.Y.; Epstein, S.S.; Joung, Y.; Han, J.; Parfenova, V.V.; Belykh, O.I.; Gladkikh, A.S.; Ahn, T.S. Application of a new cultivation technology, I-tip, for studying microbial diversity in freshwater sponges of Lake Baikal, Russia. *FEMS Microbiol. Ecol.* **2014**, *90*, 417–423. [CrossRef] [PubMed]
8. Wargasetia, T.L. Mechanisms of cancer cell killing by sea cucumber-derived compounds. *Investig. New Drugs* **2017**, *35*, 820–826. [CrossRef] [PubMed]
9. Taiyeb-Ali, T.B.; Zainuddin, S.L.; Swaminathan, D.; Yaacob, H. Efficacy of 'Gamadent' toothpaste on the healing of gingival tissues: A preliminary report. *J. Oral Sci.* **2003**, *45*, 153–159. [CrossRef]
10. Shi, S.; Feng, W.; Hu, S.; Liang, S.; An, N.; Mao, Y. Bioactive compounds of sea cucumbers and their therapeutic effects. *Chin. J. Oceanol. Limnol.* **2016**, *34*, 549–558. [CrossRef]
11. Hossain, A.; Dave, D.; Shahidi, F. Northern sea cucumber (*Cucumaria frondosa*): A potential candidate for functional food, nutraceutical, and pharmaceutical sector. *Mar. Drugs* **2020**, *18*, 274. [CrossRef]
12. Chari, A.; Mazumder, A.; Lau, K.; Catamero, D.; Galitzeck, Z.; Jagannath, S. A phase II trial of TBL-12 sea cucumber extract in patients with untreated asymptomatic myeloma. *Br. J. Haematol.* **2018**, *180*, 296–298. [CrossRef] [PubMed]
13. Schmidt, E.W. The secret to a successful relationship: Lasting chemistry between ascidians and their symbiotic bacteria. *Invertebr. Biol.* **2015**, *134*, 88–102. [CrossRef] [PubMed]
14. Bewley, C.A.; Holland, N.D.; Faulkner, D.J. Two classes of metabolites from *Theonella swinhoei* are localized in distinct populations of bacterial symbionts. *Experientia* **1996**, *52*, 716–722. [CrossRef]

15. Bewley, C.A.; Faulkner, D.J. Lithistid sponges: Star performers or hosts to the stars. *Angew. Chem. Int. Ed.* **1998**, *37*, 2162–2178. [CrossRef]

16. Piel, J. Metabolites from symbiotic bacteria. *Nat. Prod. Rep.* **2004**, *21*, 519–538. [CrossRef]

17. Zhang, X.; Nakahara, T.; Murase, S.; Nakata, H.; Inoue, T.; Kudo, T. Physiological characterization of aerobic culturable bacteria in the intestine of the sea cucumber *Apostichopus japonicus*. *J. Gen. Appl. Microbiol.* **2013**, *59*, 1–10. [CrossRef]

18. Kurahashi, M.; Fukunaga, Y.; Sakiyama, Y.; Harayama, S.; Yokota, A. *Iamia majanohamensis* gen. nov., sp. nov., an actinobacterium isolated from sea cucumber *Holothuria edulis*, and proposal of *Iamiaceae* fam. nov. *Int. J. Syst. Evol. Microbiol.* **2009**, *59*, 869–873. [CrossRef]

19. Gozari, M.; Bahador, N.; Jassbi, A.R.; Mortazavi, M.S.; Eftekhar, E. Antioxidant and cytotoxic activities of metabolites produced by a new marine *Streptomyces* sp. isolated from the sea cucumber *Holothuria leucospilota*. *Iran. J. Fish. Sci.* **2018**, *17*, 413–426. [CrossRef]

20. Pivkin, M.V. Filamentous fungi associated with holothurians from the Sea of Japan, off the primorye coast of Russia. *Biol. Bull.* **2000**, *198*, 101–109. [CrossRef]

21. Afiyatullov, S.S.; Kuznetsova, T.A.; Isakov, V.V.; Pivkin, M.V.; Prokof'eva, N.G.; Elyakov, G.B. New diterpenic altrosides of the fungus *Acremonium striatisporum* isolated from a sea cucumber. *J. Nat. Prod.* **2000**, *63*, 848–850. [CrossRef]

22. Marchese, P.; Garzoli, L.; Gnavi, G.; O'Connell, E.; Bouraoui, A.; Mehiri, M.; Murphy, J.M.; Varese, G.C. Diversity and bioactivity of fungi associated with the marine sea cucumber *Holothuria poli*: Disclosing the strains potential for biomedical applications. *J. Appl. Microbiol.* **2020**, *129*, 612–625. [CrossRef]

23. Gong, J.; Tang, H.; Geng, W.L.; Liu, B.S.; Sun, P.; Li, L.; Li, Z.Y.; Zhang, W. Cyclic dipeptides in actinomycete *Brevibacterium* sp. associated with sea cucumber *Apostichopus japonicus* Selenka: Isolation and identification. *Acad. J. Second Mil. Med. Univ.* **2012**, *33*, 1284–1287. (In Chinese) [CrossRef]

24. Enomoto, M.; Nakagawa, S.; Sawabe, T. Microbial communities associated with Holothurians: Presence of unique bacteria in the coelomic fluid. *Microbes Environ.* **2012**, *27*, 300–305. [CrossRef] [PubMed]

25. Wibowo, J.T.; Kellermann, M.Y.; Versluis, D.; Putra, M.Y.; Murniasih, T.; Mohr, K.I.; Wink, J.; Engelmann, M.; Praditya, D.F.; Steinmann, E.; et al. Biotechnological potential of bacteria isolated from the sea cucumber *Holothuria leucospilota* and *Stichopus vastus* from Lampung, Indonesia. *Mar. Drugs* **2019**, *17*, 635. [CrossRef]

26. Alipiah, N.M.; Ramli, N.H.S.; Low, C.F.; Shamsudin, M.N.; Yusoff, F.M. Protective effects of sea cucumber surface-associated bacteria against *Vibrio harveyi* in brown-marbled grouper fingerlings. *Aquacult. Environ. Interact.* **2016**, *8*, 147–155. [CrossRef]

27. Xia, X.; Qi, J.; Wei, F.; Jia, A.; Yuan, W.; Meng, X.; Zhang, M.; Liu, C.; Wang, C. Isolation and characterization of a new Benzofuran from the fungus *Alternaria* sp. (HS-3) associated with a sea cucumber. *Nat. Prod. Commun.* **2011**, *6*, 1913–1914. [CrossRef] [PubMed]

28. Liu, C.H.; Xia, X.K.; Qi, J.; Zhang, Y.G.; Yuan, W.P.; Meng, X.M.; Jia, A.R.; Sun, Y.J.; Hu, W. The secondary metabolites of the HS-3 *Alternaria* sp. fungus associated with holothurians. *J. Chin. Med. Mater.* **2010**, *33*, 1875–1877. (In Chinese) [CrossRef]

29. Qi, J.; Jiang, L.; Zhao, P.; Chen, H.; Jia, X.; Zhao, L.; Dai, H.; Hu, J.; Liu, C.; Shim, S.H.; et al. Chaetoglobosins and azaphilones from *Chaetomium globosum* associated with *Apostichopus japonicus*. *Appl. Microbiol. Biotechnol.* **2020**, *104*, 1545–1553. [CrossRef]

30. Farouk, A.E.; Ghouse, F.A.H.; Ridzwan, B.H. New bacterial species isolated from malaysia sea cucumbers with optimized secreted antibacterial activity. *Am. J. Biochem. Biotechnol.* **2007**, *3*, 60–65.

31. Kamarudin, K.R. Microbial population in the coelomic fluid of *Stichopus chloronotus* and *Holothuria* (*Mertensiothuria*) *leucospilota* collected from Malaysian waters. *Sains Malays.* **2014**, *43*, 1013–1021.

32. Li, Z.; Huang, X.S.; Zheng, B.D.; Deng, K.B. Biodiversity analysis and pathogen inhibition mechanism of the endogenous bacteria in Fujian *Apostichopus japonicus*. *Sci. Technol. Food Ind.* **2018**, *39*, 137–141, 170. (In Chinese) [CrossRef]

33. Bogatyrenko, E.A.; Buzoleva, L.S. Characterization of the gut bacterial community of the Japanese sea cucumber *Apostichopus japonicus*. *Microbiology* **2016**, *85*, 116–123. [CrossRef]

34. Jo, J.; Choi, H.; Lee, S.-G.; Oh, J.; Lee, H.-G.; Park, C. Draft genome sequences of *Pseudoalteromonas tetraodonis* CSB01KR and *Pseudoalteromonas lipolytica* CSB02KR, isolated from the gut of the sea cucumber *Apostichopus japonicus*. *Genome Announc.* **2017**, *5*, e00627-17. [CrossRef]

35. Tan, J.J.; Liu, X.Y.; Yang, Y.; Li, F.H.; Tan, C.H.; Li, Y.M. Aspergillolide, a new 12-membered macrolide from sea cucumber-derived fungus *Aspergillus* sp. S-3-75. *Nat. Prod. Res.* **2020**, *34*, 1131–1137. [CrossRef]

36. Hu, Y.; Yang, M.; Zhao, J.; Liao, Z.; Qi, J.; Wang, X.; Jiang, W.; Xia, X. A Meroterpenoid isolated from the fungus *Aspergillus* sp. *Nat. Prod. Commun.* **2019**, *14*, 1–3. [CrossRef]

37. Sun, P.; Xu, D.X.; Mandi, A.; Kurtan, T.; Li, T.J.; Schulz, B.; Zhang, W. Structure, absolute configuration, and conformational study of 12-membered macrolides from the fungus *Dendrodochium* sp. associated with the sea cucumber *Holothuria nobilis* Selenka. *J. Org. Chem.* **2013**, *78*, 7030–7047. [CrossRef] [PubMed]

38. Wang, X.D.; Sun, P.; Xu, D.X.; Tang, H.; Liu, B.S.; Zhang, W. Identification of secondary metabolites of the fungus *Phialemonium* sp. associated with the South China sea cucumber *Holothuria nobilis* Selenka. *Acad. J. Second Mil. Med. Univ.* **2014**, *35*, 988–991. (In Chinese) [CrossRef]

39. Xia, X.K.; Qi, J.; Liu, C.H.; Zhang, Y.G.; Jia, A.R.; Yuan, W.P.; Liu, X.; Zhang, M.S. Polyketones from *Aspergillus terreus* associated with *Apostichopus japonicus*. *Mod. Food Sci. Technol.* **2014**, *30*, 10–14, 62. (In Chinese) [CrossRef]

40. Xia, X.; Zhang, J.; Zhang, Y.; Wei, F.; Liu, X.; Jia, A.; Liu, C.; Li, W.; She, Z.; Lin, Y. Pimarane diterpenes from the fungus *Epicoccum* sp. HS-1 associated with *Apostichopus japonicus*. *Bioorg. Med. Chem. Lett.* **2012**, *22*, 3017–3019. [CrossRef] [PubMed]

41. Wang, F.; Fang, Y.; Zhu, T.; Zhang, M.; Lin, A.; Gu, Q.; Zhu, W. Seven new prenylated indole diketopiperazine alkaloids from holothurian-derived fungus *Aspergillus fumigatus*. *Tetrahedron* **2008**, *64*, 7986–7991. [CrossRef]

42. Wang, F.-Z.; Zhang, M.; Sun, W.; Gu, Q.-Q.; Zhu, W.-M. 5a-Hydroxy-9-methoxy-11-(3-methyl-2-butenyl)-12-(2-methyl-1-propenyl)-2,3,11,12-tetrahydro-1H,5H-pyrrolo[1″,2″:4′,5′]pyrazino[1′,2′:1,6]pyrido[3,4-b]indole-5,6,14(5aH,14aH)-trione. *Acta Crystallogr. Sect. E Struct. Rep. Online* **2007**, *63*, o1859–o1860. [CrossRef]

43. Xia, X.K.; Liu, X.; Zhang, Y.G.; Yuan, W.P.; Zhang, M.S.; Wan, X.J.; Meng, X.M.; Liu, C.H. Study on the second metabolisms from fungus HS-1 *Epicoccum* spp. from the sea cucumber in Yellow Sea. *J. Chin. Med. Mater.* **2010**, *33*, 1577–1579. (In Chinese) [CrossRef]

44. Qi, J.; Zhao, P.; Zhao, L.; Jia, A.; Liu, C.; Zhang, L.; Xia, X. Anthraquinone derivatives from a sea cucumber-derived *Trichoderma* sp. fungus with antibacterial activities. *Chem. Nat. Compd.* **2020**, *56*, 112–114. [CrossRef]

45. Xia, X.; Liu, X.; Koo, D.C.; Sun, Z.; Shim, S. Chemical constituents of *Fusarium* sp. fungus associated with sea cucumbers. *Chem. Nat. Compd.* **2014**, *50*, 1103–1105. [CrossRef]

46. Afiyatullov, S.S.; Kalinovsky, A.I.; Antonov, A.S.; Zhuravleva, O.I.; Khudyakova, Y.V.; Aminin, D.L.; Yurchenko, A.N.; Pivkin, M.V. Isolation and structures of virescenosides from the marine-derived fungus *Acremonium striatisporum*. *Phytochem. Lett.* **2016**, *15*, 66–71. [CrossRef]

47. Qi, J.; Zhao, B.; Zhao, P.; Jia, A.; Zhang, Y.; Liu, X.; Liu, C.; Zhang, L.; Xia, X. Isolation and characterization of antiangiogenesis compounds from the fungus *Aspergillus terreus* associated with *Apostichopus japonicus* using zebrafish assay. *Nat. Prod. Commun.* **2017**, *12*, 261–262. [CrossRef]

48. Tae, H.; Sohng, J.K.; Park, K. MapsiDB: An integrated web database for type I polyketide synthases. *Bioprocess. Biosyst. Eng.* **2009**, *32*, 723–727. [CrossRef]

49. Risdian, C.; Mozef, T.; Wink, J. Biosynthesis of polyketides in *Streptomyces*. *Microorganisms* **2019**, *7*, 124. [CrossRef]

50. Hertweck, C. The biosynthetic logic of polyketide diversity. *Angew. Chem. Int. Ed.* **2009**, *48*, 4688–4716. [CrossRef]

51. Hussain, H.; Al-Sadi, A.M.; Schulz, B.; Steinert, M.; Khan, A.; Green, I.R.; Ahmed, I. A fruitful decade for fungal polyketides from 2007 to 2016: Antimicrobial activity, chemotaxonomy and chemodiversity. *Future Med. Chem.* **2017**, *9*, 1631–1648. [CrossRef]

52. Staunton, J.; Weissman, K.J. Polyketide biosynthesis: A millennium review. *Nat. Prod. Rep.* **2001**, *18*, 380–416. [CrossRef]

53. Xu, D.X.; Sun, P.; Kurtan, T.; Mandi, A.; Tang, H.; Liu, B.; Gerwick, W.H.; Wang, Z.W.; Zhang, W. Polyhydroxy cyclohexanols from a *Dendrodochium* sp. fungus associated with the sea cucumber *Holothuria nobilis* Selenka. *J. Nat. Prod.* **2014**, *77*, 1179–1184. [CrossRef] [PubMed]

54. Pavesi, C.; Flon, V.; Mann, S.; Leleu, S.; Prado, S.; Franck, X. Biosynthesis of azaphilones: A review. *Nat. Prod. Rep.* **2021**, *38*, 1058–1071. [CrossRef] [PubMed]

55. Schlager, S.; Drager, B. Exploiting plant alkaloids. *Curr. Opin. Biotechnol.* **2016**, *37*, 155–164. [CrossRef] [PubMed]

56. Desgagné-Penix, I. Distribution of alkaloids in woody plants. *Plant Sci. Today* **2017**, *4*, 137–142. [CrossRef]

57. Zotchev, S.B. Alkaloids from marine bacteria. *Adv. Bot. Res.* **2013**, *68*, 301–333. [CrossRef]

58. Souza, C.R.M.; Bezerra, W.P.; Souto, J.T. Marine alkaloids with anti-inflammatory activity: Current knowledge and future perspectives. *Mar. Drugs* **2020**, *18*, 147. [CrossRef] [PubMed]

59. Chen, J.; Zhang, W.; Guo, Q.; Yu, W.; Zhang, Y.; He, B. Bioactivities and future perspectives of Chaetoglobosins. *J. Evid. Based Complement. Altern. Med.* **2020**, *2020*, 8574084. [CrossRef]

60. Wang, F.-Z.; Li, D.-H.; Zhu, T.-J.; Zhang, M.; Gu, Q.-Q. Pseurotin A_1 and A_2, two new 1-oxa-7-azaspiro[4.4]non-2-ene-4,6-diones from the holothurian-derived fungus *Aspergillus fumigatus* WFZ-25. *Can. J. Chem.* **2011**, *89*, 72–76. [CrossRef]

61. Kim, S.K.; Li, Y.X. Biological activities and health effects of terpenoids from marine fungi. *Adv. Food Nutr. Res.* **2012**, *65*, 409–413. [CrossRef] [PubMed]

62. Yang, W.; Chen, X.; Li, Y.; Guo, S.; Wang, Z.; Yu, X. Advances in pharmacological activities of terpenoids. *Nat. Prod. Commun.* **2020**, *15*, 1–13. [CrossRef]

63. Gozari, M.; Alborz, M.; El-Seedi, H.R.; Jassbi, A.R. Chemistry, biosynthesis and biological activity of terpenoids and meroterpenoids in bacteria and fungi isolated from different marine habitats. *Eur. J. Med. Chem.* **2021**, *210*, 112957. [CrossRef] [PubMed]

64. Qi, J.; Xia, X.K.; Jia, A.R.; Liu, X.; Zhang, M.S.; Liu, C.H. Separation and preparation of chemical components from sea cucmber-derived fungus *Epicoccum* sp. by two-dimensional high-throughput chromatography. *Shandong Sci.* **2015**, *28*, 14–18, 24. (In Chinese) [CrossRef]

65. Xia, X.; Qi, J.; Liu, Y.; Jia, A.; Zhang, Y.; Liu, C.; Gao, C.; She, Z. Bioactive isopimarane diterpenes from the fungus, *Epicoccum* sp. HS-1, associated with *Apostichopus japonicus*. *Mar. Drugs* **2015**, *13*, 1124–1132. [CrossRef]

66. Berrue, F.; McCulloch, M.W.; Kerr, R.G. Marine diterpene glycosides. *Bioorg. Med. Chem.* **2011**, *19*, 6702–6719. [CrossRef] [PubMed]

67. Afiyatullov, S.S.; Kalinovsky, A.I.; Kuznetsova, T.A.; Isakov, V.V.; Pivkin, M.V.; Dmitrenok, P.S.; Elyakov, G.B. New diterpene glycosides of the fungus *Acremonium striatisporum* isolated from a sea cucumber. *J. Nat. Prod.* **2002**, *65*, 641–644. [CrossRef]

68. Afiyatullov, S.S.; Kalinovsky, A.I.; Kuznetsova, T.A.; Pivkin, M.V.; Prokof'eva, N.G.; Dmitrenok, P.S.; Elyakov, G.B. New glycosides of the fungus *Acremonium striatisporum* isolated from a sea cucumber. *J. Nat. Prod.* **2004**, *67*, 1047–1051. [CrossRef]

69. Afiyatullov, S.S.; Kalinovsky, A.I.; Pivkin, M.V.; Dmitrenok, P.S.; Kuznetsova, T.A. New diterpene glycosides of the fungus *Acremonium striatisporum* isolated from a sea cucumber. *Nat. Prod. Res.* **2006**, *20*, 902–908. [CrossRef]
70. Afiyatullov, S.S.; Kalinovsky, A.I.; Antonov, A.S. New Virescenosides from the marine-derived fungus *Acremonium striatisporum*. *Nat. Prod. Commun.* **2011**, *6*, 1063–1068. [CrossRef]
71. Zhuravleva, O.I.; Antonov, A.S.; Oleinikova, G.K.; Khudyakova, Y.V.; Popov, R.S.; Denisenko, V.A.; Pislyagin, E.A.; Chingizova, E.A.; Afiyatullov, S.S. Virescenosides from the Holothurian-associated fungus *Acremonium striatisporum* Kmm 4401. *Mar. Drugs* **2019**, *17*, 616. [CrossRef] [PubMed]
72. Wibowo, J.T.; Kellermann, M.Y.; Kock, M.; Putra, M.Y.; Murniasih, T.; Mohr, K.I.; Wink, J.; Praditya, D.F.; Steinmann, E.; Schupp, P.J. Anti-Infective and antiviral activity of valinomycin and its analogues from a sea cucumber-associated bacterium, *Streptomyces* sp. SV 21. *Mar. Drugs* **2021**, *19*, 81. [CrossRef] [PubMed]

 *marine drugs*

Article

Cytotoxic Indole-Diterpenoids from the Marine-Derived Fungus *Penicillium* sp. KFD28

Lu-Ting Dai [1], Li Yang [2], Fan-Dong Kong [3], Qing-Yun Ma [2], Qing-Yi Xie [2], Hao-Fu Dai [4], Zhi-Fang Yu [1],* and You-Xing Zhao [2],*

[1] College of Food Science and Technology, Nanjing Agricultural University, Nanjing 210095, China; dailuting121@163.com
[2] Haikou Key Laboratory for Research and Utilization of Tropical Natural Products, Institute of Tropical Bioscience and Biotechnology, CATAS, Haikou 571101, China; yangli@itbb.org.cn (L.Y.); maqingyun@itbb.org.cn (Q.-Y.M.); xieqingyi@itbb.org.cn (Q.-Y.X.)
[3] Key Laboratory of Chemistry and Engineering of Forest Products, State Ethnic Affairs Commission, Guangxi Key Laboratory of Chemistry and Engineering of Forest Products, Guangxi Collaborative Innovation Center for Chemistry and Engineering of Forest Products, School of Chemistry and Chemical Engineering, Guangxi University for Nationalities, Nanning 530006, China; kongfandong501@126.com
[4] Hainan Institute for Tropical Agricultural Resources, CATAS, Haikou 571101, China; daihaofu@itbb.org.cn
* Correspondence: yuzhifang@njau.edu.cn (Z.-F.Y.); zhaoyouxing@itbb.org.cn (Y.-X.Z.); Tel.: +86-139-5169-2350 (Z.-F.Y.); +86-898-6698-9095 (Y.-X.Z.)

Abstract: Four new indole-diterpenoids, named penerpenes K–N (**1–4**), along with twelve known ones (**5–16**), were isolated from the fermentation broth produced by adding L-tryptophan to the culture medium of the marine-derived fungus *Penicillium* sp. KFD28. The structures of the new compounds were elucidated extensively by 1D and 2D NMR, HRESIMS data spectroscopic analyses and ECD calculations. Compound **4** represents the second example of paxilline-type indole diterpene bearing a 1,3-dioxepane ring. Three compounds (**4**, **9**, and **15**) were cytotoxic to cancer cell lines, of which compound **9** was the most active and showed cytotoxic activity against the human liver cancer cell line BeL-7402 with an IC_{50} value of 5.3 µM. Moreover, six compounds (**5, 7, 10, 12, 14,** and **15**) showed antibacterial activities against *Staphylococcus aureus* ATCC 6538 and *Bacillus subtilis* ATCC 6633.

Keywords: marine-derived fungus; *Penicillium* sp.; indole-diterpenoids; cytotoxicity; antibacterial activity

Citation: Dai, L.-T.; Yang, L.; Kong, F.-D.; Ma, Q.-Y.; Xie, Q.-Y.; Dai, H.-F.; Yu, Z.-F.; Zhao, Y.-X. Cytotoxic Indole-Diterpenoids from the Marine-Derived Fungus *Penicillium* sp. KFD28. *Mar. Drugs* **2021**, *19*, 613. https://doi.org/10.3390/md19110613

Academic Editor: Khaled A. Shaaban

Received: 28 September 2021
Accepted: 24 October 2021
Published: 28 October 2021

Publisher's Note: MDPI stays neutral with regard to jurisdictional claims in published maps and institutional affiliations.

1. Introduction

Marine fungi have formed different metabolic pathways and adaptation mechanisms within the peculiar marine environment. Hence, marine fungi can produce natural secondary metabolites that are characterized by unique chemical structures and high biological activities [1,2]. Alkaloids derived from marine-derived compounds have received extensive attention in recent years. Indole alkaloids, as an important class of secondary metabolites produced by marine-derived fungi [3], showed excellent biological activities, including cytotoxic [4], antibacterial [5], quorum sensing inhibitory [6], anti-Zika virus [7], and protein tyrosine phosphatase inhibitory activities [8,9].

Marine-derived fungus *Penicillium* sp. KFD28 was isolated and identified from bivalve shellfish, *Meretrix lusoria*, collected from Haikou Bay, China. Our previous study on the secondary metabolites of this fungus discovered a series of indole alkaloids with novel structures and intriguing bioactivities, e.g., protein tyrosine phosphatase inhibitory activity [8,9]. The OSMAC (one strain, many compounds) approach is highly efficient for inducing structural diversity by the variation of cultivation conditions [10]. Most of the indole alkaloids precursors are related to L-tryptophan [11]. To find more new alkaloids from marine fungi, adding amino acids to the culture media is becoming a viable strategy [12,13]. In order to explore the metabolic potential of the fungus *Penicillium* sp. KFD28, we recultured this fungus by adding L-tryptophan to the solid rice culture and found that the HPLC profiles of the extract are different from that of the first time obtained from the liquid medium. Later chemical investigation of the fermentation broth led to the isolation of four new indole-diterpenoids, penpenes K–N (**1–4**), along with 12 known ones, including paxilline (**5**) [14], dehydroxypaxilline (**6**) [15], 7-hydroxyl-13-dehydroxypaxilline (**7**) [16], 3-deoxo-4b-deoxypaxilline (**8**) [17], epipaxilline (**9**) [18], 7-hydroxypaxilline-13-ene (**10**) [19], paspaline (**11**) [20], 4a-demethylpaspaline-4a-carboxylic acid (**12**) [17], paspalinine (**13**) [21], PC-M6 (**14**) [22], emindole SB (**15**) [23], emeniveol (**16**) [24] (Figure 1). Herein, the isolation, structure elucidation, and bioactivities of these compounds are reported.

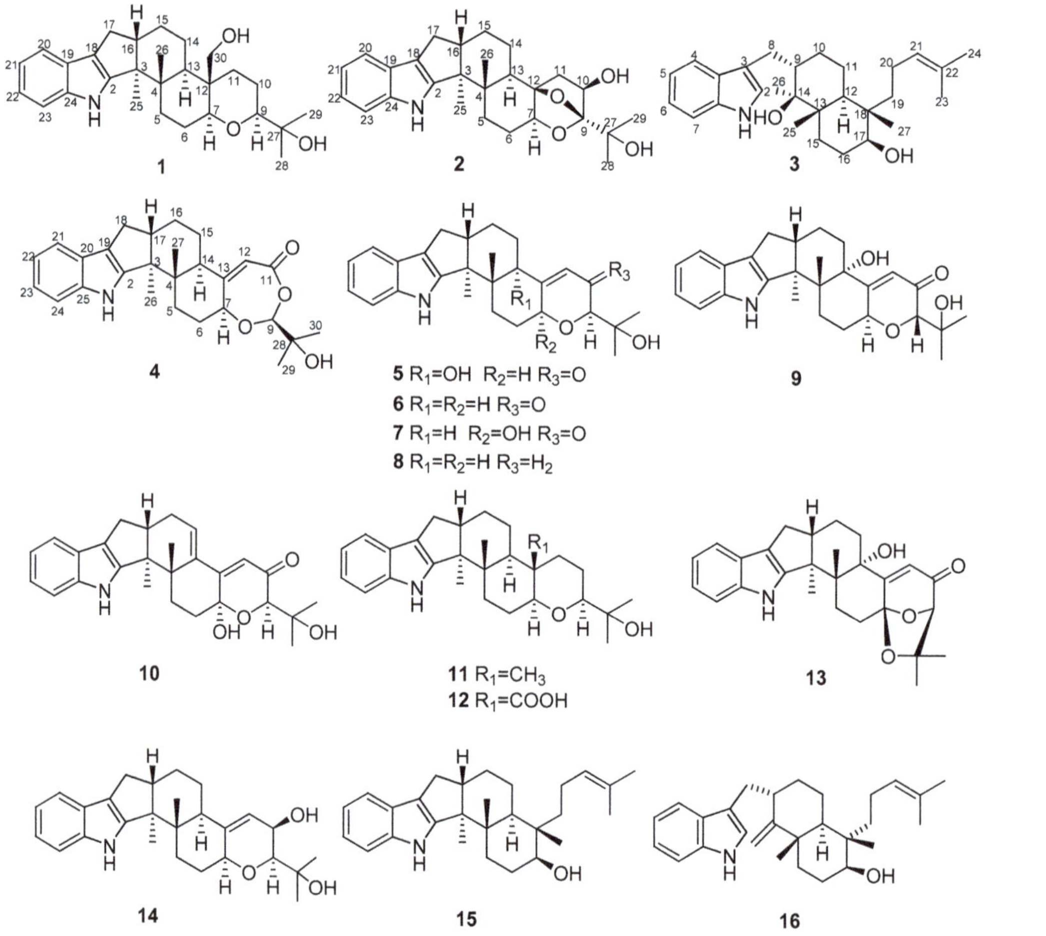

Figure 1. The chemical structures of compounds **1–16**.

2. Results and Discussions

2.1. Structure Elucidation

Compound **1** was obtained as a yellow oil. Its formula was determined as $C_{28}H_{39}NO_3$ on the basis of HRESIMS data (m/z 476.2565 for $[M+K]^+$), indicating ten degrees of unsaturation. The ^1H and ^{13}C NMR data of **1**, with the aid of its HSQC spectrum (Supplementary Materials, Figures S1–S4), showed a total of 28 carbon signals comprising eight olefinic or aromatic carbons (four protonated) for a 2,3-disubstituted indole moiety, four methyls, eight sp^3 methylenes with one oxygenated, four sp^3 methines with two oxygenated, and four non-protonated sp^3 carbons with one oxygenated. These NMR data allowed the construction of the carbon skeleton of indole-diterpenoid. Careful contrast of the similar NMR spectral data between **1** and paspaline (**11**) [20] revealed that they had the same planar structure except that the methyl group at C-12 (δ_C 36.6) in paspaline was oxidized to oxymethylene in **1**. Consistent with the introduction of the hydroxyl at C-30 (δ_C 58.5), it showed a distinctive deshielding of the C-12 signal (δ_C 48.8) in **1** compared to those of paspaline (δ_C 36.6). The linkage of a hydroxyl at C-30 was further supported by the COSY correlation of 30-OH/H_2-30 (Figure 2; Supplementary Material, Figure S8) and ROESY correlation of H_3-26/H_2-30 (Supplementary Material, Figure S9). Detailed analysis of 2D NMR data further confirmed that **1** and paspaline share the same indole-diterpenoid skeleton. The relative configuration of 6/6/6 tricyclic rings in indole-diterpenoid were determined by the ROESY spectrum (Figure 3), in which the sequential correlations of H-16/H_3-26/H_2-30/H-10β/H_3-28 suggested the same face of H-16, CH_3-26, CH_2-30, and C-27 in the 6/6/6 tricyclic ring system, while the correlations of H_3-25/H-13/H-7/H-9 suggested CH_3-25, H-13, H-7, and H-9 were on the opposite face of this system. It has been reported that the strong Cotton effect (CE) around 220 nm was related to the absolute configurations of the chiral carbons around the indole chromophore in the paxilline-type indole-diterpene [17]. Thus, the strong negative CE at 223 nm in the experimental ECD spectrum of **1** (Figure 4) suggested its (3*S*,4*S*,7*S*,9*S*,12*S*,13*R*,16*S*)-**1** absolute configuration [17]. As for the absolute configuration of **1**, the ECD spectrum of (3*S*,4*S*,7*S*,9*S*,12*S*,13*R*,16*S*)-**1** was calculated; this deduction was further supported, establishing the absolute configuration of **1** as presented in Figure 1.

Compound **2** was isolated as a yellow oil. Its formula was determined as $C_{27}H_{35}NO_4$ on the basis of HRESIMS data, indicating 11 degrees of unsaturation. Its ^{13}C NMR data showed a total of 27 carbon signals comprising eight aromatic or olefinic carbons for an indole moiety, four methyls, six sp^3 methylenes, four sp^3 methines with two oxygenated, and five non-protonated sp^3 carbons with three oxygenated. Analysis of the NMR spectra (Tables 1 and 2, Supplementary Materials, Figures S12–S18) of **2** suggested that its structure was related to that of 4a-demethylpaspaline-3,4,4a-triol [17], and the main difference being C-9 (δ_C 109.8), C-11 (δ_C 43.7), and C-12 (δ_C 84.6) in **2** instead of C-9 (δ_C 82.6), C-11 (δ_C 73.5), and C-12 (δ_C 76.1) in 4a-demethylpaspaline-3,4,4a-triol, indicating the replacement of two oxygenated carbons by a methylene (C-11) and an acetal carbon (C-9) in **2**. The HMBC correlations (Supplementary Material, Figure S16) from H-11, H-7, H-29, and H-28 to C-9 confirmed this deduction. The presence of an indole moiety, together with the HRESIMS data (Supplementary Material, Figure S19), indicated that **2** has a heptacyclic ring system. At this point, one more ring was required to fulfill the 11 double-bond equivalents, and an oxygen bridge was proposed according to the molecular formula. The oxygen bridge was assigned to connect C-9 and C-12 as deduced from distinctive deshielding of the C-12 (δ_C 84.6) and C-9 (δ_C 109.8) in **2** compared to corresponding C-12 (δ_C 76.1) and C-9(δ_C 96.4-96.5) [25,26] with a free hydroxyl group. In the ROESY spectrum (Supplementary Material, Figure S18), correlations of H-16/H_3-26/H-5β, and H-5α/H_3-25/H-13/H-7/H-10/H_3-28 determined the relative configuration of **2**, as shown in Figure 3. In addition, the β orientation of the oxygen bridge was also proved by the ROESY correlation of H-7/H_2-11 with the aid of the 3D ball-and-stick molecular model. Thus, the planar structure of compound **2** was established and named penerpene L. The ECD curve (Figure 4) of compound **2** is similar to **1**, indicating that the absolute configurations for the chiral carbons C-3, C-4, C-16, and C-13 in **1** were the same as those of **2**. The ECD calculation experiment also confirmed the above deduction (Figure 4), establishing the (3*S*,4*S*,7*S*,9*S*,10*R*,12*R*,13*R*,16*S*)-**2** absolute configuration.

Compound **3** was assigned the molecular formula of $C_{28}H_{41}NO_2$ by HRESIMS, indicating nine degrees of unsaturation. The ^1H NMR spectrum (Supplementary Material, Figure S21) displayed the typical pattern of a 3-substituted indole moiety with five aromatic protons at δ_H 7.04 (s, H-2), 7.30 (d, J = 8.0, H-7), 6.93 (t, J = 7.6, H-5), 7.02 (t, J = 7.6, H-6) and 7.55 (d, J = 8.0, H-4), as well as one clear olefinic proton at δ_H 5.05 (t, H-21), one oxygenated proton δ_H 3.27 (1H, H-17), and five methyls at δ_H 1.62 (s, 24-Me), 1.55 (s, 23-Me), 0.62 (s, 27-Me), 1.11 (s, 26-Me), 0.94 (s, 25-Me). The ^{13}C NMR data of **3** (Supplementary Materials, Figures S22 and S23) displayed 28 carbon signals, including ten aromatic or olefinic carbons (six protonated), five methyls, seven sp^3 methylenes, three sp^3 methines with one oxygenated, three non-protonated sp^3 carbons with one oxygenated. These NMR spectra data indicated that **3** was very similar to emeniveol (**16**) [24] except that a methyl in C-26 ($\delta_{C/H}$ 16.3/1.11) and an oxygenated non-protonated sp^3 carbon C-14 (δ_C 76.4) in **3** replaced two olefinic carbons in emeniveol, suggesting that the exocyclic double bond in emeniveol changed to a methyl and an oxygenated non-protonated sp^3 carbon. This obvious difference was supported by the HMBC correlations (Supplementary Material, Figure S25) from the H$_3$-26 to C-14, C-13 (δ_C 42.0) and C-9 (δ_C 42.9). The relative configuration of **3** was determined by the ROESY spectrum (Figure 3), and the correlations of 14-OH/H$_3$-25/H$_3$-27/17-OH and H-9/H$_3$-25 suggested the same face of these protons, while the correlation of H-12/H-17 indicated that these protons were on the opposite face to 17-OH. Thus, the planar structure of **3** was assigned, as shown in Figure 1. The absolute configuration of **3** was established as (9*S*,12*S*,13*S*,14*S*,17*S*,18*S*)-**3** by comparison of its experimental ECD spectrum with the calculated ECD curves (Figure 4).

The molecular formula of compound **4** was established as C$_{27}$H$_{33}$NO$_4$ on the basis of HRESIMS data, indicating 12 degrees of unsaturation. Analysis of the ^{1}H and ^{13}C NMR spectra (Tables 1 and 2, Supplementary Materials, Figures S30 and S31) of **4** suggested that its structure was closely related to that of penerpene G [9], a previously reported indole diterpene with an unusual 6/5/5/6/6/7 hexacyclic ring system bearing a 1,3-dioxepane ring. The main difference between them was the replacement of an oxygenated non-protonated sp^3 carbon C-14 (δ_C 78.3) in penerpene G by that of a sp^3 methine (δ_C 42.8) in **4**. This assignment was confirmed by HMBC (Figure 2) correlations from H-12 (δ_H 5.54) and H-27 (δ_H 0.82) to C-14 and the COSY correction of H-14/H-15. The ROESY corrections (Figure 3) of H$_3$-27/H-17/H-16β and H-16α/H$_3$-26/H-14/H-7/H-9 assigned the same relative configuration of **4** as that of penerpene G. The ECD curves of compound **4** (Supplementary Material, Figure S43) are similar to that reported for penerpene G, containing strong positive CEs at a shorter wavelength (219 nm in **4** and 220 nm in penerpene G) and strong negative CEs at a longer wavelength (238 nm in **4** and 234 nm in penerpene G), leading to the determination of the absolute configuration of **4** as (3*S*,4*S*,7*S*,9*S*,14*R*,17*S*)-**4**.

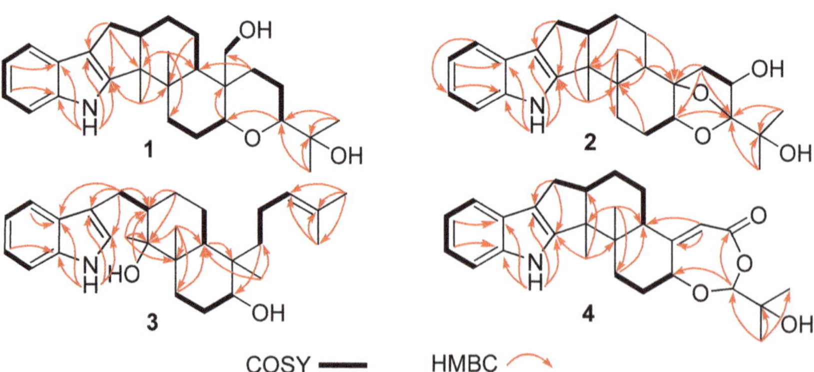

Figure 2. Key COSY and HMBC correlations of compounds **1–4**.

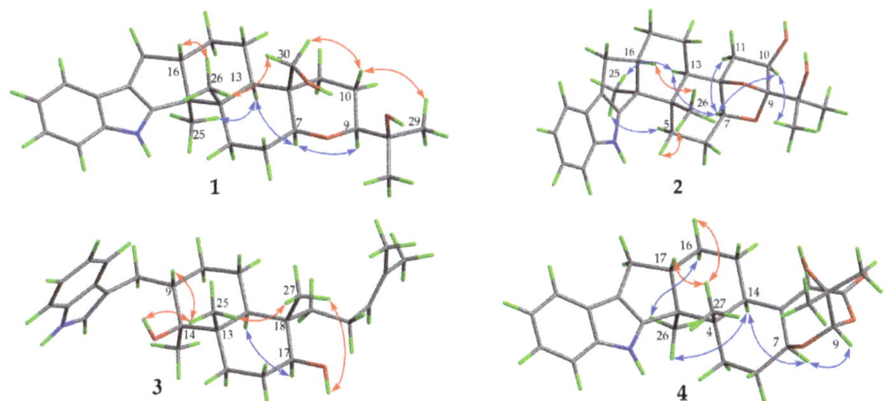

Figure 3. Key ROESY correlations of compounds **1–4**.

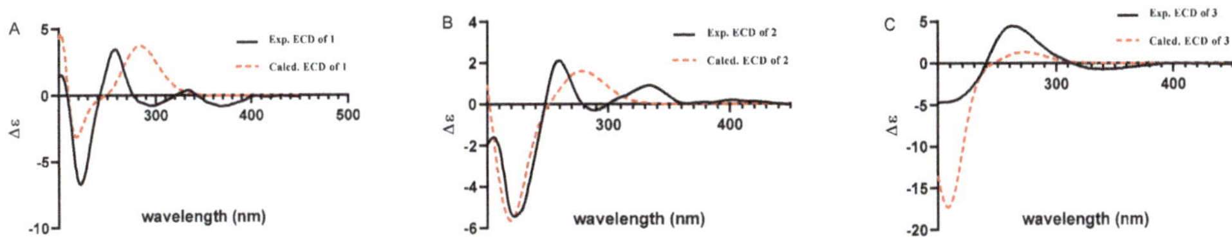

Figure 4. Experimental and calculated ECD curves for compound **1** (**A**); compound **2** (**B**); compound **3** (**C**).

Table 1. ^{1}H NMR (500 and 600 MHz) data of **1–4** in DMSO-d_6.

Position	1 δ_H (*J* in Hz)	2 δ_H (*J* in Hz)	3 δ_H (*J* in Hz)	4 δ_H (*J* in Hz)
2			7.04 (1H, s)	
4			7.55 (1H, d, 8.0)	
5	1.92 (1H, m) 1.80 (1H, m)	1.88 (1H, overlap) 1.60 (1H, overlap)	6.93 (1H, t, 8.0)	2.01 (1H, m) 1.94 (1H, m)
6	1.51 (1H, m) 1.62 (1H, overlap)	1.90 (1H, m) 1.48 (1H, m)	7.02 (1H, t, 8.0)	1.72 (1H, overlap) 2.20 (1H, m)
7	3.03 (1H, dd, 12.1, 3.8)	3.49 (1H, dd, 9.7, 7.0)	7.30 (1H, d, 8.0)	4.46 (1H, m)
8			2.07 (1H, m) 3.15 (1H, d, 13.3)	
9	3.10 (1H, dd, 12.0, 2.7)		1.81 (1H, m)	5.05 (1H, s)
10	1.41 (1H, m) 1.62 (1H, overlap)	4.04 (1H, dd, 2.1, 6.7)	1.10 (1H, m) 1.65 (1H, m)	
11	1.74 (1H, m) 1.69 (1H, s)	1.88 (1H, overlap) 1.70 (1H, m)	1.26 (2H, m)	
12			1.81 (1H, m)	5.54 (1H, s)
13	1.39 (1H, m)	2.06 (1H, dd, 12.7, 3.1)		
14	0.83 (1H, m) 2.35 (1H, m)	1.74 (1H, m) 1.64 (1H, m)		2.55 (1H, m)
15	1.44 (1H, m) 1.62 (1H, m)	1.78 (1H, m) 1.60 (1H, overlap)	1.42 (2H, overlap)	1.53 (1H, m) 1.62 (1H, overlap)
16	2.61 (1H, m)	2.86 (1H, m)	1.24 (1H, overlap) 1.54 (1H, m)	1.72 (1H, overlap) 1.62 (1H, overlap)

Table 1. *Cont.*

Position	1 δ_H (*J* in Hz)	2 δ_H (*J* in Hz)	3 δ_H (*J* in Hz)	4 δ_H (*J* in Hz)
17	2.54 (1H, m) 2.21 (1H, dd, 12.3, 10.9)	2.61 (1H, m) 2.29 (1H, dd, 11.2, 12.6)	3.27 (1H, m)	2.68 (1H, m)
18				2.33 (1H, m) 2.61 (1H, m)
19			1.02 (1H, m) 1.47 (1H, m)	
20	7.27 (1H, d, 7.5)	7.27 (1H, d, 7.5)	1.84 (2H, m)	
21	6.91 (1H, t, 7.5)	6.90 (1H, m)	5.05 (1H, t, 7.6)	7.27 (1H, m)
22	6.91 (1H, t, 7.5)	6.94 (1H, m)		6.92 (1H, t, 7.9)
23	7.24 (1H, d, 7.5)	7.26 (1H, d, 7.5)	1.55 (3H, s)	6.96 (1H, t, 7.9)
24			1.62 (3H, s)	7.27 (1H, m)
25	0.94 (3H, s)	0.94 (3H, s)	0.94 (3H, s)	
26	1.13 (3H, s)	1.17 (3H, s)	1.11 (3H, s)	1.01 (3H, s)
27			0.62 (3H, s)	0.82 (3H, s)
28	1.08 (3H, s)	1.26 (3H, s)		
29	1.03 (3H, s)	1.24 (3H, s)		1.14 (3H, s)
30	3.69 (1H, m) 3.86 (1H, m)			1.13 (3H, s)
1-NH	10.55 (1H, s)	10.68 (1H, s)	10.67 (1H, s)	10.72(1H, s)
14-OH			4.04 (1H, s)	
17-OH			4.22 (1H, d, 4.8)	
10-OH		4.88 (1H, s)		
27-OH	4.16 (1H, s)			
28-OH				4.77 (1H, s)
30-OH	4.07 (1H, s)			

Compound **1** was measured with 500 MHz ^{1}H NMR. Compounds **2**, **3**, and **4** were measured with 600 MHz ^{1}H NMR.

Table 2. ^{13}C NMR (125 and 150 MHz) data of **1–4** in DMSO-d_6.

Position	1 δ_C	2 δ_C	3 δ_C	4 δ_C
2	151.5, C	150.9, C	122.6, CH	150.2, C
3	52.9, C	50.8, C	114.4, C	42.5, C
3a			127.6, C	
4	39.9, C	38.5, C	118.7, CH	50.0, C
5	32.7, CH$_2$	30.6, CH$_2$	117.9, CH	30.8, CH$_2$
6	24.5, CH$_2$	23.2, CH$_2$	120.6, CH	30.0, CH$_2$
7	85.2, CH	78.2, CH	111.2, CH	83.0, CH
7a			136.3, C	
8			25.5, CH$_2$	
9	84.9, CH	109.8, C	42.9, CH	102.3, CH
10	25.7, CH$_2$	73.1, CH	27.2, CH$_2$	
11	23.8, CH$_2$	43.7, CH$_2$	20.7, CH$_2$	166.8, C
12	48.8, C	84.6, C	41.1, CH	113.1, CH
13	46.9, CH	38.8, CH	42.0, C	163.1, C
14	31.8, CH$_2$	26.2, CH$_2$	76.4, C	42.8, CH
15	21.3, CH$_2$	25.2, CH$_2$	29.9, CH$_2$	25.6, CH$_2$
16	49.0, CH	49.0, CH	28.5, CH$_2$	23.9, CH$_2$
17	27.3, CH$_2$	27.1, CH$_2$	71.5, CH	48.7, CH
18	116.1, C	115.9, C	40.6, C	26.9, CH$_2$
19	124.6, C	124.4, C	37.0, CH$_2$	115.8, C
20	117.8 CH	117.7, CH	21.2, CH$_2$	124.3, C
21	118.6, CH	118.5, CH	125.3, CH	117.7, CH

Table 2. *Cont.*

Position	1	2	3	4
	δ_C	δ_C	δ_C	δ_C
22	119.3, CH	119.4, CH	129.8, C	118.5, CH
23	112.0, CH	111.9, CH	17.5, CH$_3$	119.5, CH
24	140.4, C	140.1, C	25.6, CH$_3$	111.9, CH
25	14.8, CH$_3$	14.4, CH$_3$	14.6, CH$_3$	140.1, C
26	19.5, CH$_3$	16.7, CH$_3$	16.3, CH$_3$	14.7, CH$_3$
27	70.7, C	70.9, C	17.2, CH$_3$	15.4, CH$_3$
28	26.9, CH$_3$	26.4, CH$_3$		69.9, C
29	24.9, CH$_3$	24.7, CH$_3$		24.0, CH$_3$
30	58.5, CH$_2$			23.8, CH$_3$

Compound **1** was measured with 125 MHz ^{13}C NMR. Compounds **2**, **3**, and **4** were measured with 150 MHz ^{13}C NMR.

2.2. Biological Assay

The cytotoxic activities of compounds **1–16** were conducted by the MTT assay method [27] using cisplatin as the positive control. All compounds were tested against the human cervical cancer cell line HeLa, human gastric cancer cell line SGC-7901, human lung carcinoma cell line A549, and human liver cancer cell line BeL-7402. The results (Table 3) indicated that compound **9** exhibited the most pronounced activity against BeL-7402 with an IC$_{50}$ value of 5.3 μM and was comparable to that of positive control cisplatin (IC$_{50}$ 4.1 μM). Compound **9** also displayed moderate cytotoxic activity against A549. While compounds **4** showed low cytotoxicity against HeLa. Compound **15** displayed mild inhibitory activity against HeLa, A549, and BeL-7402 (IC$_{50}$ = 24.4–40.6 μM). The remaining compounds were found to be inactive against the three cell lines. All the tested compounds **1–16** were inactive against the cell line SGC-7901. The loss of a hydroxyl at C-14 suggested being a determinant of cytotoxicity shown by compound **4** against HeLa (**4** vs. penerpene G [9]). It is worth mentioning that it was the first-time report of the cytotoxicity of epipaxilline (**9**) [18]. Arintari et al. [19] and Sallam et al. [28] demonstrated the cytotoxicities of emindole SB (**15**) against human breast cell line MCF-7, murine lymphoma cell line L5178Y, and the human embryonic kidney cell line HEK-293, which provides emindole SB (**15**) a meaningful pharmacophore for further biological studies.

Table 3. Cytotoxicity of compounds **4**, **9**, and **15**.

Compound	IC$_{50}$ (μM)		
	HeLa	A549	BeL-7402
4	36.3	>50	>50
9	>50	28.4	5.3
15	33.1	24.4	40.6
Cisplatin [a]	8.6	4.5	4.1

[a] Positive control.

Compounds **1–16** were also tested for their antibacterial activity against *Escherichia coli* ATCC 25922, *Staphylococcus aureus* ATCC 6538, *Listeria monocytogenes* ATCC 1911, and *Bacillus subtilis* ATCC 6633 using the 96-well microtiter plates method [29] reported previously and using ampicillin as a positive control. The results (Table 4) revealed that six compounds **5**, **7**, **10**, **12**, **14**, and **15** showed moderate inhibitory activity against *S. aureus* ATCC 6538. Emindole SB (**15**) displayed reported selectivity toward *S. aureus* ATCC 33591 (MIC = 6.25 μg/mL) [21]. Compound **7** showed reasonable antibacterial activity against *B. subtilis* ATCC 6633 (MIC = 16 μg/mL), but compounds **5**, **10**, and **12** exhibited lower inhibitory. The results of the rest ten compounds (**1–4**, **6**, **8–9**, **11**, **13**, and **16**) did not show remarkable antibacterial activities (MIC > 128 μg/mL) against *S. aureus* and *B. subtilis*. In this assay, none of these compounds showed inhibitory activity against *E. coli* ATCC 25922 and *L. monocytogenes* ATCC 1911 (MIC > 128 μg/mL).

Table 4. Antibacterial activities of compounds **5**, **7**, **10**, **12**, **14**, and **15**.

Compound	MIC(µg/mL)	
	Staphylococcus aureus ATCC 6538	*Bacillus subtilis* ATCC 6633
5	128	32
7	64	16
10	64	64
12	64	128
14	64	>128
15	32	>128
Ampicillin [a]	<1	<1

[a] Positive control.

3. Experimental

3.1. General Experimental Procedures

NMR spectra were recorded on Bruker AV-500 and Bruker AV-600 spectrometers (Bruker, Bremen, Germany) with TMS as an internal standard. The mass spectrometric (HRESIMS) data were acquired using an API QSTAR Pulsar mass spectrometer (Bruker, Bremen, Germany) and an AB SCIEX Trip TOF 5600+ mass spectrometer (SCIEX, Framingham, MA, USA). Optical rotations were measured with a JASCO P-1020 digital polarimeter (Jasco, Tokyo, Japan). IR spectra were recorded on a Shimadzu UV2550 spectrophotometer (Shimadzu, Kyoto, Japan). UV spectra and ECD data were collected using a JASCO J-715 spectropolarimeter (Jasco, Tokyo, Japan). Semipreparative HPLC was carried out using an ODS column (YMC-pack ODS-A, 10 × 250 mm, 5 µm, 4 mL/min, YMC, Kyoto, Japan).

3.2. Fungus Material

The fungus *Penicillium* sp. KFD28 (GenBank accession No. MK934323) was isolated from a bivalve mollusk, *Meretrix lusoria*, collected from Haikou Bay, Hainan province, in China. A reference culture of *Penicillium* sp. KFD28 is deposited in our laboratory and maintained at −80 °C.

3.3. Culture Conditions

The fungus *Penicillium* sp. KFD28 was cultured in 200 × 1000 mL Erlenmeyer flasks containing 100 g rice and 100 mL of water (33 g sea salt, 5 g L-tryptophan per liter pure water). The fungus was cultured in the medium and incubated at room temperature for thirty days.

3.4. Extraction and Isolation

The fermented material was extracted three times with EtOAc to obtain 300 g crude extract. The extract was extracted between petroleum ether and 90% methanol (1:1) to remove the oil. The secondary metabolites extract (46 g) was subjected to a silica gel VLC column, eluting with a stepwise gradient of petroleum ether-EtOAc (10:1, 8:1, 6:1, 4:1, 2:1, 1:1, 1:2, *v/v*) to yield ten subfractions (Fr. 1–Fr. 10).

Fr. 3 (4.2 g) was applied to ODS silica gel with gradient elution of MeOH-H$_2$O (1:4, 3:7, 2:3, 1:1, 3:2, 7:3, 4:1, 9:1, 0:1, *v/v*) and afforded nine subfractions (Fr. 3-1–Fr. 3-9). Fr. 3-9 (182 mg) was applied to semipreparative HPLC (YMC-pack ODS-A, 5 µm; 10 × 250 mm; 85% MeOH/H$_2$O; 4 mL/min) to obtain compounds **16** (t_R 20.0 min, 1.0 mg), **11** (t_R 23.0 min, 9.7 mg), and **8** (t_R 32.0 min, 20.8 mg). Fr. 5 (3.8 g) was applied to ODS silica gel with gradient elution of MeOH-H$_2$O (1:9, 1:4, 3:7, 2:3, 1:1, 3:2, 7:3, 4:1, 9:1, 0:1, *v/v*) to get ten subfractions. Fr. 5-10 (323 mg) was purified by semipreparative HPLC (YMC-pack ODS-A, 5 µm; 10 × 250 mm; 80% MeOH/H$_2$O; 4 mL/min) to give compound **15** (t_R 28.5 min, 53.4 mg). Fr. 6 (5.1 g) was separated by ODS silica gel with gradient elution of MeOH-H$_2$O (1:9, 1:4, 3:7, 2:3, 1:1, 3:2, 7:3, 4:1, 9:1, 0:1, *v/v*) to yield ten subfractions (Fr. 6-1–Fr. 6-10). Fr. 6-10 (1.1 g) was subjected to semipreparative HPLC (YMC-pack ODS-A, 5 µm;

10 × 250 mm; 65% MeCN/H$_2$O; 4 mL/min) to give compounds **5** (t_R 11.5 min, 182.6 mg) and **6** (t_R 18.0 min, 25.8 mg). Fr. 7 (4.7 g) was applied to ODS silica gel with gradient elution of MeOH-H$_2$O (1:9, 1:4, 3:7, 2:3, 1:1, 3:2, 7:3, 4:1, 9:1, 0:1, *v/v*), ten subfractions (Fr. 7-1–7-10) were obtained. Fr. 7-7 (587 mg) was eluted with MeCN-H$_2$O through ODS silica gel (3:7, 2:3, 1:1, 3:2, 7:3, *v/v*) to afford five subfractions (Fr. 7-7-1–7-7-5). Fr. 7-7-4 (117 mg) was separated by semipreparative HPLC (YMC-pack ODS-A, 5 μm; 10 × 250 mm; 70% MeCN/H$_2$O; 4 mL/min) to give compounds **2** (t_R 8.6 min, 0.6 mg), **3** (t_R 9.4 min, 1.8 mg), **1** (t_R 10.4 min, 2.0 mg), **13** (t_R 11.6 min, 0.6 mg), **14** (t_R 13.5 min, 9.9 mg), and **7** (t_R 22.0 min, 8.1 mg). Fr. 7-7-5 (137 mg) was applied to semipreparative HPLC (YMC-pack ODS-A, 5 μm; 10 × 250 mm; 70% MeOH/H$_2$O; 4 mL/min) to get compounds **9** (t_R 14.0 min, 2.0 mg) and **10** (t_R 18.4min, 3.4 mg). Fr. 8 (5.8 g) was subjected to ODS silica gel with gradient elution of MeOH-H$_2$O (1:4, 3:7, 2:3, 1:1, 3:2, 7:3, 4:1, 9:1, 0:1, *v/v*) to get nine subfractions. Then, Fr. 8-7 (925 mg) was eluted through ODS silica gel column of MeCN-H$_2$O (1:4, 3:7, 2:3, 1:1, 3:2, *v/v*) to get five subfractions (Fr. 8-7-1–8-7-5). Fr. 8-7-5 (29 mg) was separated by a semipreparative HPLC (YMC-pack ODS-A, 5 μm; 10 × 250 mm; 60% MeCN/H$_2$O; 4 mL/min) to yield compound **4** (t_R 13.0 min, 2.2 mg). Fr. 8-9 (812 mg) was eluted by a semipreparative HPLC (YMC-pack ODS-A, 5 μm; 10 × 250 mm; 80% MeOH/H$_2$O; 4 mL/min) to afford compound **12** (t_R 30.0 min, 157.9 mg).

Penerpene K (**1**): Yellow oil; $[\alpha]_D^{25}$ −23.0 (*c* 0.01, MeOH); UV (MeOH) λ_{max} (logε): 231 (3.41), 280 (2.77) nm; ECD (0.38 mM, MeOH) λ_{max} ($\Delta\varepsilon$): 204 (2.46), 223 (−6.08), 255 (2.89); 294 (−0.62) nm; IR (KBr) ν_{max}: 3412, 2933, 1640, 1606, 1458, 1382, 1307, 1257, 1212, 1160, and 1088 cm^{-1}; ^{1}H and ^{13}C NMR spectral data, Tables 1 and 2; HRESIMS *m/z* 476.2565 ([M+K]$^+$ calcd 476.2562).

Penerpene L (**2**): Yellow oil; $[\alpha]_D^{25}$ −22.0 (*c* 0.01, MeOH); UV (MeOH) λ_{max} (logε): 231 (3.15), 282 (2.41) nm; ECD (1.10 mM, MeOH) λ_{max} ($\Delta\varepsilon$): 203 (0.14), 224 (−1.63), 260 (0.58), 292 (−0.08), 330 (0.24) nm; IR (KBr) ν_{max}: 3411, 2935, 1645, 1604, 1452, 1381, 1307, 1256, 1158, 1074, and 1018 cm^{-1}; ^{1}H and ^{13}C NMR spectral data, Tables 1 and 2; HRESIMS *m/z* 476.2210 ([M+K]$^+$ calcd 476.2198).

Penerpene M (**3**): White powder; $[\alpha]_D^{25}$ +15.0 (*c* 0.01, MeOH); UV (MeOH) λ_{max} (logε): 226 (2.84), 261 (2.32) nm; ECD (1.15 mM, MeOH) λ_{max} ($\Delta\varepsilon$): 198 (1.32), 202 (−1.27), 221 (−0.92), 263 (0.94) nm; IR (KBr) ν_{max}: 3383, 2931, 1648, 1609, 1575, 1357, 1316, 1257, 1208, 1160, 1118, and 1044 cm^{-1}; ^{1}H and ^{13}C NMR spectral data, Tables 1 and 2; HRESIMS *m/z* 446.3020 ([M+Na]$^+$ calcd 446.3030).

Penerpene N (**4**): Yellow oil; $[\alpha]_D^{25}$ +21.0 (*c* 0.01, MeOH); UV (MeOH) λ_{max} (logε): 231 (3.05), 277 (2.44) nm; ECD (0.77mM, MeOH) λ_{max} ($\Delta\varepsilon$): 197 (−3.23), 219 (4.27), 238 (−6.39), 257 (3.21) nm; IR (KBr) ν_{max}: 3415, 2937, 1641, 1408, 1110, 1408, 1110, 1042, 922, 852, 688, and 565 cm^{-1}; ^{1}H and ^{13}C NMR spectral data, Tables 1 and 2; HRESIMS *m/z* 434.2338 ([M-H]$^-$ calcd 434.2331).

4. Conclusions

To summarize, based on the OSMAC culture strategy, four new indole-diterpenoids were isolated from the marine-derived fungus *Penicillium* sp. KFD28 secondary metabolites. The absolute configurations of new compounds **1**–**3** were determined by spectroscopic methods coupled with experimental and calculated ECD. New compound **4** showed mild cytotoxicity against the human cervical cancer cell line HeLa. Notably, compound **9** exhibited strong cytotoxic activity against the human liver cancer cell line BeL-7402 with IC$_{50}$ values of 5.3 μM, indicating that compound **9** deserves further study for its therapeutic potential to develop new anti-hepatoma drugs. Compound **7** showed pronounced antibacterial activity against *Bacillus subtilis* with MIC values of 16 μg/mL, which had the potential to become an antibiotic. However, the mechanisms causing cytotoxicity and bacteria restraint were unknown and require further study. In general, this study expanded the application of the OSMAC method to increase the chemical and biological diversity of natural products isolated from the *Penicillium* sp. KFD28.

Supplementary Materials: The following are available online at https://www.mdpi.com/article/10.3390/md19110613/s1, Figures S1–S38: 1D NMR, 2D NMR, IR, and HRESIMS spectra of the new compounds **1–4**; Figure S39: the picture of strain *Penicillium* sp. KFD28; Figure S40: Separation flow chart of fungus *Penicillium* sp. KFD28 secondary metabolites extract; Figure S41: HPLC chromatograms of the EtOAc extracts monitored at wavelengths of 230 nm and 274 nm; Figure S42: HPLC chromatograms of the EtOAc extracts using Rice solid medium with L-tryptophan, and compounds **6**, **12**, **14**, **8**, and **11** monitored at wavelengths of 230 nm and 274 nm; Figure S43: ECD curve for compound **4** and computation section of compounds **1–3**.

Author Contributions: L.-T.D. contributed to fungal fermentation, compounds purification, and structural elucidation. L.Y. and F.-D.K. were responsible for chemical computation and preparation of the paper. Q.-Y.M. and Q.-Y.X. contributed to the bioassays and identified the fungal strain. H.-F.D. revised the paper. Z.-F.Y. and Y.-X.Z. designed the work and revised the paper. All authors have read and agreed to the published version of the manuscript.

Funding: This research was supported by the Natural Science Foundation of Hainan Province (2019CXTD411), Financial Fund of the Ministry of Agriculture and Rural Affairs, P.R. of China (NFZX2021), China Agriculture Research System of MOF and MARA (CARS-21), Central Public-interest Scientific Institution Basal Research Fund for Chinese Academy of Tropical Agricultural Sciences (1630052017002, 1630052021019).

Institutional Review Board Statement: Not applicable.

Conflicts of Interest: The authors declare no conflict of interest.

References

1. Lindequist, U. Marine-derived pharmaceuticals-challenges and opportunities. *Biomol. Ther.* **2016**, *24*, 561–571. [CrossRef] [PubMed]
2. Carroll, A.R.; Copp, B.R.; Davis, R.A.; Keyzers, R.A.; Prinsep, M.R. Marine natural products. *Nat. Prod. Rep.* **2019**, *36*, 122–173. [CrossRef]
3. Hu, Y.W.; Chen, J.H.; Hu, G.P.; Yu, J.C.; Zhu, X.; Lin, Y.C.; Chen, S.P.; Yuan, J. Statistical research on the bioactivity of new marine natural products discovered during the 28 years from 1985 to 2012. *Mar. Drugs* **2015**, *13*, 202–221. [CrossRef]
4. Li, T.T.; Wang, Y.; Li, L.; Tang, M.Y.; Meng, Q.H.; Zhang, C.; Hua, E.B.; Pei, Y.H.; Sun, Y. New cytotoxic cytochalasans from a plant-associated fungus *Chaetomium globosum* kz-19. *Mar. Drugs* **2021**, *19*, 438. [CrossRef]
5. Kubota, T.; Nakamura, K.; Kurimoto, S.I.; Sakai, K.; Fromont, J.; Gonoi, T.; Kobayashi, J. Zamamidine D, a manzamine alkaloid from an okinawan *Amphimedon* sp. marine sponge. *J. Nat. Prod.* **2017**, *80*, 1196–1199. [CrossRef]
6. Kong, F.D.; Zhang, S.L.; Zhou, S.Q.; Ma, Q.Y.; Xie, Q.Y.; Chen, J.P.; Li, J.H.; Zhou, L.M.; Yuan, J.Z.; Hu, Z.; et al. Quinazoline-containing indole alkaloids from the marine-derived fungus *Aspergillus* sp. HNMF114. *J. Nat. Prod.* **2019**, *82*, 3456–3463. [CrossRef]
7. Guo, Y.W.; Liu, X.J.; Yuan, J.; Li, H.J.; Mahmud, T.; Hong, M.T.; Yu, J.C.; Lan, W.J. L-tryptophan induces a marine-derived *Fusarium* sp. to produce indole alkaloids with activity against the Zika virus. *J. Nat. Prod.* **2020**, *83*, 3372–3380. [CrossRef] [PubMed]
8. Kong, F.D.; Fan, P.; Zhou, L.M.; Ma, Q.Y.; Xie, Q.Y.; Zheng, H.Z.; Zhang, R.S.; Yuan, J.Z.; Dai, H.F.; Luo, D.Q.; et al. Penerpenes A-D, four indole terpenoids with potent protein tyrosine phosphatase inhibitory activity from the marine-derived fungus *Penicillium* sp. KFD28. *Org. Lett.* **2019**, *21*, 4864–4867. [CrossRef]
9. Zhou, L.M.; Kong, F.D.; Fan, P.; Ma, Q.Y.; Xie, Q.Y.; Li, J.H.; Zheng, H.Z.; Zheng, Z.H.; Yuan, J.Z.; Dai, H.F.; et al. Indole-diterpenoids with protein tyrosine phosphatase inhibitory activities from the marine-derived fungus *Penicillium* sp. KFD28. *J. Nat. Prod.* **2019**, *82*, 2638–2644. [CrossRef] [PubMed]
10. Bode, H.B.; Bethe, B.; Höfs, R.; Zeeck, A. Big effects from small changes: Possible ways to explore nature's chemical diversity. *ChemBioChem* **2002**, *3*, 619–627. [CrossRef]
11. Xu, W.; Gavia, D.J.; Tang, Y. Biosynthesis of fungal indole alkaloids. *Nat. Prod. Rep.* **2014**, *31*, 1474–1487. [CrossRef] [PubMed]
12. Huang, L.H.; Xu, M.Y.; Li, H.J.; Li, J.Q.; Chen, Y.X.; Ma, W.Z.; Li, Y.P.; Xu, J.; Yang, D.P.; Lan, W.J. Amino acid-directed strategy for inducing the marine-derived fungus *Scedosporium apiospermum* F41–1 to maximize alkaloid diversity. *Org. Lett.* **2017**, *19*, 4888–4891. [CrossRef] [PubMed]
13. Liu, S.S.; Yang, L.; Kong, F.D.; Zhao, J.H.; Yao, L.; Yuchi, Z.G.; Ma, Q.Y.; Xie, Q.Y.; Zhou, L.M.; Guo, M.F.; et al. Three new quinazoline-containing indole alkaloids from the marine-derived fungus *Aspergillus* sp. HNMF114. *Front. Microbiol.* **2021**, *12*. [CrossRef]
14. Hosoe, T.; Nozawa, K.; Udagawa, S.I.; Nakajima, S.; Kawai, K.I. Structures of new indoloditerpenes, possible biosynthetic precursors of the tremorgenic mycotoxins, penitrems, from *Penicillium crustosum*. *Chem. Phar. Bull.* **2008**, *38*, 3473–3475. [CrossRef]
15. Nozawa, K.; Horie, Y.; Udagawa, S.I.; Kawai, K.I.; Yamazaki, M. Isolation of a new tremorgenic indologiterpene, 1'-O-Acetylpaxilline, from *Emericella striata* and distribution of paxilline in *Emericella* spp. *Chem. Phar. Bull.* **1989**, *37*, 1387–1389. [CrossRef]

16. Belofsky, G.N.; Gloer, J.B.; Wicklow, D.T.; Dowd, P.F. Antiinsectan alkaloids: Shearinines A-C and a new paxilline derivative from the Ascostromata of *Eupenicillium Shearii*. *Tetrahedron* **1995**, *51*, 3959–3968. [CrossRef]

17. Fan, Y.Q.; Wang, Y.; Liu, P.P.; Fu, P.; Zhu, T.H.; Wang, W.; Zhu, W.M. Indole-diterpenoids with anti-H1N1 activity from the aciduric fungus *Penicillium camemberti* OUCMDZ-1492. *J. Nat. Prod.* **2013**, *76*, 1328–1336. [CrossRef]

18. Chen, M.Y.; Xie, Q.Y.; Kong, F.D.; Ma, Q.Y.; Zhou, L.M.; Yuan, J.Z.; Dai, H.F.; Wu, Y.G.; Zhao, Y.X. Two new indole-diterpenoids from the marine-derived fungus *Penicillium* sp. KFD28. *J. Asian Nat. Prod. Res.* **2020**, 1–8. [CrossRef]

19. Ariantari, N.P.; Ancheeva, E.; Wang, C.Y.; Mándi, A.; Knedel, T.O.; Kurtán, T.; Chaidir, C.; Müller, W.E.G.; Kassack, M.U.; Janiak, C.; et al. Indole diterpenoids from an endophytic *Penicillium* sp. *J. Nat. Prod.* **2019**, *82*, 1412–1423. [CrossRef]

20. Munday-Finch, S.C.; Wilkins, A.L.; Miles, C.O. Isolation of paspaline B, an indole-diterpenoid from *Penicillium paxilli*. *Phytochemistry* **1996**, *41*, 327–332. [CrossRef]

21. Liang, Z.Y.; Shen, N.X.; Zhou, X.J.; Zheng, Y.Y.; Chen, M.; Wang, C.Y. Bioactive indole diterpenoids and polyketides from the marine-derived fungus *penicillium javanicum*. *Chem. Nat. Comp.* **2020**, *56*, 379–382. [CrossRef]

22. Yamaguchi, T.; Nozawa, K.; Hosoe, T.; Nakajima, S.; Kawai, K.I. Indoloditerpenes related to tremorgenic mycotoxins, penitrems, from *Penicillium crustosum*. *Phytochemistry* **1993**, *32*, 1177–1181. [CrossRef]

23. Nozawa, K.; Nakajima, S.; Kawai, K.I.; Udagawa, S.I. Isolation and structures of indoloditerpenes, possible biosynthetic intermediates to the tremorgenic mycotoxin, paxilline, from *Emericella striata*. *J. Chem. Soc. Perkin Trans.* **1988**, *9*, 2607–2610. [CrossRef]

24. Kimura, Y.; Nishibe, M.; Nakajima, H.; Hamasaki, T.; Shigemitsu, N.; Sugawara, F.; Stout, T.J.; Clardy, J. Emeniveol: A new pollen growth inhibitor from the fungus, *Emericella nivea*. *Tetrahedr. Lett.* **1992**, *33*, 6987–6990. [CrossRef]

25. Yang, M.H.; Wang, J.S.; Luo, J.G.; Wang, X.B.; Kong, L.Y. Chisopanins A-K, 11 new protolimonoids from *chisocheton paniculatus* and their anti-inflammatory activities. *Bioorg. Med. Chem.* **2011**, *19*, 1409–1417. [CrossRef] [PubMed]

26. Luo, X.D.; Wu, S.H.; Wu, D.G.; Ma, Y.B.; Qi, S.H. Three new apo-tirucallols with six-membered hemiacetal from meliaceae. *Tetrahedron* **2002**, *58*, 6691–6695. [CrossRef]

27. Chen, C.; Liang, F.; Chen, B.; Sun, Z.Y.; Xue, T.D.; Yang, R.L.; Luo, D.Q. Identification of demethylincisterol A_3 as a selective inhibitor of protein tyrosine phosphatase Shp2. *Eur. J. Phar.* **2016**, *795*, 124–133. [CrossRef] [PubMed]

28. Sallam, A.A.; Houssen, W.E.; Gissendanner, C.R.; Orabi, K.Y.; Foudah, A.I.; Sayed, K.A.E. Bioguided discovery and pharmacophore modeling of the mycotoxic indole diterpene alkaloids penitrems as breast cancer proliferation, migration, and invasion inhibitors. *Med. Chem. Commun.* **2013**, *4*, 1360–1369. [CrossRef] [PubMed]

29. Guo, J.J.; Dai, B.L.; Chen, N.P.; Jin, L.X.; Jiang, F.S.; Ding, Z.S.; Qian, C.D. The anti-staphylococcus aureus activity of the phenanthrene fraction from fibrous roots of *Bletilla striata*. *BMC Complement. Altern. Med.* **2016**, *16*, 491–497. [CrossRef] [PubMed]

Article

Metabolomics Tools Assisting Classic Screening Methods in Discovering New Antibiotics from Mangrove Actinomycetia in Leizhou Peninsula

Qin-Pei Lu [1,2,†], Yong-Mei Huang [3,4,†], Shao-Wei Liu [1,2], Gang Wu [1,2], Qin Yang [1,2], Li-Fang Liu [1,2], Hai-Tao Zhang [3], Yi Qi [3,4], Ting Wang [1,2], Zhong-Ke Jiang [1,2], Jun-Jie Li [3], Hao Cai [5], Xiu-Jun Liu [5], Hui Luo [3,4,*] and Cheng-Hang Sun [1,2,*]

1 Department of Microbial Chemistry, Institute of Medicinal Biotechnology, Chinese Academy of Medical Sciences & Peking Union Medical College, Beijing 100050, China; qinpei89@hotmail.com (Q.-P.L.); liushaowei3535@163.com (S.-W.L.); gangwu@aliyun.com (G.W.); yangqin@imb.pumc.edu.cn (Q.Y.); LiuLiFang@imb.pumc.edu.cn (L.-F.L.); tingwang0707@imb.pumc.edu.cn (T.W.); jiangzhongke@126.com (Z.-K.J.)
2 Beijing Key Laboratory of Antimicrobial Agents, Institute of Medicinal Biotechnology, Chinese Academy of Medical Sciences & Peking Union Medical College, Beijing 100050, China
3 The Key Lab of Zhanjiang for R&D Marine Microbial Resources in the Beibu Gulf Rim, Marine Biomedical Research Institute, Guangdong Medical University, Zhanjiang 524023, China; huangym@gdmu.edu.cn (Y.-M.H.); taohaizhang33@163.com (H.-T.Z.); qiyi7272@gdmu.edu.cn (Y.Q.); jjleeee@gdmu.edu.cn (J.-J.L.)
4 Marine Biomedical Research Institute of Guangdong Zhanjiang, Zhanjiang 524023, China
5 Department of Oncology, Institute of Medicinal Biotechnology, Chinese Academy of Medical Sciences & Peking Union Medical College, Beijing 100050, China; caihao@imb.pumc.edu.cn (H.C.); Liuxiujun2000@imb.pumc.edu.cn (X.-J.L.)
* Correspondence: luohui@gdmu.edu.cn (H.L.); chenghangsun@hotmail.com or sunchenghang@imb.pumc.edu.cn (C.-H.S.); Tel.: +86-10-63165272 (C.-H.S.)
† These authors contributed equally to this work.

Citation: Lu, Q.-P.; Huang, Y.-M.; Liu, S.-W.; Wu, G.; Yang, Q.; Liu, L.-F.; Zhang, H.-T.; Qi, Y.; Wang, T.; Jiang, Z.-K.; et al. Metabolomics Tools Assisting Classic Screening Methods in Discovering New Antibiotics from Mangrove Actinomycetia in Leizhou Peninsula. *Mar. Drugs* **2021**, *19*, 688. https://doi.org/10.3390/md19120688

Academic Editors: Khaled A. Shaaban and Bill J. Baker

Received: 5 September 2021
Accepted: 28 November 2021
Published: 1 December 2021

Publisher's Note: MDPI stays neutral with regard to jurisdictional claims in published maps and institutional affiliations.

Abstract: Mangrove actinomycetia are considered one of the promising sources for discovering novel biologically active compounds. Traditional bioactivity- and/or taxonomy-based methods are inefficient and usually result in the re-discovery of known metabolites. Thus, improving selection efficiency among strain candidates is of interest especially in the early stage of the antibiotic discovery program. In this study, an integrated strategy of combining phylogenetic data and bioactivity tests with a metabolomics-based dereplication approach was applied to fast track the selection process. A total of 521 actinomycetial strains affiliated to 40 genera in 23 families were isolated from 13 different mangrove soil samples by the culture-dependent method. A total of 179 strains affiliated to 40 different genera with a unique colony morphology were selected to evaluate antibacterial activity against 12 indicator bacteria. Of the 179 tested isolates, 47 showed activities against at least one of the tested pathogens. Analysis of 23 out of 47 active isolates using UPLC-HRMS-PCA revealed six outliers. Further analysis using the OPLS-DA model identified five compounds from two outliers contributing to the bioactivity against drug-sensitive *A. baumannii*. Molecular networking was used to determine the relationship of significant metabolites in six outliers and to find their potentially new congeners. Finally, two *Streptomyces* strains (M22, H37) producing potentially new compounds were rapidly prioritized on the basis of their distinct chemistry profiles, dereplication results, and antibacterial activities, as well as taxonomical information. Two new trioxacarcins with keto-reduced trioxacarcinose B, gutingimycin B (**16**) and trioxacarcin G (**20**), together with known gutingimycin (**12**), were isolated from the scale-up fermentation broth of *Streptomyces* sp. M22. Our study demonstrated that metabolomics tools could greatly assist classic antibiotic discovery methods in strain prioritization to improve efficiency in discovering novel antibiotics from those highly productive and rich diversity ecosystems.

Keywords: Leizhou Peninsula; mangrove soil; actinomycetia; diversity; antimicrobial activity; secondary metabolites; dereplication; metabolomics tools; trioxacarcins

1. Introduction

Highly effective antibiotics have been waning since they were introduced into the clinic to treat infectious diseases more than 75 years ago. Antibacterial-resistant bacteria account for approximately 700,000 deaths each year worldwide [1]. The near-empty pharmaceutical pipeline for new antibiotics has now reached alarming levels. In 2017, the World Health Organization (WHO) released a list of antibiotic-resistant "priority pathogens" that posed the greatest threat to human health for the first time, reminding that we were stepping into the post-antibiotic era [2]. Therefore, the urgent need for new antimicrobial agents has augmented scientists' interest in discovering new antibiotics.

Natural products have a profound role in drug discovery and development [3]. Between 1981 and 2019, natural products and their derivatives accounted for 55% of the 162 new antibacterial chemical entities [4]. The strains in the class *Actinomycetia* (former class *Actinobacteria*) [5], especially species of the genus *Streptomyces*, are known as a rich source of novel antibiotics [6–9]. However, finding novel antibiotics from in-depth investigated terrestrial microorganisms has become more challenging. Researchers have been exploring untapped sources of biodiversity for novel pharmaceutical compounds, such as the deep sea [10], deserts [11], polar areas [12], and mangroves [13,14]. Because of its high moisture, high salinity, and hypoxia-tolerant ecosystem in the intertidal zones, the mangrove ecosystem is unique and contains a wealth of undiscovered bacteria and natural products, making it a focus area for bioactive natural product discovery [15]. To the best of our knowledge, very little research has been carried out on bioprospecting of Leizhou Peninsula mangrove actinomycetia. Only a few reports describing actinomycetia isolated from this ecosystem have been published so far, and no research on prioritizing the actinomycetial strains and discovering their secondary metabolites has been carried out [16–18].

The selection of strain candidates from large strain collections using traditional bioactivity screening and/or taxonomy-based methods usually result in the reisolation of known compounds [19,20]. Nowadays, exploring new bioactive compounds from microbial strains has moved toward integrated strategies, which combine phylogenetic data and bioactivity tests with dereplication approaches for rapid identification of known bioactive metabolites. Dereplication using chromatographic and spectroscopic methods and database searches can save time and avoid repetitive work during natural product discovery programs. Recently, a new dereplication method was developed by Tim Bugni using chemical dereplication coupled with metabolomics tools [21]. By incorporating metabolomics approaches, dereplication can focus on chemically diverse bacterial extracts from the bioactive strains, and this new method has shown to be effective in discovering putative new bioactive compounds and is frequently employed in microbial drug discovery programs.

Metabolomics is the comprehensive analysis of small molecule metabolites in a biological system to reflect the phenotype of its genomic, transcriptomic, and proteomic networks, providing insight into the biological functions [22]. In the past, metabolomics was mainly used to investigate primary metabolites, such as nucleotides, amino acids, and lipids. However, the majority of modern metabolomics coupled with molecular networking is readily applied to secondary metabolites discovery for new natural products. Combined principal component analysis (PCA) with LC/MS-based metabolomics is an efficient analytical tool to differentiate the bacterial strains based on their LC/MS profiles [23,24]. The strains producing similar secondary metabolites are clustered together, whereas those with different metabolites are separated. This new approach could significantly accelerate the bioassay-guided selection for chemically distinct strains that might yield novel bioactive secondary metabolites [25–27]. In addition, orthogonal partial least squares-discriminant analysis (OPLS-DA) is a supervised multivariate analysis that targets bioactive metabolites between the active and inactive groups to give information about the chemical composition of selected active extracts before isolation [19,20,28,29]. Molecular networking, one of the main analysis tools in the GNPS platform, creates structured networking based on MS/MS similarity to reflect the molecular diversity in the extracts. It has been proven to be successful for effective chemical dereplication and novel metabolite discovery [30–32].

In the present study, we employed integrated strategies in prioritizing the actinomycetial strains from the underexplored mangrove soil in Leizhou Peninsula. Actinomycetial strains were isolated using the culture-dependent method and phylogenetically characterized based on 16S rRNA gene sequencing. The selected actinomycetial strains were further subjected to antibacterial assays followed by metabolomics analysis, such as PCA, OPLS-DA, and molecular networking. The obtained data were integrated to prioritize the strains for follow-up chemical isolation and structural identification work of putative new compounds.

2. Results

2.1. Strain Isolation and Phylogenetic Identification

A total of 521 actinomycetial strains were isolated by 12 different agar media (Table S1). Partial 16S rRNA gene sequence analysis identified these strains as the following genera: *Micromonospora* (121 strains), *Streptomyces* (116), *Microbacterium* (36), *Rhodococcus* (35), *Brachybacterium* (28), *Isoptericola* (25), *Cellulosimicrobium* (21), *Brevibacterium* (17), *Serinibacter* (13), *Mycolicibacterium* (10), *Micrococcus* (10), *Agromyces* (10), *Kocuria* (9), *Mycobacterium* (6), *Gordonia* (6), *Aeromicrobium* (5), *Arthrobacter* (5), *Citricoccus* (5), *Janibacter* (5), *Nocardia* (5), *Corynebacterium* (4), *Glutamicibacter* (4), *Agrococcus* (3), *Intrasporangium* (2), *Kineococcus* (2), *Phycicoccus* (2), *Sinomonas* (2), *Serinicoccus* (2), *Actinomadura* (1), *Actinopolymorpha* (1), *Blastococcus* (1), *Demequina* (1), *Georgenia* (1), *Gulosibacter* (1), *Jonesia* (1), *Leucobacter* (1), *Motilibacter* (1), *Mumia* (1), *Salinibacterium* (1), and *Streptacidiphilus* (1). These strains were affiliated to 40 different genera in 23 families (Figure 1a, Tables S2 and S3). The genera *Micromonospora* (23.2%) and *Streptomyces* (22.3%) were dominant, followed by *Microbacterium* (6.9%), *Rhodococcus* (6.7%), *Brachybacterium* (5.4%), and *Isoptericola* (4.8%). Additionally, 12 genera were represented by only one isolate. The distribution of the 521 actinomycetial strains from 13 different mangrove soil samples is shown in Figure 1b and Table S2. Sample 1 gave the highest diversity and abundance (74 isolates in 17 genera). Samples 3, 8, and 9 shared the second-highest diversity (15 genera), followed closely by sample 13 (14 genera) and sample 2 (13 genera). Two samples collected from Dongsong island showed the lowest diversity (sample 6, 4 genera; sample 7, 5 genera). With respect to the medium composition, M1 yielded the highest recovery rate (12.5%), with 65 isolates representing 18 different genera (Figure 1c and Table S3). A similar result showing higher diversity (63 isolates in 23 genera) was obtained from the modified M1 medium (M11) by the addition of kelp (Table S1). M5 displayed the third-best recovery rates (10.9%) with 57 isolates in 15 genera and showed the best recovery rate of *Streptomyces* (21 isolates). M4 yielded the lowest number and diversity of isolates (13 isolates in 6 genera). Interestingly, M12 (18 isolates in 12 genera), the amended M4 medium with the addition of coconut juice (Table S1), showed a higher number and diversity of isolates than M4 (13 isolates in 6 genera).

2.2. Antibacterial Assay

The antibacterial assay was used as one effective way to exclude the inactive strains and select the candidates for metabolomics analysis. Out of 521 actinomycetial strains from 13 different mangrove soil samples, 179 strains affiliated to 40 different genera with a unique colony morphology were selected to evaluate the antibacterial activities against six sets of indicator bacteria; each set contained one drug-sensitive and one drug-resistant pathogen (*Enterococcus faecium*, *Staphylococcus aureus*, *Klebsiella pneumoniae*, *Acinetobacter baumannii*, *Pseudomonas aeruginosa*, and *Enterobacter species*). Of the 179 strains, 47 exhibited antagonistic activity against at least one of the tested pathogens. These bioactive strains were distributed in 13 genera, including *Streptomyces* (29), *Micromonospora* (4), *Micrococcus* (3), *Rhodococcus* (2), *Brevibacterium* (1), *Demequina* (1), *Actinomadura* (1), *Georgenia* (1), *Gordonia* (1), *Isoptericola* (1), *Microbacterium* (1), *Streptacidiphilus* (1), and *Serinibacter* (1) (Table S4). The antibacterial spectra of the 47 isolates against indicator bacteria are shown in Figure 2. Of the 47 strains, 35 were active only against the Gram-positive bacteria, while two isolates (*Streptomyces* sp. H124 and *Streptomyces* sp. Y129) were active only against the Gram-negative bacteria. In addition, ten strains showed inhibitory activities against both the Gram-positive and Gram-negative bacteria. Interestingly, isolate M45, a rare actinomycetial strain affiliated to genus *Demequina*, showed moderate activities against all four Gram-positive bacterial strains tested. Thirty-seven isolates showed anti-MRSA activity, accounting for the highest proportion of the total active strains (78.7%). Twenty-three strains showing high anti-MRSA activity (diameter of inhibition zone ≥ 10 mm) were selected for subsequent metabolomics analyses (Table 1).

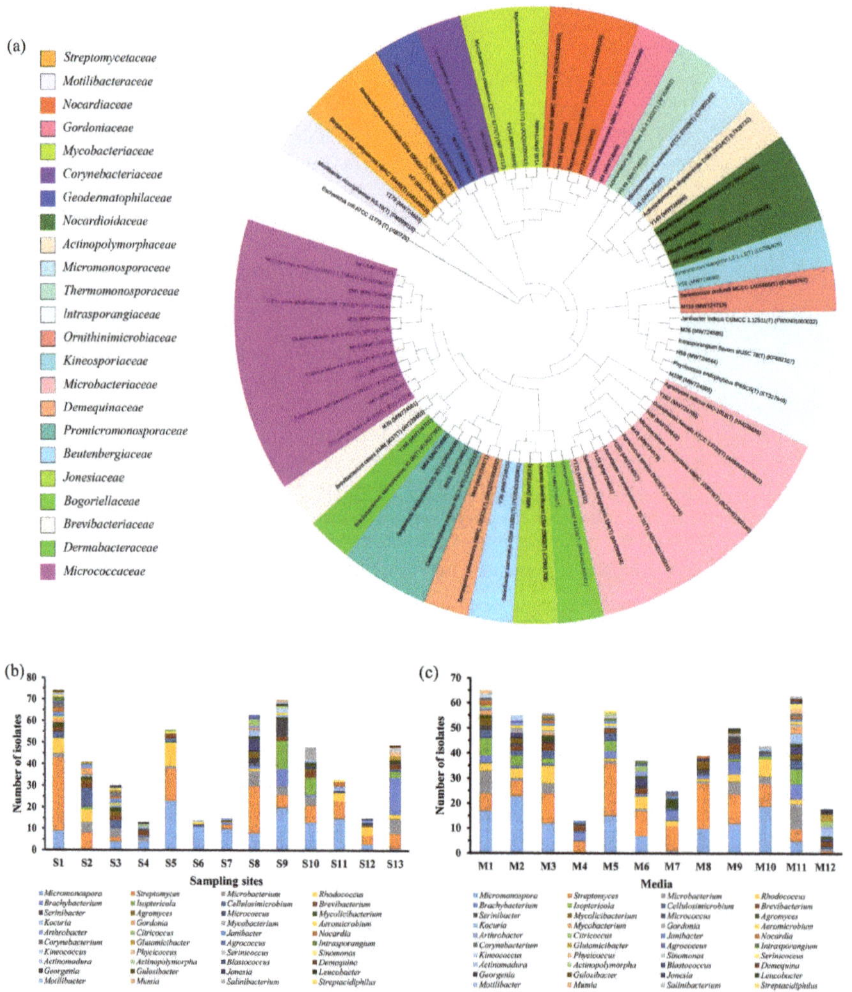

Figure 1. The distribution of cultivable actinomycetial strains isolated from mangrove soil in Leizhou Peninsula. (**a**) Phylogenetic tree (16S rRNA gene) obtained by neighbor-joining analysis of 40 isolates from each genus and their closely related type strains; *Escherichia coli* was used as an outgroup; (**b**) genera distribution in different sampling sites; (**c**) genera distribution according to the culture media used for the isolation.

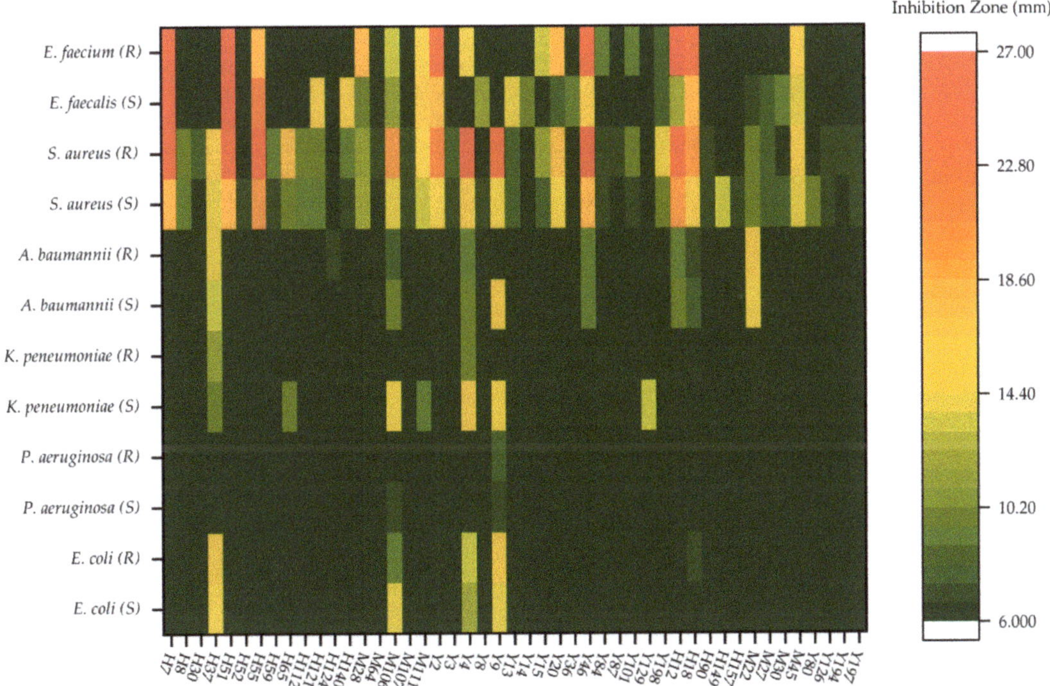

Figure 2. The antibacterial spectra of 47 strains against indicator bacteria (6 mm, no inhibitory activity; *S*, sensitive; *R*, drug resistant; *E. coli*, *Escherichia coli*; *P. aeruginosa*, *Pseudomonas aeruginosa*; *K. pneumoniae*, *Klebsiella pneumoniae*; *A. baumannii*, *Acinetobacter baumannii*; *S. aureus*, *Staphylococcus aureus*; *E. faecalis*, *Enterococcus faecalis*; *E. faecium*, *Enterococcus faecium*).

2.3. Strain Prioritization by Metabolomics

The ethyl acetate extracts of 23 isolates shown in Table 1 were subjected to metabolomics analysis using UPLC-HRMS-PCA. PCA, an unsupervised statistical analysis method, was used to identify differing chemical features found in the outlying strains and to prioritize the strains that produced unique secondary metabolites. After UPLC-MS/MS data acquisition and processing, a total of 6544 features (RT-*m/z* pairs) and 6 principal components were generated by PCA analysis, giving the R^2 (goodness of fit) and Q^2 (predictability) values of 0.82 and 0.70, respectively. After analyzing the scores plot (PC1 vs. PC2), the quality control (QC) samples (purple circle) clustered and were close to the center of the scores plot (Figure 3a), indicating good reproducibility and stability of the system. Four predominant outliers, *Streptomyces* sp. H7, *Streptomyces* sp. Y2, *Streptomyces* sp. H12, and *Streptomyces* sp. Y46, were observed, suggesting their chemical uniqueness from the main groups of samples in the scores plot. *Streptomyces* sp. H7 and *Streptomyces* sp. Y2 were located in the same quadrant, indicating similarities in their secondary metabolite profiles. Similarly, the metabolite profiles of *Streptomyces* sp. H12 and *Streptomyces* sp. Y46 were also alike. The loadings plot (Figure 3b) was geometrically related to the scores plot and described the variance observed in the scores plot, so it could be used to identify compounds that caused a clustered group to separate. From the loadings plot, three outlying metabolites, 12.88_1268.6095n (n: neutral mass) (**1**), 12.69_1254.6282n (**2**), and 11.11_1270.6216n (**3**), were significant contributors to the group of *Streptomyces* sp. H12 and *Streptomyces* sp. Y46. They were putatively assigned as actinomycin-type antibiotics through dereplication with the Natural Products Atlas (NPAtlas) v19_12 database [33] and StreptomeDB v3.0 database [34] and further confirmed as the actinomycins by comparisons of their ultraviolet (UV) spectra with previously published data (Figure 3c, Table 2, and Figure S1) [35,36]. Comparison with the other extracts revealed that the outlying feature 16.71_724.4749n (**4**) was abundant in *Streptomyces* sp. H7 and present in *Streptomyces* sp. Y2 with lower peak intensities (Figure S2). The predicted molecular formula $C_{40}H_{68}O_{11}$ was tentatively identified as nigericin, epinigericin, or abierixin by dereplication. Abierixin was first excluded because it had UV absorptions at 200–400 nm but the feature 16.71_724.4749n (**4**) lacked UV absorptions [37]. Meanwhile, the MS/MS fragment pattern (Figure S3) and UV characterization of 16.71_724.4749n (**4**) strongly suggested that it was most likely to be either nigericin or epinigericin [38,39]. Another feature 18.21_708.4796n (**6**) was putatively elucidated as 22-member macrolides (ushikulide A or B) or nigericin-type polyethers (grisorixin or epigrisorixin); all were previously isolated from *Streptomyces* spp. (Table 2, Figure 3c) [40–42].

Table 1. Information of 23 isolates selected for metabolomics analyses and their antimicrobial activity against drug-resistant *S. aureus* (MRSA) and drug-sensitive *A. baumannii*.

Isolates ID	GenBank Accession No.	Top-Hit Taxon (Pairwise Similarity, %)	Diameter of Inhibition Zone (mm)	
			*A. baumannii*_S	*S. aureus*_R
H7	MW724538	*Streptomyces malaysiensis* NBRC 16446^T (100.00)	-	24
H8	MW724539	*Streptomyces bungoensis* DSM 41781^T (99.62)	-	10
H37	MW724543	*Streptomyces thermoviolaceus* subsp. *Thermoviolaceus* DSM 40443^T (98.50)	13	17
H51	MW724550	*Streptomyces griseochromogenes* ATCC 14511^T (99.38)	-	23
H55	MW724552	*Streptomyces galbus* DSM 40089^T (99.50)	-	24
H65	MW724554	*Streptomyces griseoincarnatus* LMG 19316^T (100.00)	-	18
H112	MW724556	*Gordonia polyisoprenivorans* NBRC 16320^T (98.54)	-	10
H121	MW724559	*Rhodococcus zopfii* NBRC 100606^T (100.00)	-	10
M28	MW724573	*Streptomyces sundarbansensis* MS1/7^T (99.76)	-	11
M106	MW724593	*Streptomyces geysiriensis* NBRC 15413^T (99.75)	10	20
M111	MW724597	*Streptomyces smyrnaeus* SM3501^T (99.74)	-	16
Y2	MW724603	*Streptomyces hygroscopicus* subsp. *hygroscopicus* NBRC 13472^T (99.63)	-	21
Y4	MW724605	*Streptomyces albogriseolus* NRRL B-1305^T (100.00)	10	26
Y9	MW724607	*Streptomyces pluripotens* MUSC 135^T (99.87)	17	23
Y15	MW724610	*Streptomyces cellulosae* NBRC 13027^T (100.00)	-	11
Y20	MW724611	*Streptomyces levis* NBRC 15423^T (99.75)	-	18
Y46	MW724616	*Streptomyces antibioticus* NBRC 12838^T (100.00)	9	25
Y101	MW724624	*Micromonospora humi* DSM 45647^T (99.61)	-	10
Y198	MW724634	*Micromonospora pallida* DSM 43817^T (99.13)	-	14
H12	MW724637	*Streptomyces similanensis* KC-106^T (99.13)	10	23
H18	MW724638	*Streptomyces seoulensis* NRRL B-24310^T (100.00)	8	20
M22	MW724664	*Streptomyces shenzhenensis* 172115^T (99.88)	16	10
M45	MW724671	*Demequina salsinemoris* NBRC 105323^T (99.63)	-	17

Paper disk diameter, 6 mm; -, no inhibitory zone; *A. baumannii*, *Acinetobacter baumannii*; *S. aureus*, *Staphylococcus aureus*; S, sensitive; R, drug resistant.

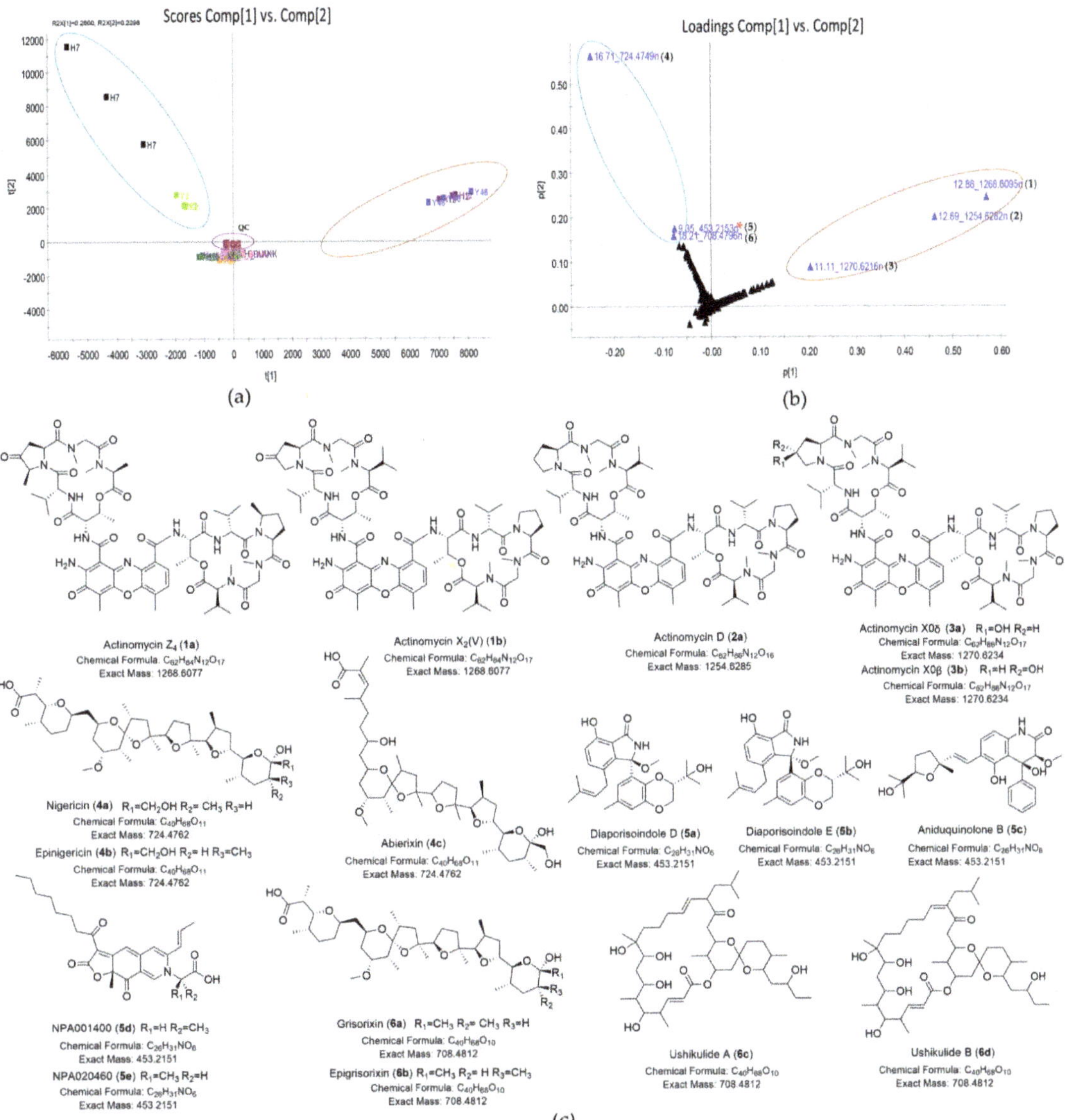

Figure 3. Discovery of natural products unique to four outlier strains (*Streptomyces* sp. Y46, *Streptomyces* sp. H12, *Streptomyces* sp. H7, and *Streptomyces* sp. Y2). (**a**) Scores plot (PC1 vs. PC2) of 23 extracts; (**b**) loadings plot (PC1 vs. PC2) of 23 extracts; the loadings plot showed compounds **1–3** were unique to *Streptomyces* sp. H12 and *Streptomyces* sp. Y46, and compounds **4–6** were special to *Streptomyces* sp. H7 and *Streptomyces* sp. Y2; 9.35_453.2153n* (**5**): false positive; (**c**) The structures of putative metabolites for compounds **1–6** from four outlier strains by database dereplication.

Table 2. The putative metabolites for compounds **1–9** against the NPAtlas (Hit 1) and StreptomeDB (Hit 2) databases in the Progenesis QI v3.0 software.

Peak ID	Compound	Isolate No. [a]	R_t (min)	Neutral Mass	Observed m/z	UV-Vis (nm) [b]	Hit 1	Hit 2	Predicted MF [e]
1	12.88_1268.6095n	H12,Y46	12.88	1268.6095	1269.6168	245, 442	Actinomycin Z4 (**1a**)		$C_{62}H_{84}N_{12}O_{17}$
								Actinomycin X2(V) (**1b**)	$C_{62}H_{84}N_{12}O_{17}$
2	12.69_1254.6282n	H12,Y46	12.69	1254.6282	1255.6355	245, 442	Actinomycin D (**2a**)	Actinomycin D (**2a**)	$C_{62}H_{86}N_{12}O_{16}$
3	11.11_1270.6216n	H12,Y46	11.11	1270.6216	1271.6289	245, 442	Actinomycin X0δ (**3a**)		$C_{62}H_{86}N_{12}O_{17}$
								Actinomycin X0β (**3b**)	$C_{62}H_{86}N_{12}O_{17}$
4	16.71_724.4749n	H7,Y2	16.71	724.4749	747.4654	- [c]	Nigericin (**4a**)	Nigericin (**4a**)	$C_{40}H_{68}O_{11}$
							Epinigericin (**4b**)		$C_{40}H_{68}O_{11}$
							Abierixin (**4c**)		$C_{40}H_{68}O_{11}$
5	9.35_453.2153n [d]	H7,Y2	9.35	453.2153	454.2226	257, 308	Diaporisoindole D (**5a**)		$C_{26}H_{31}NO_6$
							Diaporisoindole E (**5b**)		$C_{26}H_{31}NO_6$
							Aniduquinolone B (**5c**)		$C_{26}H_{31}NO_6$
							NPA001400 (**5d**)		$C_{26}H_{31}NO_6$
							NPA020460 (**5e**)		$C_{26}H_{31}NO_6$
6	18.21_708.4796n	H7,Y2	18.21	708.4796	731.4806	- [c]	Grisorixin (**6a**)	Grisorixin (**6a**)	$C_{40}H_{68}O_{10}$
								Epigrisorixin (**6b**)	$C_{40}H_{68}O_{10}$
							Ushikulide A (**6c**)	Ushikulide A (**6c**)	$C_{40}H_{68}O_{10}$
							Ushikulide B (**6d**)	Ushikulide B (**6d**)	$C_{40}H_{68}O_{10}$
7	9.35_546.2571n	H7,Y2	9.35	546.2571	564.2910	257, 308		17-O-demethyl-geldanamycin (**7**)	$C_{28}H_{38}N_2O_9$
8	9.46_548.2717n	H7,Y2	9.46	548.2717	566.3055	257, 308	4,5-dihydro-17-O-demethyl-geldanamycin (**8a**)	4,5-dihydro-17-O-demethyl-geldanamycin (**8a**)	$C_{28}H_{40}N_2O_9$
							17-O-demethylgeldanamycin hydroquinone (**8b**)		
							Herbimycin F (**8c**)		
							Antimycin A1a (**8d**)	Antimycin A1a (**8d**)	
								Antimycin A1 (**8e**)	
							Antimycin A12 (**8f**)		
							Antimycin A13 (**8g**)		
							Antimycin A19 (**8h**)	Antimycin A19 (**8h**)	
								Deformylantimycin A2a (**8i**)	
9	10.91_560.2719n	H7,Y2	10.91	560.2719	578.3062	257, 308		Geldanamycin (**9a**)	$C_{29}H_{40}N_2O_9$
							17-hydroxymethyl-17-demethoxygeldanamycin (**9b**)	17-hydroxymethyl-17-demethoxygeldanamycin (**9b**)	$C_{29}H_{40}N_2O_9$
							17-formyl-17-demethoxy18-O,21-O-dihydrogeldanamycin (**9c**)	17-formyl-17-demethoxy18-O,21-O-dihydrogeldanamycin (**9c**)	$C_{29}H_{40}N_2O_9$
							Herbimycin C (**9d**)	Herbimycin C (**9d**)	$C_{29}H_{40}N_2O_9$

[a], the isolate with the high intensity; [b], re-checked in the LC conditions with ACN-H_2O; [c], no or end absorption; [d], false positive; [e], molecular formula.

After rapid structure dereplication using high-resolution MS data of the feature 9.35_453.2153n (**5**), it was putatively assigned to diaporisoindole D, E, aniduquinolone B, NPA001400, or NPA020460, all described as fungal metabolites by the NPAtlas database (Table 2 and Figure 3c). However, none of them matched the UV spectrum of this feature [43–45]. When searching its UV maximum (308 nm) in the UPLC-UV-HRMS chromatogram of sample *Streptomyces* sp. H7, two analogs of compound **5**, 9.53_456.4081*m/z* and 10.90_468.4110*m/z*, were found and they did not get any hits from the databases. The dereplication results of three homologs (9.35_453.2153n (**5**), 9.53_456.4081*m/z*, and 10.90_468.4110*m/z*) indicated they might be the putative new metabolites. However, when we reacquired the MS/MS data in the data-dependent acquisition (DDA) method, all of them showed poor MS/MS fragmentations and high background noise (Figure S4), indicating that they probably were in-source fragments (false positives) rather than molecular ion peaks when acquiring in MSE method. As a data-independent acquisition (DIA) method, the MSE can simultaneously record exact mass precursor and fragment ion information in the full *m/z* range, but precursor and fragment spectra are aligned mainly according to retention times. Therefore, the MSE acquisition might mismatch the product ions with its parent ion in the analysis of complex samples [46–48], leading to the misidentification of compounds when processing the data in software, such as Progenesis QI. By examination of the raw MS and MS/MS data in almost the same elution time as three in-source fragments (9.35_453.2153n (**5**), 9.53_456.4081*m/z* and 10.90_468.4110*m/z*), three other compounds, 9.35_546.2571n (**7**), 9.46_548.2717n (**8**), and 10.91_560.2719n (**9**), had a high intensity of MS/MS fragments in the DDA acquisition, and their pseudomolecular ion peaks were also identified in MS spectra of UPLC-HRMS and HPLC-MS chromatograms (Figures 4, S5 and S6). Therefore, the corresponding compounds in the same elution time should be revised as compounds **7–9**, respectively. By searching databases and comparing their UV spectral data (Figures 4 and S7) with the literature data, compounds **7–9** were most likely to be benzoquinoid ansamycin-type compounds as shown in Figure 4 [49–51]. In summary, four predominant outliers, *Streptomyces* sp. H7, *Streptomyces* sp. Y2, *Streptomyces* sp. H12, and *Streptomyces* sp. Y46, were preliminarily excluded for prospecting new antibiotics because differing features of those strains could be matched well with known antibiotics.

Identifying the compounds responsible for groups near the center of the scores plot, such as sample *Streptomyces* sp. H37 or sample *Streptomyces* sp. M22 in the PC1 vs. PC2 scores plot (Figure 3a) was usually not straightforward. However, these groups could be separated by observing different PC planes [21]. This approach could lead to identifying more significant compounds that we might be interested in. As shown in Figure 5a, sample *Streptomyces* sp. H37 was separated in the PC1 vs. PC4 scores plot. Compounds unique to this sample were identified in the corresponding position in the loadings plot. Therefore, the major compounds unique to *Streptomyces* sp. H37 were shown as 10.64_900.5435n (**10**) and 11.08_928.5742n (**11**) (Figure 5b). The outlying feature (compound **10**) with neutral mass ion peak at 900.5435 Da in the predicted molecular formula $C_{47}H_{80}O_{16}$ was putatively identified as either cytovaricin or W341C (Figure 5c, Table 3). The MS/MS fragment spectrum showed a series of peaks with loss of H_2O (Figure S8), and a characteristic fragmentation pattern of polyether ionophores [39] representing compound **10** was likely identified as the polyether ionophore compound W341C. Database searches revealed no hit for feature 11.08_928.5742n (**11**), suggesting that it might be a putative new compound. Two features (**10–11**) showed similarity in their MS/MS spectra, demonstrating they should be the structural analogs (Figure S8).

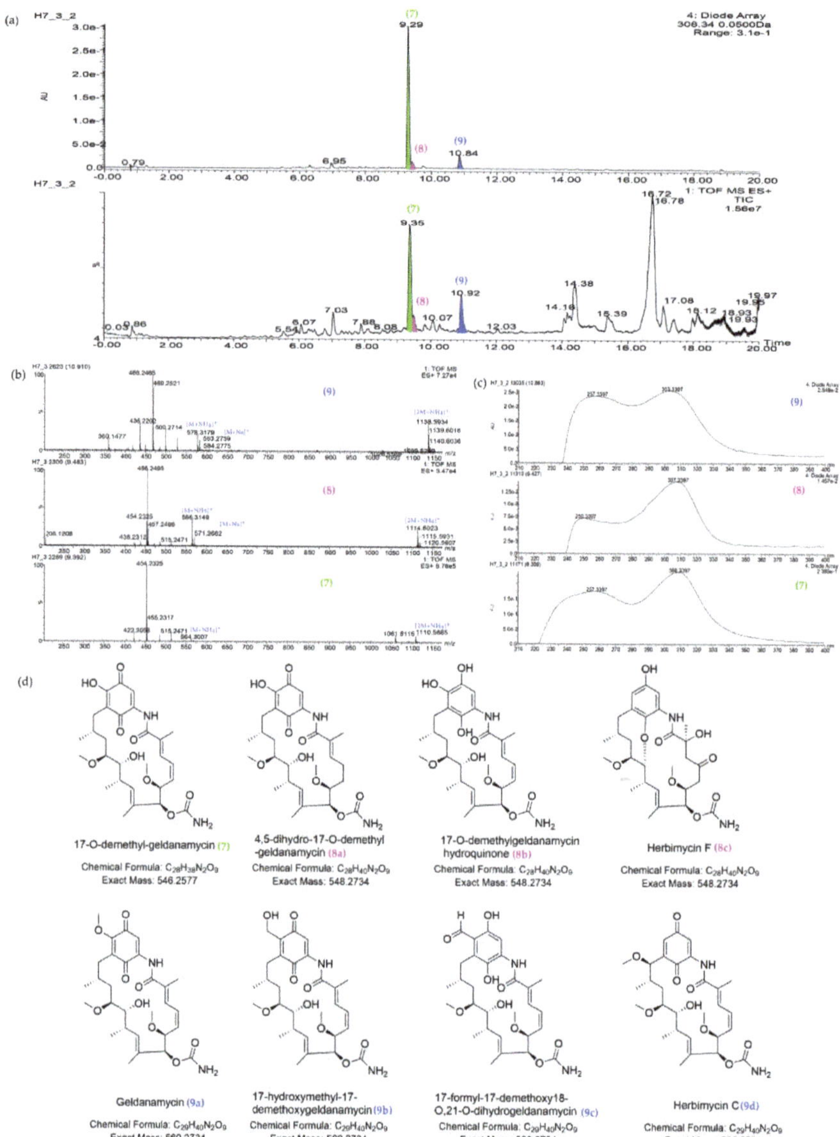

Figure 4. The characteristic of major benzoquinoid ansamycin-type compounds in the UPLC-UV-HRMS chromatogram of sample *Streptomyces* sp. H7 (**7**: 9.35_546.2571n, 17-*O*-demethyl-geldanamycin; **8**: 9.46_548.2717n, 4,5-dihydro-17-O-demethyl-geldanamycin (**8a**) or 17-O-demethylgeldanamycin hydroquinone (**8b**), or herbimycin F (**8c**); **9**: 10.91_560.2719n, geldanamycin (**9a**) or 17-hydroxymethyl-17-demethoxygeldanamycin (**9b**), or 17-formyl-17-demethoxy18-O,21-O-dihydrogeldanamycin (**9c**)). (**a**) UV spectra at 308 nm and TIC plot. (**b**) MS spectra of three compounds, **7–9**. (**c**) UV spectra of three compounds, **7–9**. (**d**) The structures of putative metabolites for compounds **7–9** after comparison with the databases and literatures.

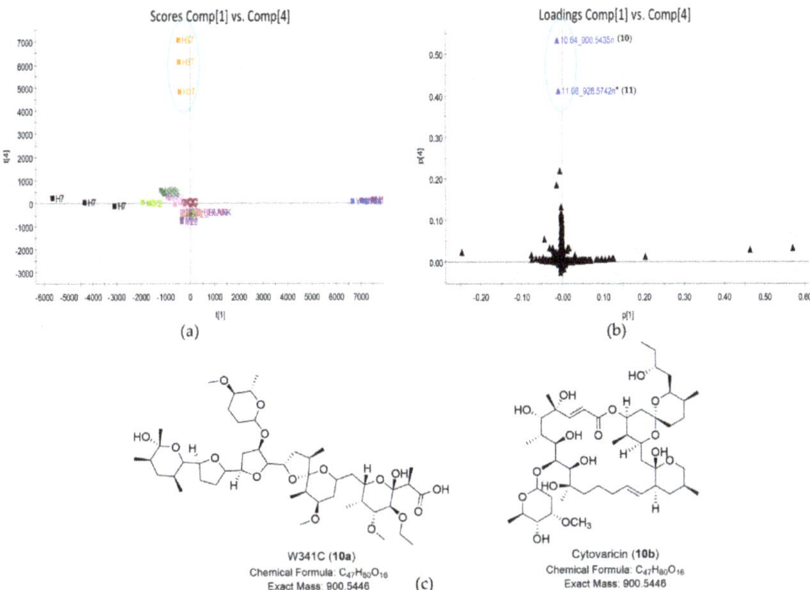

Figure 5. Discovery of natural products unique to strain *Streptomyces* sp. H37. (**a**) Scores plot (PC1 vs. PC4) of 23 extracts; the PC planes were adjusted to separate *Streptomyces* sp. H37; (**b**) loadings plot (PC1 vs. PC4) of 23 extracts; the loadings plot showed compounds **10–11** that were unique to *Streptomyces* sp. H37; *, no hit. (**c**) The structures of putative metabolites for compound **10** from *Streptomyces* sp. H37 by database dereplication.

In the scores plot (PC1 vs. PC6) (Figure 6a), strain *Streptomyces* sp. M22 was clearly distinct from other strains. The major compounds unique to *Streptomyces* sp. M22 were 7.16_1028.3600m/z (**12**), 15.40_566.4171n (**13**), 7.47_1028.3592m/z (**14**), and 9.55_876.2968n (**15**). The dereplicated results of these outlying metabolites are shown in Table 3. The UV spectra of compounds **12**, **14**, and **15** confirmed their identities as trioxacarcin-type compounds (Figure S9), corresponding to putative gutingimycin, gutingimycin, and trioxacarcin A, respectively [52,53]. Both compounds **12** and **14** matched gutingimycin in the database searching because they had the same MS data (1028.3592m/z), indicating that one of them should be an isomer of gutingimycin. To the best of our knowledge, no isomers of gutingimycin have been found so far (Table S5). The presence of gutingimycin isomer in 7.16 or 7.47 min (compound **12** or **14**) indicated that one of them should be a putative new compound. After examining the UPLC-UV-HRMS profile of *Streptomyces* sp. M22, trioxacarcin-type compounds, including compounds **12**, **14**, and **15**, were presented within a retention time range of 6.5–10.0 min (Figures 7, S9 and S10). Two compounds, 6.69_1030.3751m/z (**16**) and 7.94_1013.3486m/z (**17**), in the chromatogram were identified as the putative new compounds after searching the databases and reverting back to the literature (Table S5). Lastly, another differing compound (**13**) was tentatively confirmed not to be 4-ketozeinoxanthin due to its lack of characteristic maximum UV absorption at 445–470 nm [54].

Table 3. The putative metabolites for compounds **10–19** against the NPAtlas (Hit 1) and StreptomeDB (Hit 2) databases in the Progenesis QI v3.0 software.

Peak ID	Compound	Isolate No. [a]	R_t (min)	Neutral Mass	Observed m/z	UV-Vis (nm) [b]	Hit 1	Hit 2	Predicted MF [d]
10	10.64_900.5435n	H37	10.64	900.5435	901.5508	-[c]	Cytovaricin (**10b**)	W341C (**10a**) Cytovaricin (**10b**)	$C_{47}H_{80}O_{16}$ $C_{47}H_{80}O_{16}$
11	11.08_928.5742n	H37	11.08	928.5742	929.5815	-[c]	No hit	No hit	unknown
12	7.16_1028.3600m/z	M22	7.16	unknown	1028.3600	219, 230 (sh), 270, 397	Gutingimycin (**12a**) Cyanopeptolin SS (**12b**) Micropeptin HU989 (**12c**)		$C_{47}H_{57}N_5O_{21}$ $C_{40}H_{63}N_9O_{17}S_2$ $C_{41}H_{64}ClN_9O_{15}S$
13	15.40_566.4171n	M22	15.4	566.4171	549.4138	-[c]	X-14889-D (**13a**) CP-78545 (**13b**) 4-ketozeinoxanthin (**13c**)	CP-78545 (**13b**)	$C_{33}H_{58}O_7$ $C_{33}H_{58}O_7$ $C_{40}H_{54}O_2$
14	7.47_1028.3592m/z	M22	7.47	unknown	1028.3592	219, 230 (sh), 270, 397	Gutingimycin (**14a**) Cyanopeptolin SS (**14b**) Micropeptin HU989 (**14c**)		$C_{47}H_{57}N_5O_{21}$ $C_{40}H_{63}N_9O_{17}S_2$ $C_{41}H_{64}ClN_9O_{15}S$
15	9.55_876.2968n	M22	9.55	876.2968	894.3355	232, 270, 399	Nai414-B (**15b**)	Trioxacarcin A (**15a**)	$C_{42}H_{52}O_{20}$ $C_{44}H_{55}C_{13}N_2O_{10}$
16	6.69_1030.3751m/z	M22	6.69	unknown	1030.3751	219, 230 (sh), 270, 397	Stremycin A (**16a**)	Stremycin A (**16a**) Kedarcidin (**16b**)	
17	7.94_1013.3486m/z	M22	7.94	unknown	1013.3486	232, 270, 399	No hit	No hit	unknown
18	8.43_894.3132n	M22	8.43	894.3132		232, 270, 399		Trioxacarcin B (**18**)	$C_{42}H_{54}O_{21}$
19	8.89_878.3168n	M22	8.89	878.3168		232, 270, 399		Trioxacarcin C (**19**)	$C_{42}H_{54}O_{20}$

[a], the isolate with the high intensity; [b], re-checked in the LC conditions with ACN-H_2O; [c], no or end absorption; [d], molecular formula.

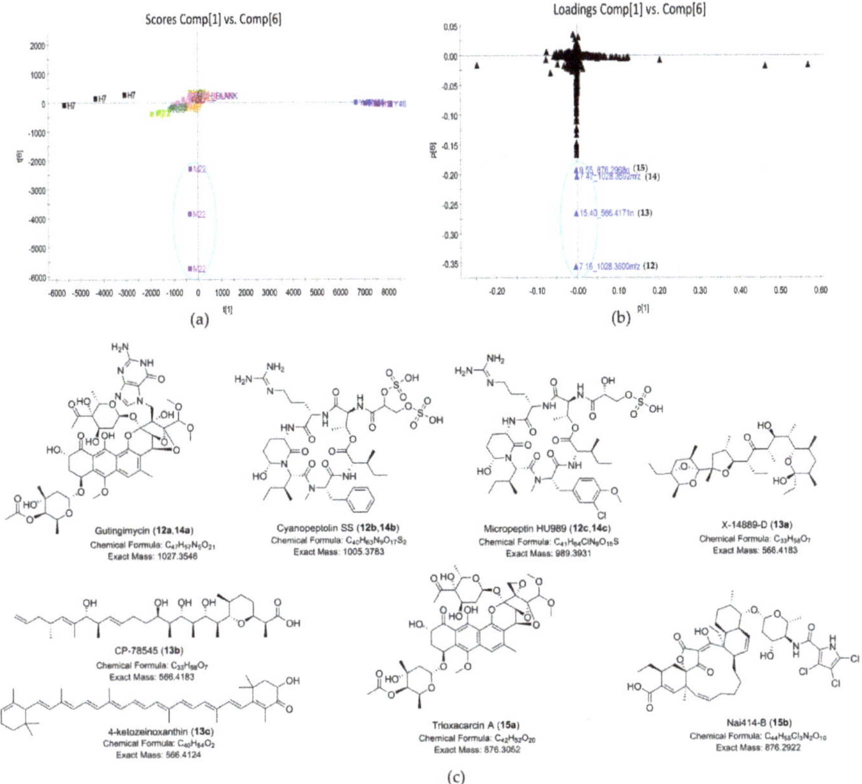

Figure 6. Discovery of natural products unique to strain *Streptomyces* sp. M22. (**a**) Scores plot (PC1 vs. PC6) of 23 extracts; the PC planes were adjusted to separate *Streptomyces* sp. M22; (**b**) loadings plot (PC1 vs. PC6) of 23 extracts; the loadings plot showed compounds **12–15** that were unique to *Streptomyces* sp. M22. (**c**) The structures of putative metabolites for compounds **12–15** from *Streptomyces* sp. M22 by database dereplication.

As shown in Table 1, among 23 strains showing anti-MRSA activity, 8 strains also displayed inhibition zones against drug-sensitive *Acinetobacter baumannii*. However, in PCA model, the groups of active or inactive strains against the *Acinetobacter baumannii* were not completely separated from each other as shown in Figure 3a. To discriminate the two classes (active vs. inactive) and identify compounds mainly contributing to the bioactivity of the eight strains, the OPLS-DA model was constructed. As shown in Figure 8a, active and inactive samples were clearly separated in the model. The R^2 value of 0.98 and Q^2 value of 0.95 suggested that the OPLS-DA model possessed reliable fitness and predictability. In the S-plot analysis, five potential marker compounds were selected to chemically distinguish the active from inactive extracts (Figure 8b). The variable importance in the projection (VIP) plot showed that all selected potential markers had high VIP values (VIP ≥ 10) (Figure S11), revealing that these marker compounds were largely responsible for the discrimination between active and inactive groups. Therefore, the main contributors to the activity were the putative three actinomycins (**1–3**) from the group of *Streptomyces* sp. H12 and *Streptomyces* sp. Y46, and the two compounds 10.64_900.5435n (**10**) and 11.08_928.5742n (**11**) from *Streptomyces* sp. H37. To the best of our knowledge, the inhibitory activity of actinomycin D against *A. baumannii* was reported [55]. Furthermore, cytovaricin and W341C, the dereplicated metabolites for compound **10**, had no antibacterial activity against Gram-negative pathogens [56,57]. In contrast to their antibacterial spectra, compound **10** showed antibacterial activity against Gram-negative pathogens, suggesting that it might be a putative new compound.

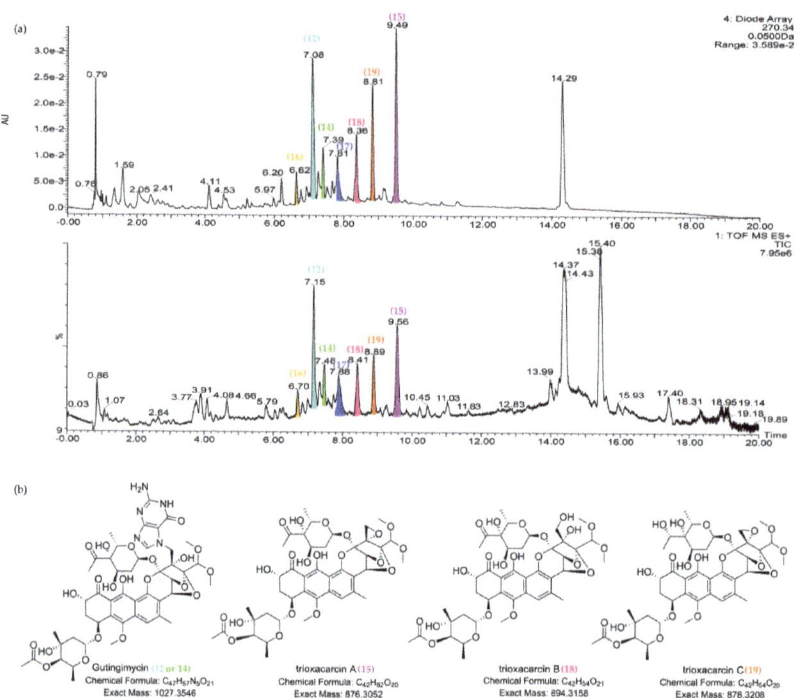

Figure 7. The characteristic of major trioxacarcin-type compounds in the UPLC-UV-HRMS chromatogram of *Streptomyces* sp. M22. (**12**, 7.16_1028.3600*m/z*; **14**, 7.47_1028.3592*m/z*; **15**, 9.55_876.2958n, trioxacarcin A; **16**, 6.69_1030.3751*m/z*; **17**, 7.94_1013.3486*m/z*; **18**, 8.43_894.3132n, trioxacarcin B; **19**, 8.89_878.3168n, trioxacarcin C). (**a**) UV spectra at 270 nm and TIC plot. (**b**) The structures of putative metabolites for compounds **12**, **14–15**, and **18–19** after comparison with the databases and literature.

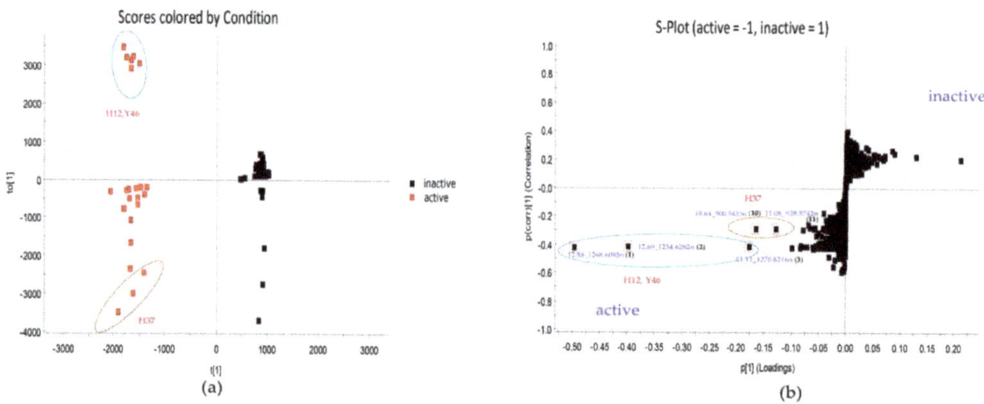

Figure 8. OPLS-DA analysis of 23 extracts. (**a**) OPLS-DA scores plot against *A. baumannii*. (**b**) The OPLS-DA loadings S-plot with the selected markers (12.88_1268.6095n (**1**), 12.69_1254.6282n (**2**), 11.11_1270.6216n (**3**), 10.64_900.5435n (**10**), and 11.08_928.5742n (**11**)).

2.4. Molecular Network Analysis of Outlier Strains

To survey the global map of the metabolites of six outliers (*Streptomyces* sp. Y46, *Streptomyces* sp. H12, *Streptomyces* sp. H7, *Streptomyces* sp. Y2, *Streptomyces* sp. H37, and *Streptomyces* sp. M22), classical molecular networking acquired by the DDA method was performed. After the removal of nodes associated with the blank medium control, the molecular network consisted of 5033 nodes connected with 5247 edges. It was noted that the number of nodes in the network did not correspond exactly to the number of metabolites, as different adducts or charges of the same compounds could generate different nodes [58]. As shown in Figure 9, six molecular families that contained spectra matching the discriminatory metabolites in the six outliers were identified. The putative actinomycin-type compounds, the trioxacarcins group, and benzoquinoid ansamycin-type compounds only existed in samples of *Streptomyces* sp. H12 and *Streptomyces* sp. Y46, *Streptomyces* sp. M22, *Streptomyces* sp. H7 and *Streptomyces* sp. Y2, respectively. They were in agreement with the results of the PCA analysis. In addition, the molecular network could clarify the relationship of the discriminatory metabolites in the outlier strains, assisting in the identification of the discriminatory metabolites. The two discriminatory metabolites 16.71_724.4749n (**4**) and 18.21_708.4796n (**6**) clustered as adjacent nodes in the nigericin family (Figure 9), indicating that compound **6** could be assigned as putative grisorixin or epigrisorixin, the analog of compound **4**. The putative metabolites (**8a** and **9a**) showed higher structural similarity with 17-*O*-demethyl-geldanamycin (**7**) than other "hit" compounds (**8b–8c**; **9b–9d**), and thus, compounds **8** and **9** in *Streptomyces* sp. H7 and *Streptomyces* sp. Y2 were further putatively deduced as 4,5-dihydro-17-*O*-demethyl-geldanamycin and geldanamycin, respectively (Figure 9). Similarly, the compounds 10.64_900.5435n (**10**) and 11.08_928.5742n (**11**) from *Streptomyces* sp. H37 were further confirmed as two congeners with a mass difference of 28 Da (CO or C_2H_4 group). In sample *Streptomyces* sp. M22, two putative new compounds, 1030.3751*m/z* (**16**) and 1013.3486*m/z* (**17**), were adjacent to gutingimycin (**12** or **14**) with mass differences of 2 and 15 Da, respectively, suggesting that the former was the putative hydrogenated gutingimycin, and the latter was the putative gutingimycin with loss of NH group (Figure 10).

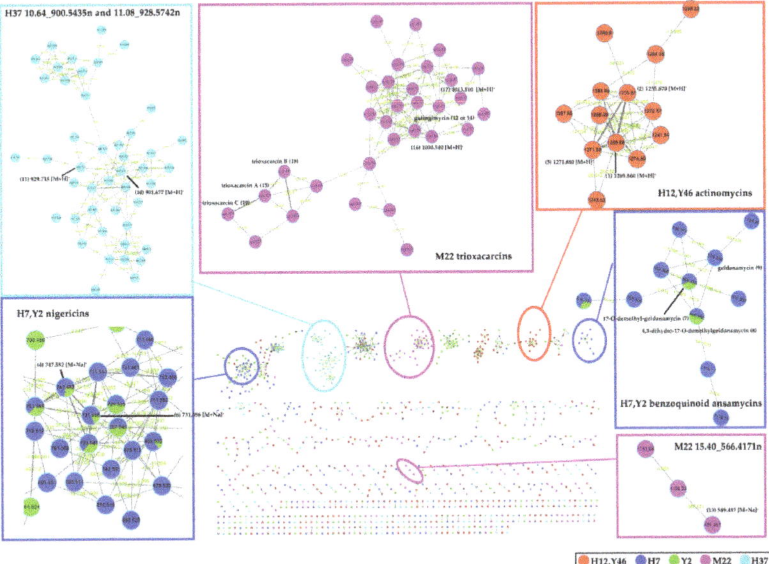

Figure 9. Molecular network of six outliers. The different colors of the nodes represented by different outliers, but the red node represented both *Streptomyces* sp. H12 and *Streptomyces* sp. Y46. Only clusters containing at least two nodes were shown.

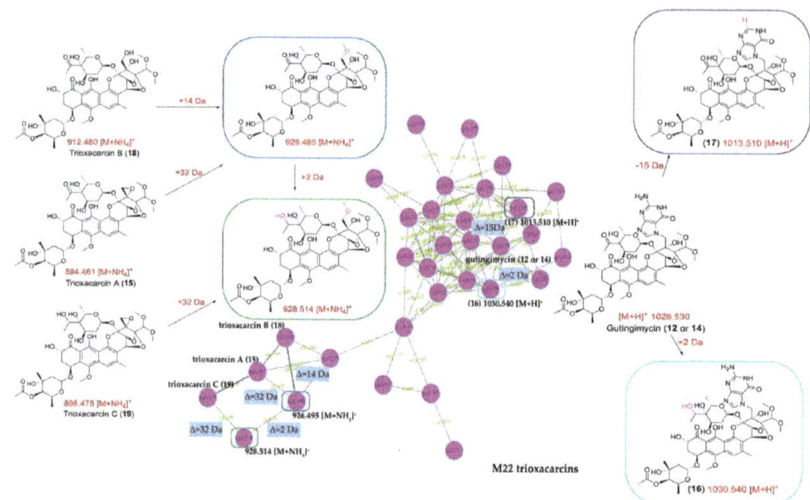

Figure 10. Molecular network analysis of putative trioxacarcins family in *Streptomyces* sp. M22 extract.

Furthermore, a large number of nodes clustered with the discriminatory metabolites suggested the presence of additional analogs in these compound classes. For example, in addition to trioxacarcin A (**15**), B (**18**), and C (**19**), the inspection of their parent mass, mass difference, and the MS/MS data implied that putative methylated trioxacarcin B (*m/z* 926.495) and its hydrogenation product (*m/z* 928.514) were present (Figure 10). Another outlier, 15.40_566.4171n (**13**), in the *Streptomyces* sp. M22 was clustered with two other compounds, whose molecular weights were almost twice as high, indicating that its dimers might exist in the extract (Figure S12). In the cluster containing the 10.64_900.5435n (**10**) and 11.08_928.5742n (**11**) of *Streptomyces* sp. H37, putative dehydrogenated, methylated, and demethylated analogs were detected (Figure S13). However, most of the putative new compounds discovered from the molecular network were presented as trace components in the extracts when checking their original UPLC-HRMS/MS data.

According to the results above, actinomycin-producing strains *Streptomyces* sp. Y46 and *Streptomyces* sp. H12, and nigericin-producing strains *Streptomyces* sp. H7 and *Streptomyces* sp. Y2 were excluded since their discriminatory metabolites were well known and frequently discovered from the actinomycetial strains, even though they were chemically unique. Samples of *Streptomyces* sp. H37 and *Streptomyces* sp. M22 were prioritized for scale-up and further isolation work. The strain *Streptomyces* sp. H37 was selected because the PCA analysis revealed that it contained the putative novel compounds. Further analysis using the OPLS-DA model identified these putative novel compounds contributing to significant inhibitory activity against drug-sensitive *A. baumannii*. Lastly, a variety of their potential new analogs were found through molecular network analysis. The purification and structural elucidation of new putative compounds in *Streptomyces* sp. H37 is still underway. The strain *Streptomyces* sp. M22 was selected because the putative novel trioxacarcin-type compounds were identified by the metabolomics-based dereplication approach (PCA, metabolic profile, molecular network, and database searching). Furthermore, the structures of some putative novel compounds were deduced from molecular networking. As a result, two novel trioxacarcin analogs, gutingimycin B (**16**) and trioxacarcin G (**20**) (Figure 11), along with gutingimycin (**12**) from the strain *Streptomyces* sp. M22, were isolated and structurally identified after large-scale fermentation.

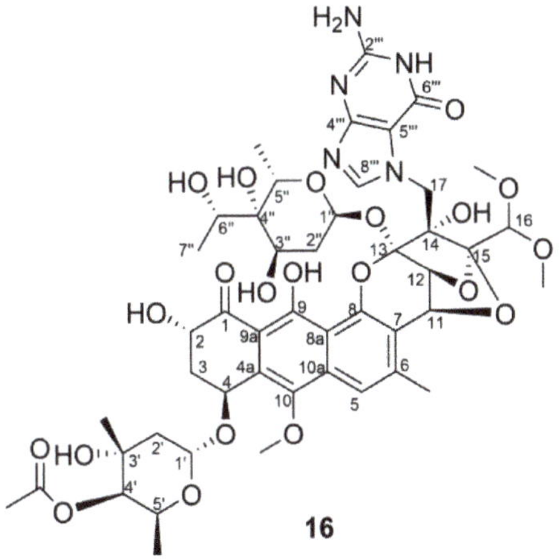

Figure 11. Chemical structures of gutingimycin B (**16**) and trioxacarcin G (**20**).

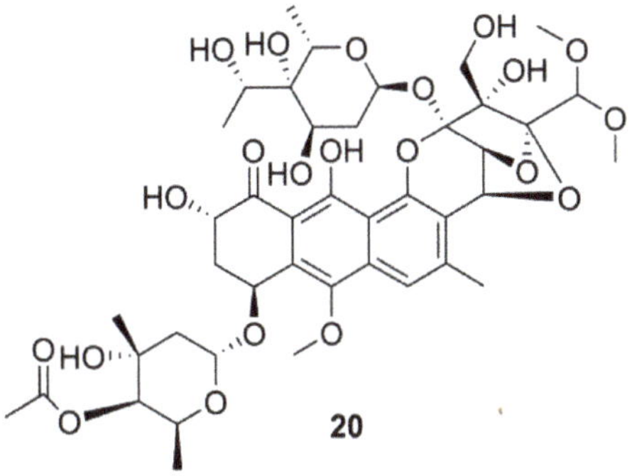

2.5. Structure Elucidation of Trioxacarcin Compounds from Strain Streptomyces sp. M22

Gutingimycin B (**16**) was obtained as a yellow powder. It fluoresced green in solution upon irradiation with 365 nm light and showed the UV absorption maxima at 274 nm and 408 nm (Figure S14), typical for trioxacarcin-type compounds. Its molecular formula was assigned as $C_{47}H_{59}N_5O_{21}$ on the basis of HRESIMS peak at m/z 1030.3778 [M+H]$^+$ (calcd for $C_{47}H_{60}N_5O_{21}$, 1030.3781) (Figure S15), 2 mass units more than gutingimycin. The ^{1}H NMR spectra (Table 4 and Figure S17) showed characteristic resonances for an exchangeable OH proton (δ_H 13.84 (s)), two aromatic protons (δ_H 8.22 (s) and 7.48 (s)), two anomeric protons (δ_H 5.64 (d, 3.6) and 5.51 (d, 4.2)), a pair of coupling oxygenated methines (δ_H 5.22 (d, 4.2) and 5.10 (d, 4.2)), a pair of geminal coupling methylenes (δ_H 5.02 (d, 15.6) and 4.33 (d,15.6)), three methoxyl singlets (δ_H 3.95 (s), 3.66 (s) and 3.53 (s)), two methyls attached to olefinic or carbonyl carbon (δ_H 2.62 (s) and 2.25 (s)), three methyl doublets (δ_H 1.37 (d, 6.6), 1.26 (d, 6.6) and 1.26 (d, 6.6)), and one methyl singlet (δ_H 1.18 (s)). In accordance with the molecular formula, the ^{13}C (Table 4 and Figure S18) and DEPT NMR spectra (Figure S19) showed the 47 carbon resonances assigned to 1 keto carbonyl (δ_C 207.6), 1 ester carbonyl (δ_C 173.3), a characteristic of the guanine group (δ_C 157.8, 153.8, 151.8, 140.6, and 108.1), 3 naphthalene carbons bearing to oxygen (δ_C 162.9, 152.7, and 144.6), 6 acetal carbons or electron-rich $sp2$ atoms (δ_C 108.8, 108.3, 103.3, 100.9, 99.6, and 92.5), 15 oxygenated methines (δ_C 56.1–84.4), 4 methylenes (δ_C 46.2, 37.7, 37.1, and 33.8), and 6 methyls (δ_C 15.7–27.3). The aforementioned NMR data of compound **16** showed very close similarities with those of gutingimycin [52]. The differences were an additional methine quartet at δ_H 3.99 (q, 6.6) and a methyl doublet at δ_H 1.37 (d, 6.6) in compound **16** instead of the methyl singlet at δ_H 2.50 (s) in gutingimycin. In the ^{13}C NMR spectra, the carbonyl at δ_C 211.1 in gutingimycin was missing, and additional oxygenated methine at δ_C 70.3 in compound **16** was detected. The missing ketone signal, together with the additional two mass units compared to gutingimycin, indicated that the ketone group in the sugar moiety of gutingimycin was reduced to an alcohol in compound **16**.

Table 4. ^{1}H (600 MHz) and ^{13}C (150 MHz) NMR data for compounds **16** and **20** in CDCl$_3$.

Position	16 δ_H (multi, J, Hz)	16 δ_C	20 δ_H (multi, J, Hz)	20 δ_C
1		207.6		203.9
2	5.50 (br s)	68.2	4.81 (dd, 12.6, 5.4)	68.0
3A	2.97 (d, 13.8)	37.1	2.77 (m)	36.7
3B	2.21 (overlapped)		2.22 (overlapped)	
4	5.34 (br s)	72.2	5.41 (t, 3.0)	66.7
4a		128.4		126.7
5	7.48 (s)	116.8	7.50 (br s)	117.1
6		142.5		142.7
6-CH$_3$	2.62 (s)	20.2	2.59 (s)	20.3
7		113.2		114.2
8		152.7		152.6
8a		114.4		114.2
9		162.9		162.5
9a		108.3		108.2
9-OH	13.84 (s)		14.61 (s)	
10		144.6		145.3
10-OCH$_3$	3.95 (s)	62.7	3.84 (s)	62.9
10a		135.7		135.5
11	5.10 (d, 4.2)	69.0	5.13 (d, 3.6)	69.3
12	5.22 (d, 4.2)	71.5	5.21 (d, 3.6)	71.2
13		103.3		105.2
14		84.4		85.0
15		108.8		108.2
16	5.08 (s)	99.6	5.04 (s)	99.9
16-OCH$_3$	3.66 (s)	56.1	3.63 (s)	56.4
16-OCH$_3$	3.53 (s)	56.4	3.54 (s)	57.3

Position	16 δ_H (multi, J, Hz)	16 δ_C	20 δ_H (multi, J, Hz)	20 δ_C
17A	5.02 (d, 15.6)	46.2	3.68 (m)	62.8
17B	4.33 (d, 15.6)		3.68 (m)	
1'	5.51 (d, 4.2)	100.9	5.35 (d, 4.2)	97.4
2'A	1.99 (overlapped)	37.7	1.93 (dd, 15.0, 4.2)	36.5
2'B	1.92 (d, 15.0)		1.62 (d, 15.0)	
3'		67.3		69.0
3'-CH$_3$	1.18 (s)	27.3	1.06 (s)	25.8
4'	5.14 (s)	75.9	4.74 (s)	74.5
4'-OCOCH$_3$		173.3		170.5
4'-OCOCH$_3$	2.25 (s)	21.1	2.12 (s)	21.0
5'	4.92 (br s)	62.2	4.53 (q, 6.6)	62.9
5'-CH$_3$	1.26 (d, 6.6)	16.9	1.23 (d, 6.6)	17.0
1''	5.64 (d, 3.6)	92.5	5.65 (d, 3.0)	93.7
2''A	2.21 (overlapped)	33.8	2.28 (overlapped)	33.0
2''B	1.76 (d, 15.6)		2.05 (d, 15.0)	
3''	3.91 (br s)	68.6	3.97 (br s)	68.4
4''		72.1		72.6
5''	4.68 (q, 6.6)	66.2	4.60 (q, 6.6)	66.3
5''-CH$_3$	1.26 (d, 6.6)	15.7	1.23 (d, 6.6)	15.5
6''	3.99 (q, 6.6)	70.3	3.91 (q, 6.6)	70.7
7''	1.37 (d, 6.6)	17.8	1.34 (d, 6.6)	18.2
2'''		151.8		
4'''		153.8		
5'''		108.1		
6'''		157.8		
8'''	8.22 (s)	140.6		

The abovementioned structural features in compound **16** were confirmed by the 2D NMR spectra (Figures S20–S22). The crucial ^{1}H-^{1}H COSY correlation of H-6'' (δ_H 3.99, q, 6.6)/H$_3$-7'' (δ_H 1.37, d, 6.6) and the key HMBC correlation from H$_3$-7'' to C-6'' (δ_C 70.3), C-4'' (72.1) (Figure 12) indicated that the hydroxyl group was linked to the C-6'' position as the keto-reduced trioxacarcinose B.

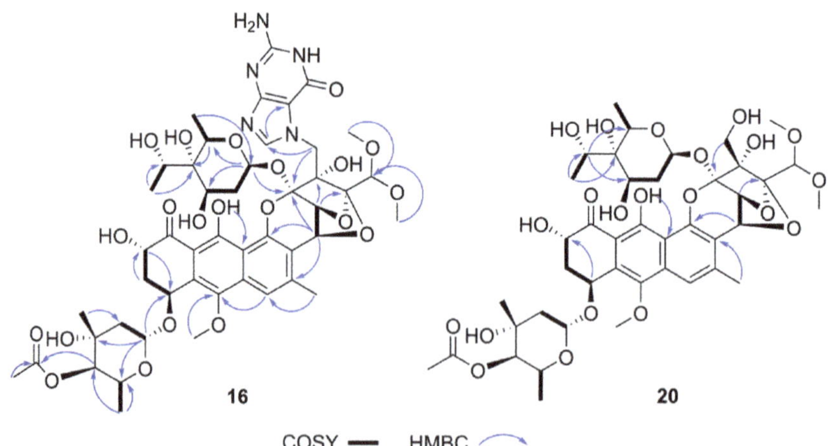

Figure 12. The key ^{1}H-^{1}H COSY and HMBC correlations of **16** and **20**.

Trioxacarcin G (**20**) was obtained as a yellow powder. The UV absorption maxima at 232, 271, and 400 nm (Figure S23) and fluorescence under 365 nm light indicated that compound **20** is a trioxacarcin-type compound. The HRESIMS suggested that its molecular formula was determined to be $C_{42}H_{56}O_{21}$ (*m/z* 914.3644 $[M+NH_4]^+$, calcd $C_{42}H_{60}NO_{21}$, 914.3658) (Figure S24), 2 mass units more than trioxacarcin B. ^1H and ^{13}C NMR of compound **20** (Table 4, Figures S26–S28) were nearly identical to those of trioxacarcin B [59,60], except for an additional methine quartet at δ_H/δ_C 3.91 (q, 6.6)/70.7 and a methyl doublet at δ_H/δ_C 1.34 (d, 6.6)/18.2 in replacement of the carbonyl at δ_C 210.9 and the methyl singlet at δ_H/δ_C 2.46 (s)/28.1. These were similar to the difference between compound **16** and gutingimycin, indicating that the ketone group in the L-trioxacarcinoses B of trioxacarcin B was also reduced to an alcohol in the keto-reduced trioxacarcinose B of compound **20**. The location of hydroxyl group was assigned at C-6″ based on the ^1H-^1H COSY correlation of H-6″ (δ_H 3.91, q, 6.6)/H_3-7″ (δ_H 1.34, d, 6.6) and the HMBC correlations from H_3-7″ to C-6″ (δ_C 70.7), C-4″ (δ_C 72.6) and from H-6″ to C-3″ (δ_C 68.4), C-5″ (δ_C 66.3) (Figures 12 and S29–S31).

The NMR data of the keto-reduced trioxacarcinose B in **16** and **20** were closely similar with those of trioxacarcin C rather than epi-6″-trioxacarcin C, indicating that the absolute configuration at C-6″of **16** and **20** was determined as 6″S, the same as trioxacarcin C [61]. According to the literature [60], the X-ray structure of gutingimycin delivered the stereochemistry of the trioxacarcin skeleton, and the sugar moieties of the trioxacarcins A–B were identified previously as L-trioxacarcinoses A and B. In addition, compounds **16** and **20** had the same specific rotation sign as the known trioxacarcin, also isolated in the present study, gutingimycin (**12**) ($[\alpha]_D^{25}$ −60.0° (c 0.02, ACN)). Therefore, the absolute configuration of compounds **16** and **20** was established, as shown in Figure 11. In addition, the structures of known compounds were identified as gutingimycin (**12**) by comparison of their physio-chemical and spectroscopic data (Figures S32–S37) with those of the literature [52].

3. Discussion

Actinomycetia derived from mangroves is a promising source for exploring novel bioactive natural products. A multitude of bioactive compounds, including the promising compounds salinosporamide A, xiamycins, and indolocarbazoles, have been isolated from mangrove actinomycetia [13,14,62,63]. Salinosporamide A, a potent 20S proteasome inhibitor, is the first mangrove-derived compound that entered phase I clinical trials for the treatment of multiple myeloma only three years after its discovery [13,64,65]. Xiamycin exhibits selective anti-HIV activity and also represents one of the first examples of indolosesquiterpenes isolated from prokaryotes [66]. The Leizhou Peninsula, located in the southernmost end of mainland China, has 9284.3 ha of mangrove distributed in over 100 sites along the coastlines, comprising approximately 79% of the total mangroves area in Guangdong province and 33% in China [67,68]. However, the mangrove from the Leizhou Peninsula is underexplored, and only a few reports on the diversity and antibacterial activity of its inhabiting actinomycetia have been published to date. Our group started exploring the diversity and bioactivity of the mangrove plant endophytic actinomycetia in this region in 2015. In total, 159 strains in 19 genera affiliated with 12 families were isolated, and 64 out of 88 tested strains exhibited activity against at least one of the tested pathogens [18]. In order to maximize the harvest of actinomycetial strains in this study, 13 soil samples from different mangrove sites were collected, and 12 isolation media were applied, leading to the isolation of 521 strains in 40 genera from 23 families. Thus, this ecosystem is proven to provide a highly productive and rich diversity of actinomycetia. The genera *Micromonospora* and *Streptomyces* were still numerically dominant, which is consistent with previous studies [63]. Interestingly, the addition of kelp or coconut juice into the medium (M11 and M12) increased the diversity of actinomycetia compared with the blank control (M1 and M4), which might be caused by trace substances from kelp or coconut juice assisting some strains growth since both seaweed and coconut trees grow in or along the sea and live in similar environments as mangroves. Lastly, 179 strains affiliated to 40 different genera with a unique colony morphology were selected to evaluate their antibacterial activities.

Bioactivity screening is, in general, the initial step to find novel antibiotics. To comprehensively analyze the antibacterial spectra of the extracts, 12 strains of drug-sensitive and drug-resistant pathogens were used as indicators in this study. In order to find the strains producing more potent antibacterial metabolites, the test volume of extracts used in the assay was reduced to 50% of the normally applied volume in the same concentration. In this assay, we found that *Streptomyces* was still the main genus producing bioactive secondary metabolites, accounting for 61.7% of all active strains. Because the ISP2 medium supports the growth of most actinomycetia and also is known to promote secondary metabolite production [24,69,70], it was the only medium used for fermentation. ISP2 medium components are also reported to reduce noise level and interference with secondary metabolites detection in the UPLC–HRMS analysis [24]. In addition, application of only one medium facilitates comparability between the different strains in metabolomics analysis [71].

Dereplication has become a key issue for the discovery of new antibiotics. An effective approach was required to maximize the detection of chemical diversity and minimize the redundancy of the samples after the bioactivity and phylogeny screening [72,73]. UPLC-HRMS/MS-based metabolomics could maximize the detection of chemical diversity among extracts in a high throughput manner [74]. UPLC-HRMS/MS gives rapid separation from the complex strain extracts and increases confidence in identifying metabolites based on mass accuracy and isotope fit. Metabolomic methods are combined with chemoinformatics approaches, such as PCA, OPLS-DA, etc. PCA, an unsupervised method without grouping, was applied to overview the differences among numerous samples, identifying strains with distinct metabolites, while excluding strains with common chemical profiles [75]. Hou Yangpeng et al. applied LC/MS-PCA to discover novel natural products from marine-derived *Streptomyces* spp.; the result was that 37% of all isolates produced a number of unique and putative new natural products, indicating that this approach could greatly improve the discovery rate from *Streptomyces* spp. [21]. However, the application of PCA was limited to the number of samples, usually between 20 and 50 strains, practically [21,24]. Therefore, 23 strains with strong anti-MRSA activity were selected for PCA analysis. The result showed that six strains with unique chemical profiles could be candidates for prioritization. After dereplicating, two *Streptomyces* strains (M22, H37) with putative novel compounds were prioritized.

OPLS-DA, a supervised method, was an effective statistic model for comparing two different sample groups [20]. A total of 8 out of the above 23 strains showing bioactivity against drug-sensitive *A. baumannii* were prioritized using the OPLS-DA model to find the bioactive compounds produced in the strains. Finally, *Streptomyces* sp. H37 with putative new compounds responsible for inhibitory activity against *A. baumannii* were prioritized. The OPLS-DA analysis effectively assists the bioactivity screening to find potential new antibiotics with great bioactivity and avoid targeting inactive metabolites in the follow-up chemical isolation. Molecular networking was a new dereplication strategy to rapidly overview the chemical family in the extracts, identify known compounds, and find the analogs with novelty. In this study, it can act as a complement for the PCA and OPLS-DA analysis to claim the relationship of the significant metabolites in the outlier strains, find the potential new congeners, and predict the structure using their MS/MS similarity. As shown in Figure 10, the predicted structure of compound **16** was consistent with the structure that was identified by spectroscopic analysis, including HRESIMS, 1D, and 2D NMR.

NPAtlas and StreptomeDB are recommended as the databases used for the dereplication of metabolites produced in microbial extracts, especially in actinomycetial extracts. These databases are manually curated by the Linington group out of Simon Fraser and Stefan Günther group from Freiburg, respectively. Unlike Antibase, MarinLit, and Dictionary of Natural Products (DNP), NPAtlas and StreptomeDB are open access, updated, and ready to be downloaded. To date, NPAtlas v19_12 contains over 25,000 microbial-produced natural products [33], and StreptomeDB v3.0 includes 6524 compounds produced by *Strep-*

tomyces [34]. As shown in Tables 2 and 3, the dereplication of metabolites from different databases produced different results sometimes. If two professional microbial databases are simultaneously applied to the dereplication process, it can increase accuracy and effectiveness in dereplication and avoid omissions caused by using one database. Hence, more than one database should be used, and loose parameters, such as precursor and fragments tolerances in 10 ppm, should be set in the search method. The limitation of dereplication for secondary metabolites is usually caused by the difficulty of obtaining an authentic standard for every "hit" from the database. Additional data, such as UV and MS/MS data, can be helpful to ensure that the identified hits in the mass ion peaks are correct.

The trioxacarcin family is a family of complex aromatic polyketides, which is produced by *Streptomyces* strains. In 1981, the trioxacarcins (trioxacarcins A–C) were first isolated from *Streptomyces bottropensis* DO-45 [76]. Subsequently, they were reisolated from a marine *Streptomyces* sp. B8652 with additional four new analogs, trioxacarcin D–F and gutingimycin, in 2004 [52,60]. The structure of the trioxacarcins was characterized as a rigid, highly oxygenated polycyclic skeleton with a fused spiro epoxide function, and one or more unusual glycosidic residues, named 'trioxacarcinoses' [77]. Trioxacarcin-type compounds display extraordinary antiproliferative effects, such as anticancer, antibacterial, and anti-malaria activities [60,76,78], which have attracted attention for chemical synthesis [61,77,79,80], mode of action, and biosynthesis studies [81–85]. It was reported that the notable biological activity of trioxacarcin A is due to its tight interaction with DNA [86,87]. In our study, several putative novel trioxacarcin-type compounds were identified by the metabolomics-based dereplication approach, such as compound **12** (7.16_1028.3600*m/z*), compound **14** (7.47_1028.3592*m/z*), compound **16** (6.69_1030.3751*m/z*), compound **17** (7.94_1013.3486*m/z*), and so on. However, some of them were unstable in scale-up fermentation. The contents of compounds **14** and **17** were decreased, making them hard to be isolated and accumulated for structural identification. Meanwhile, a low-yield compound **20** (7.86_896.3265n) was increased in the scale-up fermentation. A similar phenomenon was also reported by other researchers in metabolomics analysis of bacterial strains [88]. Finally, two new trioxacarcins, gutingimycin B (**16**) and trioxacarcin G (**20**), along with gutingimycin (**12**, 7.16_1028.3600*m/z*) were isolated from the scale-up fermentation broth of *Streptomyces* sp. M22. To the best of our knowledge, it is the second report of finding trioxacarcins with keto-reduced trioxacarcinose B moiety, except for trioxacarcin C. The new trioxacarcin-type members, compound **16** and compound **20**, together with known compound **12**, were evaluated for cytotoxicity against the H460 lung cancer cell line in this study, but no prominent cytotoxic activity against this cell line was observed in these compounds (IC_{50} > 1000 nM). The antibacterial activity of the new analogs and their cytotoxicity against other cell lines will be tested in the future. The successful isolation of novel trioxacarcin-type compounds from *Streptomyces* sp. M22 has demonstrated that our integrative strategies are effective, efficient, and suitable for seeking new antibiotics from those ecosystems inhabiting a large amount of actinomycetial strains.

4. Materials and Methods

4.1. Samples Collection

Soil samples were collected in August 2019 at different mangrove sites in Leizhou Peninsular, Guangdong province, China. The locations where samples were collected and their information are shown in Figure 13 and Table S6. All samples were collected from depth of 5–10 cm with a sterile spatula, then packed in sterile bags, and brought to the laboratory at the earliest time. Each sample was air-dried in a laminar flow hood before grinding with a mortar and pestle.

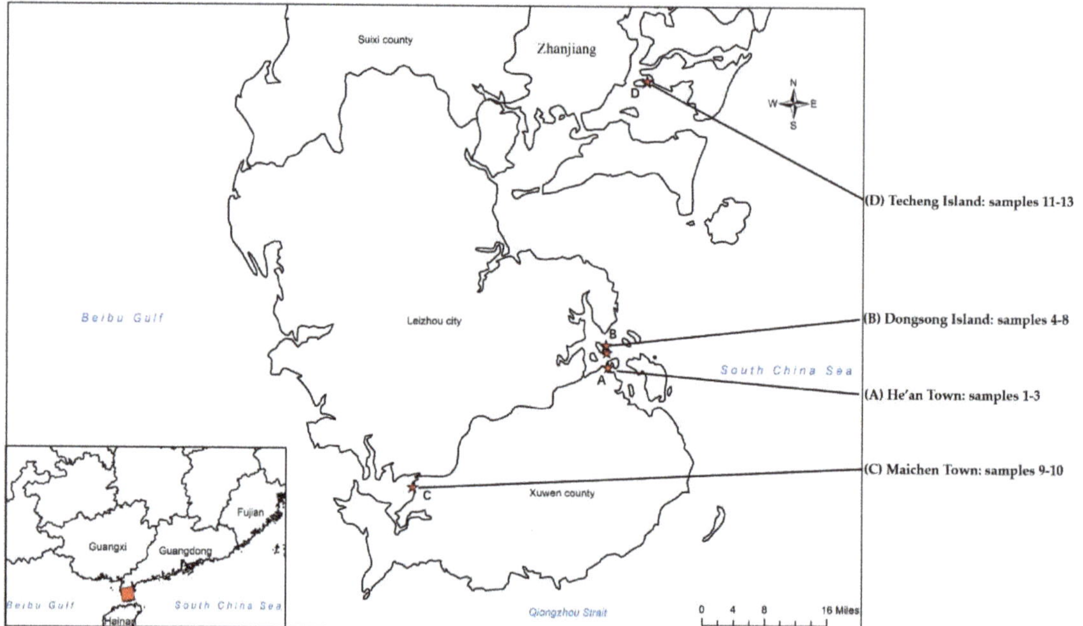

Figure 13. Locations of the sampling sites (red star) in Leizhou Peninsula, Guangdong, China.

4.2. Isolation of Actinomycetial Strains

A total of 12 cultural media were used to isolate actinomycetial strains. All media were supplemented with nalidixic acid (25 mg/L), cycloheximide (50 mg/L), and potassium dichromate (50 mg/L) to inhibit the growth of Gram-negative bacteria and fungi. The recipes for 12 media are shown in Table S1. It should be mentioned that M11 is a modified version of M1 with the addition of 15 mL kelp solution instead of 15 mL distilled water in the recipe. The kelp solution was prepared as follows: 200 g fresh kelp was cut into small pieces, added into 200 mL distilled water, and boiled for 30 min. After cooling down, the kelp soup was filtered by absorbent cotton to obtain the kelp solution. Meanwhile, M12 is a modified version of M4 with the addition of 10 mL natural fresh coconut juice instead of 10 mL distilled water in the recipe.

Strains were isolated by using the dilution plating technique. A total of 5 g of each soil sample was diluted with 45 mL of sterile 0.1% $Na_4P_2O_7$ solution, then mixed, homogenized, and shaken for 1 h at 180 rpm to release actinomycetia cells attached to the soil. Subsequently, the pretreated samples were prepared for ten-fold serial dilutions up to 10^{-4}. A total of 0.2 mL of diluted sample (10^{-2}, 10^{-3}, and 10^{-4}) from each soil sample was spread onto isolation agar plates, and plates were incubated for 7–14 days at 28 °C. Actinomycetia-like colonies depending on their morphological characters, pigment diffusion, and coloration of their mycelia were picked and streaked several times on ISP2 agar plates until pure actinomycetial colonies were isolated. The pure strains were maintained on ISP2 agar slants at 4 °C and preserved in 20% glycerol (*v/v*) at −80 °C.

4.3. Phylogenetic Analysis

Genomic DNA was extracted by using a rapid method with Chelex-100 as described previously [89]. The 16S rRNA gene amplification and sequencing analysis were performed using the universal primers 27F (5′-AGAGTTTGATCMTGGCTCAG-3′) and 1492R (5′-GGTTACCTTGTTACGACTT-3′). The PCR reaction mixture (50 μL) included 25 μL 2× supermix (TransGen Biotech, Beijing, China), 1 μL each of the primers (10 mM, Sangon

Biotech, Shanghai, China), 1.5 µL DNA, and 21.5 µL ddH$_2$O. The reaction conditions were as follows: 95 °C for 3 min, 30 cycles of 94 °C for 1 min, annealing at 60 °C for 1 min, extension at 72 °C for 1 min, followed by final extension for 10 min at 72 °C. The amplified products were sent to Shanghai Shenggong Company for sequencing. The sequencing data were BLAST analyzed using the GenBank NCBI (http://www.ncbi.nlm.nih.gov/, accessed on 11 November 2021) and the EzBioCloud database [90] to determine the similarity with type strains. Multiple alignments were generated using the Clustal_X tool in MEGA version 7.0 [91]. A phylogenetic tree based on the neighbor-joining method was constructed under Kimura's two-parameter model [92]. Bootstrap analysis with 1000 replications was performed with MEGA version 7.0 and finally visualized via the Interactive Tree of Life (iTOL) web service [93].

4.4. Extracts Preparation and Bioactivity Assay

Based on the analysis of phenotypic and phylogenetic characteristics, 179 strains were selected from the 521 isolated actinomycetial strains to examine their antibacterial potentials. The strains were inoculated into 100 mL ISP2 broth in 500 mL conical flasks and cultured for 7 days in a shaking incubator at 180 rpm at 28 °C. A total of 300 mL (3 × 100 mL) cultural broth of each strain was pooled and centrifuged at 4200 rpm for 20 min to separate the mycelium portion. The supernatants were extracted three times with ethyl acetate (1:1, *v/v*). The organic layers were combined and evaporated to obtain crude extracts. The crude extracts were dissolved in 3 mL methanol and used for antibacterial assay by the paper disc diffusion method.

The methanol sample (30 µL) was dripped on a paper disk (6 mm diameter). A total of 30 µL methanol and levofloxacin solution (10 µL, 1 mg/mL) were used as the negative and positive control, respectively. After being dried in a biosafety hood, the paper disks were transferred to agar plates seeded with pathogenic bacteria and incubated at 37 °C for 24–48 h. The antibacterial activity was evaluated by measuring the diameters of the inhibition zones with a vernier caliper. The indicator bacteria used for antimicrobial assay were six sets of indicator bacteria, including *Enterococcus* sp. (ATCC 33186 and 310682), *Staphylococcus aureus* (ATCC 29213 and ATCC 33591), *Klebsiella pneumonia* (ATCC 10031 and ATCC 700603), *Acinetobacter baumannii* (2799 and ATCC 19606), *Pseudomonas aeruginosa* (ATCC 27853 and 2774) and *Escherichia coli* (ATCC 25922 and ATCC 35218). Their drug susceptibility testing was identified and confirmed by the Beijing Key Laboratory of Antimicrobial Agents, Institute of Medicinal Biotechnology. Each set consisted of two strains, one drug-sensitive strain (the former), and one drug-resistant strain (the latter). Isolate 310682 was resistant to vancomycin. Meanwhile, isolate 2774 was resistant to aminoglycosides and carbapenems. Indicator bacteria were obtained from either the American Type Culture Collection (ATCC) or the clinic isolation from hospital in China, and they were deposited in the Institute of Medicinal Biotechnology, Chinese Academy of Medical Sciences and Peking Union Medical College.

4.5. PCA and OPLS-DA Analysis

Twenty-three strains with zones of inhibition larger than or equal to 10 mm against MRSA were selected for dereplication and microbial strain prioritization studies using UPLC-HRMS-PCA and UPLC-HRMS-OPLS-DA. Three biological replicates were prepared for each actinomycetial strain. Each strain was cultured in triplicate for 7 days in ISP2 broth medium (3 × 100 mL) as mentioned above. Only 15 mL supernatant from 100 mL cultural broth were extracted three times (3 × 15 mL) with ethyl acetate, then dried under vacuum to obtain the crude extract. The dried crude extracts were weighed and dissolved in methanol to yield a stock solution with a concentration of 2 mg/mL. The ISP2 broth medium was used as a blank medium control. After centrifugation at 14,000 rpm for 10 min, the supernatant of the stock solution was diluted 4-fold with methanol to yield the test solution (0.05 mg/mL). A quality control (QC) sample was prepared by mixing an equal volume of each test solution (including bank medium control). All test solutions were stored at 4 °C before analysis.

UPLC-HRMS/MS experiments were carried out on Waters ACQUITY UPLC I-Class system combined with Waters Xevo G2-XS Q-TOF mass spectrometer (Waters, Manchester, UK). A Waters ACQUITY UPLC BEH C18 column (2.1 × 100 mm, 1.7 μm) maintained at 25 °C was used, and the PDA scan range was 200–800 nm. The binary mobile phase consisted of solvent A (water containing 0.1% formic acid) and solvent B (acetonitrile). The gradient elution program was applied as follows: 0–1 min, 10%(B); 1–18 min, 10–95%(B); 18–20 min, 95%(B); 20–22 min, 10%(B). The flow rate was 0.3 mL/min. The injection volume was 2 μL, and the QC sample was analyzed after every six injections to evaluate system stability.

The ESI source parameters in the positive mode were set as follows: capillary, 2 kV; sampling cone, 40 V; source offset, 80 V; source temperature, 100 °C; desolvation temperature, 250 °C; cone gas, 50 L/h; and desolvation gas, 600 L/h. The desolvation and cone gases were nitrogen, and the collision gas was argon. The MSE acquisition (data-independent acquisition) was obtained in the continuum format with a mass range of 100–2000 Da in both low-energy (function 1) and high-energy (function 2) scan functions. For function 1, the collision energy was 2 V. For function 2, a collision energy ramp of 40–80 V was used. The scan time was 0.10 s. The mass accuracy was maintained by using a lock spray with leucine–enkephalin ($[M+H]^+$ = 556.2771 Da) at a concentration of 2 ng/mL and a flow rate of 5 μL/min as reference. The run sequence started with a blank solvent, then a blank medium, followed by the samples. The instrument controlling and data acquisition were performed by MassLynx V4.1 software (Waters, Milford, CT, USA).

The acquired raw data from 87 samples, including 69 test samples, six blank samples (blank solvent and blank medium), and 12 QC injections, were all imported into Progenesis QI 3.0 software (Waters, Milford, USA) to operate the chromatographic peak alignment, experimental design setup, peak picking, normalization, deconvolution, compound identification, and compound review. The imported 87 runs were aligned on the basis of an automatically selected QC sample. The retention time for peak picking was set as 0–20 min, and the limits and sensitivity were set as the default. The adduct ion forms of $[M+H-H_2O]^+$, $[M+H]^+$, $[M+NH_4]^+$, $[M+Na]^+$, $[M+K]^+$, $[2M+H]^+$, $[2M+Na]^+$ were added to deconvolute the spectral data. After performing automatic processing to all compounds in all samples, a data matrix involving sample code, RT, *m/z*, and normalized abundance was generated. In the blank samples, features with the most abundance and an abundance 20 times less than the 69 test samples were hidden manually to remove medium and blank effects for cleaner data [20,71]. The obtained data were exported into the extended statistics module EZinfo 3.0 (Umetrics, Umea, Sweden) for PCA and OPLS-DA analyses. The significant differential retention time-observed mass (RT-*m/z*) or retention time-neutral mass (RT-*m/z*) pairs in the loadings plot and S-plot were selected and imported back into Progenesis QI for compound identification. After filtering by ANOVA *p*-value $\leq$ 0.05, *q* value $\leq$ 0.05, and max fold change $\geq$ 2, the filtered pairs were identified using the search method in Progenesis QI (parameters in search method: precursor tolerance 10 ppm and theoretical fragment tolerance 10 ppm). The Natural Product Atlas v19_12 and StreptomeDB v3.0 databases as in-house libraries were use for the dereplication of the differential metabolites in the samples.

4.6. Molecular Network Analysis

The UPLC and ESI source parameters were set as same as shown above. DDA was also performed in positive ion mode. The full MS survey scan was performed for 0.2 s in the range of 100–2000 Da, and MS/MS scanned a mass range of 50–2000 Da by the same scan time. The five most intense ions were chosen for MS/MS fragmentation spectra. The gradient of collision energy was set as 20 V to 40 V for low-mass collision energy (LM CE) and 60 V to 80 V for high-mass collision energy (HM CE). Automatic switching to MS/MS mode was enabled when the TIC intensity rose above 10,000 counts and switched off when 0.4 s had elapsed, or the TIC intensity was 1,000,000 counts. The tolerance window of ± 3.0 Da was set in the deisotope peak detection mode. Dynamic peak exclusion was enabled, acquired, and then excluded for 3.0 s. Fixed peak exclusion was as follows: *m/z* 205.0877, 255.1581, 279.1591, 301.1425, 371.3183, 579.2933. Raw data files obtained from the DDA acquisition were converted to 32-bit mzxML format with MS-Convert [94] and then uploaded on the GNPS web platform (http://gnps.ucsd.edu, accessed on 11 November 2021) for dereplication and molecular networking construction.

A molecular network was created using the online workflow on the GNPS website (https://ccms-ucsd.github.io/GNPSDocumentation/, accessed on 11 November 2021). The precursor ion mass tolerance was set as 0.1 Da and an MS/MS fragment ion tolerance as 0.1 Da. A network was then created where edges were filtered to have a cosine score above 0.6 and more than 4 matched peaks. Furthermore, edges between two nodes were kept in the network only if each of the nodes appeared in each other's respective top 10 most similar nodes. Finally, the maximum size of a molecular family was set to 100, and the lowest scoring edges were removed from molecular families until the molecular family size was below this threshold. The spectra in the network were then searched against GNPS spectral libraries. The library spectra were filtered in the same manner as the input data. All matches kept between network spectra and library spectra were required to have scores above 0.6 and at least 3 matching peaks. The generated molecular network was visualized using Cytoscape 3.7.1 [95].

4.7. Scale-Up Fermentation, Extraction, and Purification of Natural Products

Streptomyces sp. M22 was grown and maintained on an ISP2 agar plate at 28 °C for 7–10 days. The spores of the strain were inoculated into 500 mL Erlenmeyer flasks contained 100 mL of the ISP2 medium, which grew at 28 °C for 2 days at 180 rpm as seed cultures. Then, each seed culture (100 mL) was inoculated into autoclaved 5 L Erlenmeyer flasks containing 1 L ISP2 medium. The flasks were incubated at 28 °C for 7 days on a rotary shaker (180 rpm). The total 18 L (18 × 1L) of fermentation broth was centrifuged at 4300 rpm for 20 min, and the supernatant was extracted three times with ethyl acetate (18 L/time) to give an organic extract. After 3 times of fermentation, the combined organic extract (5.5 g) was subjected to MPLC column chromatography eluted with MeOH-H_2O (10:90, 30:70, 50:50, 70:30, 90:10, 100:0, *v/v*) to obtain six subfractions (Fr.01–Fr.06) based on LC-MS analysis. Fraction 05 was further separated by Sephadex LH-20 (CH_2Cl_2: MeOH=1:1, *v/v*) to yield five subfractions (Fr.05a–Fr.05e). Fr.05b was subjected to semi-preparative HPLC (ACN-H_2O, 32:68, *v/v*, 0–5 min; 32:68–52:48, *v/v*, 5–40 min, 3.0 mL/min) to yield gutingimycin B (**16**, 6.0 mg), gutingimycin (**12**, 15.6 mg) and semi-pure trioxacarcin G. The semi-pure trioxacarcin G was further fractioned by semi-preparative HPLC using MeOH-H_2O (55:45, *v/v*) to yield pure trioxacarcin G (**20**, 8.0 mg).

Gutingimycin B (**16**): yellow amorphous powder. $[\alpha]_D^{25}$ −49.4° (c 0.02, ACN); UV (MeOH) λ_{max} (log ε) 274 (4.83), 408 (4.23) nm; IR v_{max}: 3366, 2932, 2853, 1689, 1628, 1386, 1223, 1089, 999 cm^{-1}; ^{1}H NMR (CDCl$_3$, 600 MHz) and ^{13}C NMR (CDCl$_3$, 150 MHz), see Table 4; HRESIMS: *m/z* 1030.3778 [M+H]$^+$ (calcd for $C_{47}H_{60}N_5O_{21}$, 1030.3781).

Trioxacarcin G (**20**): yellow amorphous powder. $[\alpha]_D^{25}$ −140.0° (c 0.02, ACN); UV (MeOH) λ_{max} (log ε) 232 (4.19), 271 (4.25), 400 (3.71) nm; IR v_{max}: 3439, 2933, 2851, 1730, 1623, 1384, 1233, 1085, 998 cm^{-1}; ^{1}H NMR (CDCl$_3$, 600 MHz) and ^{13}C NMR (CDCl$_3$, 150 MHz), see Table 4; HRESIMS: *m/z* 914.3644 [M+NH$_4$]$^+$ (calcd for $C_{42}H_{60}NO_{21}$, 914.3658).

5. Conclusions

Mangrove actinomycetia are considered one of the promising sources of novel biologically active compounds. In this study, a total of 521 actinomycetial strains were isolated from underexplored mangrove soils collected in Leizhou Peninsular, China. Our integrative strategies using taxonomical information, bioactivity, and metabolomics tools (PCA, OPLS-DA, molecular networking) for dereplication allowed us to prioritize two

Streptomyces strains (H37, M22) with the potential to produce new antibiotics. Two new trioxacarcins were isolated from the scale-up fermentation broth of *Streptomyces* sp. M22. Our study demonstrated that modern metabolomics tools greatly assist classic antibiotic discovery for strain prioritization and improve the efficiency of novel antibiotics discovery. Our data also highlighted that the mangrove in Leizhou Peninsular is an unexploited source with rich microbial diversity and bioactive actinomycetia. In summary, the new strategies presented in this research could set an example to accelerate new antibiotics discovery from mangroves and other highly productive sources, such as rainforests.

Supplementary Materials: The following are available online at https://www.mdpi.com/article/md19120688/s1, Table S1. Composition of 12 different media used to isolate actinomycetial strains from 13 mangrove soil samples; Table S2. Information on genera distribution of actinomycetial strains isolated from 13 different mangrove soil samples; Table S3. Information on genera distribution of actinomycetial strains recovered from the 12 different cultural media; Table S4. Antibacterial activities of 179 actinomycetial strains by the paper disc diffusion method; Table S5. The trioxacarcin-type antibiotics isolated from the actinomycetial strains; Table S6. The information of mangrove-derived soil samples in different sites of Leizhou Peninsula, China; Figure S1. The UV spectra of three outliers (**1–3**) of Y46 and H12. Figure S2. The feature statistic plot of 16.71_724.4749n (**4**) in all samples; Figure S3. The MS/MS fragment pattern of the outlier 16.71_724.4749n (**4**) in sample H7; Figure S4. The MS/MS spectra of three false positive compounds in H7 acquired by DDA method; Figure S5. The MS/MS spectra of three revised compounds (**7–9**) in H7 acquired by DDA method; Figure S6. The positive and negative MS spectra of two revised compounds in the LC-MS of H7; Figure S7. The UV spectra of three compounds (**7–9**) of H7 eluting with ACN and H_2O; Figure S8. The MS/MS spectra of two compounds (**10–11**) in H37 acquired by MSE method; Figure S9. The UV spectra of seven trioxacarcin-type compounds (**12, 14–19**) in the UPLC-UV-HRMS chromatograms of M22; Figure S10. The MS spectra of seven trioxacarcin-type compounds (**12, 14–19**) in the UPLC-UV-HRMS chromatograms of M22; Figure S11. The VIP score of selected markers in the OPLS-DA model; Figure S12. Molecular network of the cluster containing the compound 15.40_566.4171n (**13**) in M22 extract; Figure S13. Molecular network of the cluster containing the compounds 10.64_900.5435n (**10**) and 11.08_928.5742n (**11**) in H37 extract. Figures S14–S31. The HRESIMS, UV, IR, and NMR spectra of gutingimycin B (**16**) and trioxacarcin G (**20**); Figures S32–S37. The HRESIMS and NMR spectra of gutingimycin (**12**).

Author Contributions: Conceptualization, C.-H.S., H.L. and Q.-P.L.; methodology, Q.-P.L., Y.-M.H. and S.-W.L.; validation, Q.-P.L., Y.-M.H., G.W., Q.Y. and L.-F.L.; formal analysis, Q.-P.L., G.W., S.-W.L. and Q.Y.; investigation, Q.-P.L., Y.-M.H., S.-W.L., G.W., Q.Y., L.-F.L. and H.C.; resources, C.-H.S., H.L., H.-T.Z., Y.Q., T.W., Z.-K.J., J.-J.L. and X.-J.L.; data curation, Q.-P.L., Y.-M.H., G.W., Q.Y., L.-F.L., H.-T.Z. and Y.Q.; writing—original draft preparation, Q.-P.L., Y.-M.H.; writing—review and editing, Q.-P.L., C.-H.S., Y.-M.H. and H.L.; visualization, Q.-P.L., Y.-M.H. and S.-W.L.; supervision, C.-H.S., H.L.; project administration, C.-H.S., H.L.; funding acquisition, C.-H.S., Q.-P.L., H.L., Y.-M.H. and G.W. All authors have read and agreed to the published version of the manuscript.

Funding: This research was funded by the CAMS Innovation Fund for Medical Sciences (CIFMS 2021-I2M-1-028, CAMS 2017-I2M-B&R-08, and 2017-I2M-1-012), the National Natural Science Foundation of China (81803411 and 82011530051), PUMC Doctoral Innovation Fund Project (2018-1007-16), the Science and Technology Program of Guangdong Province (2019B090905011), the Public Service Platform of South China Sea for R&D Marine Biomedicine Resources (2017C8A), the Guangdong University Youth Innovation Talent Project (2020KQNCX023), the Scientific Research Fund of Guangdong Medical University (GDMUM202002).

Data Availability Statement: The sequencing data presented in this study are available in GenBank at NCBI (accession numbers: MW724535-MW724713).

Acknowledgments: We greatly thank Chao Zhou from Waters Corporation for his advice in MS acquisition and metabolomics analysis. We greatly thank Rongfeng Li from Department of Chemistry, Johns Hopkins University for his linguistic assistance during the preparation of this manuscript. Our deepest gratitude goes to the anonymous reviewers for their careful work and thoughtful suggestions that have helped improve this paper substantially.

Conflicts of Interest: The authors declare no conflict of interest.

References

1. Willyard, C. The drug-resistant bacteria that pose the greatest health threats. *Nature* **2017**, *543*, 15. [CrossRef] [PubMed]
2. Tacconelli, E.; Carrara, E.; Savoldi, A.; Harbarth, S.; Mendelson, M.; Monnet, D.L.; Pulcini, C.; Kahlmeter, G.; Kluytmans, J.; Carmeli, Y. Discovery, research, and development of new antibiotics: The WHO priority list of antibiotic-resistant bacteria and tuberculosis. *Lancet Infect. Dis.* **2018**, *18*, 318–327. [CrossRef]
3. Walsh, C.T.; Fischbach, M.A. Natural products version 2.0: Connecting genes to molecules. *J. Am. Chem. Soc.* **2010**, *132*, 2469–2493. [CrossRef] [PubMed]
4. Newman, D.J.; Cragg, G.M. Natural products as sources of new drugs over the nearly four decades from 01/1981 to 09/2019. *J. Nat. Prod.* **2020**, *83*, 770–803. [CrossRef]
5. Salam, N.; Jiao, J.Y.; Zhang, X.T.; Li, W.J. Update on the classification of higher ranks in the phylum *Actinobacteria*. *Int. J. Syst. Evol. Microbiol.* **2020**, *70*, 1331–1355. [CrossRef]
6. Procópio, R.E.; Silva, I.R.; Martins, M.K.; Azevedo, J.L.; Araújo, J.M. Antibiotics produced by *Streptomyces. Braz. J. Infect. Dis.* **2012**, *16*, 466–471. [CrossRef]
7. Watve, M.G.; Tickoo, R.; Jog, M.M.; Bhole, B.D. How many antibiotics are produced by the genus *Streptomyces? Arch. Microbiol.* **2001**, *176*, 386–390. [CrossRef]
8. Clardy, J.; Fischbach, M.A.; Walsh, C.T. New antibiotics from bacterial natural products. *Nat. Biotechnol.* **2006**, *24*, 1541–1550. [CrossRef]
9. Genilloud, O. Actinomycetes: Still a source of novel antibiotics. *Nat. Prod. Rep.* **2017**, *34*, 1203–1232. [CrossRef]
10. Kamjam, M.; Sivalingam, P.; Deng, Z.; Hong, K. Deep sea actinomycetes and their secondary metabolites. *Front. Microbiol.* **2017**, *8*, 760. [CrossRef]
11. Mohammadipanah, F.; Wink, J. Actinobacteria from arid and desert habitats: Diversity and biological activity. *Front. Microbiol.* **2015**, *6*, 1541. [CrossRef]
12. Tian, Y.; Li, Y.L.; Zhao, F.C. Secondary metabolites from polar organisms. *Mar. Drugs* **2017**, *15*, 28. [CrossRef]
13. Xu, D.B.; Ye, W.W.; Han, Y.; Deng, Z.X.; Hong, K. Natural products from mangrove actinomycetes. *Mar. Drugs* **2014**, *12*, 2590–2613. [CrossRef]
14. Azman, A.S.; Othman, I.; Velu, S.S.; Chan, K.G.; Lee, L.H. Mangrove rare actinobacteria: Taxonomy, natural compound, and discovery of bioactivity. *Front. Microbiol.* **2015**, *6*, 856. [CrossRef]
15. Ancheeva, E.; Daletos, G.; Proksch, P. Lead compounds from mangrove-associated microorganisms. *Mar. Drugs* **2018**, *16*, 319. [CrossRef]
16. Hong, K.; Gao, A.H.; Xie, Q.Y.; Gao, H.G.; Zhuang, L.; Lin, H.P.; Yu, H.P.; Li, J.; Yao, X.S.; Goodfellow, M.; et al. Actinomycetes for marine drug discovery isolated from mangrove soils and plants in China. *Mar. Drugs* **2009**, *7*, 24–44. [CrossRef]
17. Tong, L.Y. *Isolation and Identification of Actinomycetes from Soil of Root System of Mangrove Forest in Zhanjiang*; Guangdong Ocean University: Zhanjiang, China, 2011.
18. Xu, M.; Li, J.; Dai, S.J.; Gao, C.Y.; Liu, J.M.; Tuo, L.; Wang, F.F.; Li, X.J.; Liu, S.W.; Jiang, Z.K.; et al. Study on diversity and bioactivity of actinobacteria isolated from mangrove plants collected from Zhanjiang in Guangdong Province. *Chin. J. Antibiot.* **2016**, *41*, 26–34.
19. Betancur, L.A.; Naranjo-Gaybor, S.J.; Vinchira-Villarraga, D.M.; Moreno-Sarmiento, N.C.; Maldonado, L.A.; Suarez-Moreno, Z.R.; Acosta-González, A.; Padilla-Gonzalez, G.F.; Puyana, M.; Castellanos, L.; et al. Marine actinobacteria as a source of compounds for phytopathogen control: An integrative metabolic-profiling/bioactivity and taxonomical approach. *PLoS ONE* **2017**, *12*, e0170148. [CrossRef]
20. Sebak, M.; Saafan, A.E.; AbdelGhani, S.; Bakeer, W.; El-Gendy, A.O.; Espriu, L.C.; Duncan, K.; Edrada-Ebel, R. Bioassay- and metabolomics-guided screening of bioactive soil actinomycetes from the ancient city of Ihnasia, Egypt. *PLoS ONE* **2019**, *14*, e0226959. [CrossRef]
21. Hou, Y.P.; Braun, D.R.; Michel, C.R.; Klassen, J.L.; Adnani, N.; Wyche, T.P.; Bugni, T.S. Microbial strain prioritization using metabolomics tools for the discovery of natural products. *Anal. Chem.* **2012**, *84*, 4277–4283. [CrossRef]
22. Villas-Bôas, S.G.; Mas, S.; Åkesson, M.; Smedsgaard, J.; Nielsen, J. Mass spectrometry in metabolome analysis. *Mass Spectrom. Rev.* **2005**, *24*, 613–646. [CrossRef]
23. Samat, N.; Tan, P.J.; Shaari, K.; Abas, F.; Lee, H.B. Prioritization of natural extracts by LC-MS-PCA for the identification of new photosensitizers for photodynamic therapy. *Anal. Chem.* **2014**, *86*, 1324–1331. [CrossRef]
24. Forner, D.; Berrué, F.; Correa, H.; Duncan, K.; Kerr, R.G. Chemical dereplication of marine actinomycetes by liquid chromatography–high resolution mass spectrometry profiling and statistical analysis. *Anal. Chim. Acta* **2013**, *805*, 70–79. [CrossRef]
25. Gill, K.A.; Berrué, F.; Arens, J.C.; Kerr, R.G. Isolation and structure elucidation of cystargamide, a lipopeptide from *Kitasatospora cystarginea. J. Nat. Prod.* **2014**, *77*, 1372–1376. [CrossRef]
26. Stewart, A.K.; Ravindra, R.; Van Wagoner, R.M.; Wright, J.L.C. Metabolomics-guided discovery of microginin peptides from cultures of the cyanobacterium *Microcystis aeruginosa. J. Nat. Prod.* **2018**, *81*, 349–355. [CrossRef]
27. Gill, K.A.; Berrué, F.; Arens, J.C.; Carr, G.; Kerr, R.G. Cystargolides, 20S proteasome inhibitors isolated from *Kitasatospora cystarginea. J. Nat. Prod.* **2015**, *78*, 822–826. [CrossRef]

28. Abdelmohsen, U.R.; Cheng, C.; Viegelmann, C.; Zhang, T.; Grkovic, T.; Ahmed, S.; Quinn, R.J.; Hentschel, U.; Edrada-Ebel, R. Dereplication strategies for targeted isolation of new antitrypanosomal actinosporins A and B from a marine sponge associated-*Actinokineospora* sp. EG49. *Mar. Drugs* **2014**, *12*, 1220–1244. [CrossRef]

29. Tawfike, A.; Attia, E.Z.; Desoukey, S.Y.; Hajjar, D.; Makki, A.A.; Schupp, P.J.; Edrada-Ebel, R.; Abdelmohsen, U.R. New bioactive metabolites from the elicited marine sponge-derived bacterium *Actinokineospora spheciospongiae* sp. nov. *AMB Express* **2019**, *9*, 12. [CrossRef]

30. Yang, J.Y.; Sanchez, L.M.; Rath, C.M.; Liu, X.T.; Boudreau, P.D.; Bruns, N.; Glukhov, E.; Wodtke, A.; de Felicio, R.; Fenner, A.; et al. Molecular networking as a dereplication strategy. *J. Nat. Prod.* **2013**, *76*, 1686–1699. [CrossRef]

31. Tangerina, M.M.P.; Furtado, L.C.; Leite, V.M.B.; Bauermeister, A.; Velasco-Alzate, K.; Jimenez, P.C.; Garrido, L.M.; Padilla, G.; Lopes, N.P.; Costa-Lotufo, L.V.; et al. Metabolomic study of marine *Streptomyces* sp.: Secondary metabolites and the production of potential anticancer compounds. *PLoS ONE* **2020**, *15*, e0244385. [CrossRef]

32. Wang, M.X.; Carver, J.J.; Phelan, V.V.; Sanchez, L.M.; Garg, N.; Peng, Y.; Nguyen, D.D.; Watrous, J.; Kapono, C.A.; Luzzatto-Knaan, T.; et al. Sharing and community curation of mass spectrometry data with Global Natural Products Social Molecular Networking. *Nat. Biotechnol.* **2016**, *34*, 828–837. [CrossRef] [PubMed]

33. Van Santen, J.A.; Jacob, G.; Singh, A.L.; Aniebok, V.; Balunas, M.J.; Bunsko, D.; Neto, F.C.; Castaño-Espriu, L.; Chang, C.; Clark, T.N.; et al. The Natural products atlas: An open access knowledge base for microbial natural products discovery. *ACS Cent. Sci.* **2019**, *5*, 1824–1833. [CrossRef] [PubMed]

34. Moumbock, A.F.; Gao, M.; Qaseem, A.; Li, J.; Kirchner, P.A.; Ndingkokhar, B.; Bekono, B.D.; Simoben, C.V.; Babiaka, S.B.; Malange, Y.I.; et al. StreptomeDB 3.0: An updated compendium of streptomycetes natural products. *Nucleic Acids Res.* **2020**, *49*, D600–D604. [CrossRef] [PubMed]

35. Zhang, X.F.; Ye, X.W.; Chai, W.Y.; Lian, X.Y.; Zhang, Z.Z. New metabolites and bioactive actinomycins from marine-derived *Streptomyces* sp. ZZ338. *Mar. Drugs* **2016**, *14*, 181. [CrossRef] [PubMed]

36. Lackner, H.; Bahner, I.; Shigematsu, N.; Pannell, L.K.; Mauger, A.B. Structures of five components of the actinomycin Z complex from *Streptomyces fradiae*, two of which contain 4-chlorothreonine. *J. Nat. Prod.* **2000**, *63*, 352–356. [CrossRef] [PubMed]

37. David, L.; Ayala, H.L.; Tabet, J.C. Abierixin, a new polyether antibiotic. Production, structural determination and biological activities. *J. Antibiot.* **1985**, *38*, 1655–1663. [CrossRef]

38. Zhao, X.; Wang, B.; Xie, K.Z.; Liu, J.Y.; Zhang, Y.Y.; Wang, Y.J.; Guo, Y.W.; Zhang, G.X.; Dai, G.J.; Wang, J.Y. Development and comparison of HPLC-MS/MS and UPLC-MS/MS methods for determining eight coccidiostats in beef. *J. Chromatogr. B Analyt. Technol. Biomed. Life Sci.* **2018**, *1087–1088*, 98–107. [CrossRef]

39. Martínez-Villalba, A.; Moyano, E.; Galceran, M.T. Fast liquid chromatography/multiple-stage mass spectrometry of coccidiostats. *Rapid Commun. Mass Spectrom.* **2009**, *23*, 1255–1263. [CrossRef]

40. Takahashi, K.; Yoshihara, T.; Kurosawa, K. Ushikulides A and B, immunosuppressants produced by a strain of *Streptomyces* sp. *J. Antibiot.* **2005**, *58*, 420–424. [CrossRef]

41. Mouslim, J.; Cuer, A.; David, L.; Tabet, J.C. Epigrisorixin, a new polyether carboxylic antibiotic. *J. Antibiot.* **1993**, *46*, 201–203. [CrossRef]

42. Gachon, P.; Kergomard, A.; Veschambre, H.; Esteve, C.; Staron, T. Grisorixin, a new antibiotic related to nigericin. *J. Chem. Soc. D* **1970**, *21*, 1421–1422. [CrossRef]

43. Cui, H.; Liu, Y.N.; Li, J.; Huang, X.S.; Yan, T.; Cao, W.H.; Liu, H.J.; Long, Y.H.; She, Z.G. Diaporindenes A–D: Four unusual 2,3-dihydro-1H-indene analogues with anti-inflammatory activities from the mangrove endophytic fungus *Diaporthe* sp. SYSU-HQ3. *J. Org. Chem.* **2018**, *83*, 11804–11813. [CrossRef]

44. An, C.Y.; Li, X.M.; Luo, H.; Li, C.S.; Wang, M.H.; Xu, G.M.; Wang, B.G. 4-Phenyl-3,4-dihydroquinolone derivatives from *Aspergillus nidulans* MA-143, an endophytic fungus isolated from the mangrove plant *Rhizophora stylosa*. *J. Nat. Prod.* **2013**, *76*, 1896–1901. [CrossRef]

45. Sato, K.; Goda, Y.; Sasaki, S.S.; Shibata, H.; Maitani, T.; Yamada, T. Identification of major pigments containing D-amino acid units in commercial Monascus pigments. *Chem. Pharm. Bull.* **1997**, *45*, 227–229. [CrossRef]

46. Guo, J.; Huan, T. Comparison of Full-Scan, Data-Dependent, and Data-Independent Acquisition modes in liquid chromatography–mass spectrometry based untargeted metabolomics. *Anal. Chem.* **2020**, *92*, 8072–8080. [CrossRef]

47. Nikolskiy, I.; Mahieu, N.G.; Chen, Y., Jr.; Tautenhahn, R.; Patti, G.J. An untargeted metabolomic workflow to improve structural characterization of metabolites. *Anal. Chem.* **2013**, *85*, 7713–7719. [CrossRef]

48. Helm, S.; Baginsky, S. MSE for label-free absolute protein quantification in complex proteomes. *Methods Mol. Biol.* **2018**, *1696*, 235–247.

49. Baksh, A.; Kepplinger, B.; Isah, H.A.; Probert, M.R.; Clegg, W.; Wills, C.; Goodfellow, M.; Errington, J.; Allenby, N.; Hall, M.J. Production of 17-O-demethyl-geldanamycin, a cytotoxic ansamycin polyketide, by *Streptomyces hygroscopicus* DEM20745. *Nat. Prod. Res.* **2017**, *31*, 1895–1900. [CrossRef]

50. Lin, H.N.; Wang, K.L.; Wu, Z.H.; Tian, R.M.; Liu, G.Z.; Xu, Y. Biological and chemical diversity of bacteria associated with a marine flatworm. *Mar. Drugs* **2017**, *15*, 281. [CrossRef]

51. DeBoer, C.; Meulman, P.A.; Wnuk, R.J.; Peterson, D.H. Geldanamycin, a new antibiotic. *J. Antibiot.* **1970**, *23*, 442–447. [CrossRef]

52. Maskey, R.P.; Sevvana, M.; Usón, I.; Helmke, E.; Laatsch, H. Gutingimycin: A highly complex metabolite from a marine *Streptomycete*. *Angew. Chem. Int. Ed.* **2004**, *43*, 1281–1283. [CrossRef]

53. Tamaoki, T.; Shirahata, K.; Iida, T.; Tomita, F. Trioxacarcins, novel antitumor antibiotics. II. Isolation, physico-chemical properties and mode of action. *J. Antibiot.* **1981**, *34*, 1525–1530. [CrossRef]

54. Maoka, T.; Takemura, M.; Tokuda, H.; Suzuki, N.; Misawa, N. 4-Ketozeinoxanthin, a novel carotenoid produced in *Escherichia coli* through metabolic engineering using carotenogenic genes of bacterium and liverwort. *Tetrahedron Lett.* **2014**, *55*, 6708–6710. [CrossRef]

55. Wong, W.R.; Oliver, A.G.; Linington, R.G. Development of antibiotic activity profile screening for the classification and discovery of natural product antibiotics. *Chem. Biol.* **2012**, *19*, 1483–1495. [CrossRef]

56. Kihara, T.; Kusakabe, H.; Nakamura, G.; Sakurai, T.; Isono, K. Cytovaricin, a novel antibiotic. *J. Antibiot.* **1981**, *34*, 1073–1074. [CrossRef] [PubMed]

57. Cai, R.S.; Wu, J.M. Biochemical and biological characterization of ionophorous antibiotic W341. *Chin. J. Antibiot.* **1985**, *10*, 342–347.

58. Crusemann, M.; O'Neill, E.C.; Larson, C.B.; Melnik, A.V.; Floros, D.J.; da Silva, R.R.; Jensen, P.R.; Dorrestein, P.C.; Moore, B.S. Prioritizing natural product diversity in a collection of 146 bacterial strains based on growth and extraction protocols. *J. Nat. Prod.* **2016**, *80*, 588–597. [CrossRef]

59. Tomita, F.; Tamaoki, T.; Shirahata, K.; Iida, T.; Morimoto, M.; Fujimoto, K. Antibiotic substances DC-45, and their use as medicaments. U.S. Patent No. 4511560, 16 April 1985.

60. Maskey, R.P.; Helmke, E.; Kayser, O.; Fiebig, H.H.; Maier, A.; Busche, A.; Laatsch, H. Anti-cancer and antibacterial trioxacarcins with high anti-malaria activity from a marine Streptomycete and their absolute stereochemistry. *J. Antibiot.* **2004**, *57*, 771–779. [CrossRef]

61. Nicolaou, K.C.; Cai, Q.; Sun, H.; Qin, B.; Zhu, S. Total synthesis of trioxacarcins DC-45-A1, A, D, C, and C7″-epi-C and full structural assignment of trioxacarcin C. *J. Am. Chem. Soc.* **2016**, *138*, 3118–3124. [CrossRef]

62. Li, F.N.; Liu, S.W.; Lu, Q.P.; Zheng, H.Y.; Osterman, I.A.; Lukyanov, D.A.; Sergiev, P.V.; Dontsova, O.A.; Liu, S.S.; Ye, J.J.; et al. Studies on antibacterial activity and diversity of cultivable actinobacteria isolated from mangrove soil in futian and maoweihai of China. *Evid. Based Complement. Alternat. Med.* **2019**, *2019*, 3476567. [CrossRef]

63. Lu, Q.P.; Ye, J.J.; Huang, Y.M.; Liu, D.; Liu, L.F.; Dong, K.; Razumova, E.A.; Osterman, I.A.; Sergiev, P.V.; Dontsova, O.A.; et al. Exploitation of potentially new antibiotics from mangrove actinobacteria in Maowei Sea by combination of multiple discovery strategies. *Antibiotics* **2019**, *8*, 236. [CrossRef] [PubMed]

64. Feling, R.H.; Buchanan, G.O.; Mincer, T.J.; Kauffman, C.A.; Jensen, P.R.; Fenical, W. Salinosporamide A: A Highly Cytotoxic Proteasome Inhibitor from a Novel Microbial Source, a Marine Bacterium of the New Genus *Salinospora*. *Angew. Chem. Int. Ed.* **2003**, *42*, 355–357. [CrossRef] [PubMed]

65. Chauhan, D.; Catley, L.; Li, G.; Podar, K.; Hideshima, T.; Velankar, M.; Mitsiades, C.; Mitsiades, N.; Yasui, H.; Letai, A.; et al. A novel orally active proteasome inhibitor induces apoptosis in multiple myeloma cells with mechanisms distinct from Bortezomib. *Cancer Cell* **2005**, *8*, 407–419. [CrossRef] [PubMed]

66. Ding, L.; Münch, J.; Goerls, H.; Maier, A.; Fiebig, H.-H.; Lin, W.-H.; Hertweck, C. Xiamycin, a pentacyclic indolosesquiterpene with selective anti-HIV activity from a bacterial mangrove endophyte. *Bioorg. Med. Chem. Lett.* **2010**, *20*, 6685–6687. [CrossRef]

67. Gao, X.M.; Han, W.D.; Liu, S.Q. The mangrove and its conservation in Leizhou Peninsula, China. *J. For. Res.* **2009**, *20*, 174–178. [CrossRef]

68. Li, M.S.; Mao, L.J.; Shen, W.J.; Liu, S.Q.; Wei, A.S. Change and fragmentation trends of Zhanjiang mangrove forests in southern China using multi-temporal Landsat imagery (1977–2010). *Estuar. Coast. Shelf Sci.* **2013**, *130*, 111–120. [CrossRef]

69. Hazarika, S.N.; Thakur, D. Chapter 21—Actinobacteria. In *Beneficial Microbes in Agro-Ecology*; Amaresan, N., Kumar, M.S., Annapurna, K., Kumar, K., Sankaranarayanan, A., Eds.; Academic Press: Cambridge, MA, USA, 2020; pp. 443–476. [CrossRef]

70. Harikrishnan, H.; Shanmugaiah, V.; Nithya, K.; Balasubramanian, N.; Sharma, M.P.; Gachomo, E.W.; Kotchoni, S.O. Enhanced production of phenazine-like metabolite produced by *Streptomyces aurantiogriseus* VSMGT1014 against rice pathogen, Rhizoctonia solani. *J. Basic Microbiol.* **2016**, *56*, 153–161. [CrossRef]

71. Cheng, C.; MacIntyre, L.; Abdelmohsen, U.R.; Horn, H.; Polymenakou, P.N.; Edrada-Ebel, R.; Hentschel, U. Biodiversity, anti-trypanosomal activity screening, and metabolomic profiling of actinomycetes isolated from mediterranean sponges. *PLoS ONE* **2015**, *10*, e0138528. [CrossRef]

72. Genilloud, O.; González, I.; Salazar, O.; Martín, J.; Tormo, J.R.; Vicente, F. Current approaches to exploit actinomycetes as a source of novel natural products. *J. Ind. Microbiol. Biotechnol.* **2011**, *38*, 375–389. [CrossRef]

73. Liu, X.Y.; Ashforth, E.; Ren, B.; Song, F.H.; Dai, H.Q.; Liu, M.; Wang, J.; Xie, Q.; Zhang, L.X. Bioprospecting microbial natural product libraries from the marine environment for drug discovery. *J. Antibiot.* **2010**, *63*, 415–422. [CrossRef]

74. Stuart, K.A.; Welsh, K.; Walker, M.C.; Edrada-Ebel, R. Metabolomic tools used in marine natural product drug discovery. *Expert Opin. Drug Discov.* **2020**, *15*, 499–522. [CrossRef]

75. Gu, R.H.; Rybalov, L.; Negrin, A.; Morcol, T.; Long, W.W.; Myers, A.K.; Isaac, G.; Yuk, J.; Kennelly, E.J.; Long, C.L. Metabolic profiling of different parts of *Acer truncatum* from the Mongolian Plateau using UPLC-QTOF-MS with comparative bioactivity assays. *J. Agric. Food. Chem.* **2019**, *67*, 1585–1597. [CrossRef]

76. Tomita, F.; Tamaoki, T.; Morimoto, M.; Fujimoto, K. Trioxacarcins, novel antitumor antibiotics. I. Producing organism, fermentation and biological activities. *J. Antibiot.* **1981**, *34*, 1519–1524. [CrossRef]

77. Magauer, T.; Smaltz, D.J.; Myers, A.G. Component-based syntheses of trioxacarcin A, DC-45-A1 and structural analogues. *Nat. Chem.* **2013**, *5*, 886–893. [CrossRef]

78. Arshad, M.; Sharif, A.; Ahmed, E.; Bariyah, S. Trioxacarcins as a promising class of anticancer drugs. *World J. Pharm. Pharm. Sci.* **2019**, *8*, 81–107.

79. Nicolaou, K.C.; Chen, P.; Zhu, S.; Cai, Q.; Erande, R.D.; Li, R.; Sun, H.; Pulukuri, K.K.; Rigol, S.; Aujay, M.; et al. Streamlined total synthesis of trioxacarcins and its application to the design, synthesis, and biological evaluation of analogues thereof. discovery of simpler designed and potent trioxacarcin analogues. *J. Am. Chem. Soc.* **2017**, *139*, 15467–15478. [CrossRef]

80. Nicolaou, K.C.; Cai, Q.; Qin, B.; Petersen, M.T.; Mikkelsen, R.J.; Heretsch, P. Total synthesis of trioxacarcin DC-45-A2. *Angew. Chem. Int. Ed. Engl.* **2015**, *54*, 3074–3078. [CrossRef]

81. Zhang, M.; Hou, X.-F.; Qi, L.-H.; Yin, Y.; Li, Q.; Pan, H.-X.; Chen, X.-Y.; Tang, G.-L. Biosynthesis of trioxacarcin revealing a different starter unit and complex tailoring steps for type II polyketide synthase. *Chem. Sci.* **2015**, *6*, 3440–3447. [CrossRef]

82. Shen, Y.; Nie, Q.-Y.; Yin, Y.; Pan, H.-X.; Xu, B.; Tang, G.-L. Production of a trioxacarcin analog by introducing a C-3 dehydratase into deoxysugar biosynthesis. *Acta Biochim. Biophys. Sin.* **2019**, *51*, 539–541. [CrossRef]

83. Qi, L.-H.; Zhang, M.; Pan, H.-X.; Chen, X.-D.; Tang, G.-L. Production of a trioxacarcin analogue by engineering of its biosynthetic pathway. *Chin. J. Org. Chem.* **2014**, *34*, 1376. [CrossRef]

84. Yin, Y.; Shen, Y.; Meng, S.; Zhang, M.; Pan, H.-X.; Tang, G.-L. Characterization of a membrane-bound O-acetyltransferase involved in trioxacarcin biosynthesis offers insights into its catalytic mechanism. *Chin. J. Chem* **2020**, *38*, 1607–1611. [CrossRef]

85. Yang, K.; Qi, L.H.; Zhang, M.; Hou, X.F.; Pan, H.X.; Tang, G.L.; Wang, W.; Yuan, H. The SARP family regulator Txn9 and two-component response regulator Txn11 are key activators for trioxacarcin biosynthesis in *Streptomyces bottropensis*. *Curr. Microbiol.* **2015**, *71*, 458–464. [CrossRef]

86. Pfoh, R.; Laatsch, H.; Sheldrick, G.M. Crystal structure of trioxacarcin A covalently bound to DNA. *Nucleic Acids Res.* **2008**, *36*, 3508–3514. [CrossRef]

87. Pröpper, K.; Dittrich, B.; Smaltz, D.J.; Magauer, T.; Myers, A.G. Crystalline guanine adducts of natural and synthetic trioxacarcins suggest a common biological mechanism and reveal a basis for the instability of trioxacarcin A. *Bioorg. Med. Chem. Lett.* **2014**, *24*, 4410–4413. [CrossRef]

88. Macintyre, L.; Zhang, T.; Viegelmann, C.; Martinez, I.J.; Cheng, C.; Dowdells, C.; Abdelmohsen, U.R.; Gernert, C.; Hentschel, U.; Edrada-Ebel, R.A. Metabolomic tools for secondary metabolite discovery from marine microbial symbionts. *Mar. Drugs* **2014**, *12*, 3416–3448. [CrossRef]

89. Zhou, S.Q.; Huang, X.L.; Huang, D.Y.; Hu, X.W.; Chen, J.L. A rapid method for extracting DNA from actinomycetes by Chelex-100. *Shengwu Jishu Tongbao* **2010**, *2*, 123–125.

90. Yoon, S.H.; Ha, S.M.; Kwon, S.; Lim, J.; Kim, Y.; Seo, H.; Chun, J. Introducing EzBioCloud: A taxonomically united database of 16S rRNA gene sequences and whole-genome assemblies. *Int. J. Syst. Evol. Microbiol.* **2017**, *67*, 1613–1617. [CrossRef]

91. Kumar, S.; Stecher, G.; Tamura, K. MEGA7: Molecular evolutionary genetics analysis version 7.0 for bigger datasets. *Mol. Biol. Evol.* **2016**, *33*, 1870–1874. [CrossRef]

92. Saitou, N.; Nei, M. The neighbor-joining method: A new method for reconstructing phylogenetic trees. *Mol. Biol. Evol.* **1987**, *4*, 406–425.

93. Letunic, I.; Bork, P. Interactive Tree Of Life (iTOL) v4: Recent updates and new developments. *Nucleic Acids Res.* **2019**, *47*, W256–W259. [CrossRef]

94. Chambers, M.C.; Maclean, B.; Burke, R.; Amodei, D.; Ruderman, D.L.; Neumann, S.; Gatto, L.; Fischer, B.; Pratt, B.; Egertson, J.; et al. A cross-platform toolkit for mass spectrometry and proteomics. *Nat. Biotechnol.* **2012**, *30*, 918–920. [CrossRef] [PubMed]

95. Shannon, P.; Markiel, A.; Ozier, O.; Baliga, N.S.; Wang, J.T.; Ramage, D.; Amin, N.; Schwikowski, B.; Ideker, T. Cytoscape: A software environment for integrated models of biomolecular interaction networks. *Genome Res.* **2003**, *13*, 2498–2504. [CrossRef] [PubMed]

marine drugs

MDPI

Article

Mansouramycins E–G, Cytotoxic Isoquinolinequinones from Marine Streptomycetes

Mohamed Shaaban [1,2,†], Khaled A. Shaaban [1,†,‡], Gerhard Kelter [3], Heinz Herbert Fiebig [4] and Hartmut Laatsch [1,*]

[1] Institute of Organic and Biomolecular Chemistry, University of Göttingen, Tammannstrasse 2, D-37077 Göttingen, Germany; mshaaba@gmail.com or ms.attia@nrc.sci.eg (M.S.); khaled_shaaban@uky.edu (K.A.S.)

[2] National Research Centre, Chemistry of Natural Compounds Department, Pharmaceutical and Drug Industries Research Institute, El-Behoos St. 33, Giza 12622, Egypt

[3] Oncotest GmbH, Charles River Discovery Germany, Am Flughafen 14, D-79108 Freiburg, Germany; Gerhard.Kelter@crl.com

[4] Oncotest GmbH, Biotec GmbH, Am Flughafen 14, D-79108 Freiburg, Germany; fiebig@4hf.eu

* Correspondence: hlaatsc@gwdg.de; Tel.: +49-551-393-211; Fax: +49-551-399-660

† Authors contributed equally to this work.

‡ Current address: Center for Pharmaceutical Research and Innovation, Department of Pharmaceutical Sciences, College of Pharmacy, University of Kentucky, Lexington, KY 40536, USA.

Abstract: Chemical investigation of the ethyl acetate extract from the marine-derived *Streptomyces* sp. isolate B1848 resulted in three new isoquinolinequinone derivatives, the mansouramycins E–G (**1a–3a**), in addition to the previously reported mansouramycins A (**5**) and D (**6**). Their structures were elucidated by computer-assisted interpretation of 1D and 2D NMR spectra, high-resolution mass spectrometry, and by comparison with related compounds. Cytotoxicity profiling of the mansouramycins in a panel of up to 36 tumor cell lines indicated a significant cytotoxicity and good tumor selectivity for mansouramycin F (**2a**), while the activity profile of E (**1a**) was less attractive.

Keywords: mansouramycins; isoquinolinequinones; marine-derived *Streptomyces* sp.; cytotoxicity

Citation: Shaaban, M.; Shaaban, K.A.; Kelter, G.; Fiebig, H.H.; Laatsch, H. Mansouramycins E–G, Cytotoxic Isoquinolinequinones from Marine Streptomycetes. *Mar. Drugs* **2021**, *19*, 715. https://doi.org/10.3390/md19120715

Academic Editor: Ipek Kurtboke

Received: 11 November 2021
Accepted: 8 December 2021
Published: 20 December 2021

Publisher's Note: MDPI stays neutral with regard to jurisdictional claims in published maps and institutional affiliations.

1. Introduction

The first natural isoquinolinequinone isolated from bacteria were reported by Fukum et al. in 1977 [1] and then by Kubo et al. in 1988 [2]. A few others were isolated from porifera, including cribrostatins (produced by the blue marine sponge *Cribrochalina* sp.) [3], renierones (from *Reniera*, *Petrosia*, and *Haliclona* spp.) [4,5], and caulibugulones (found in the marine bryozoon *Caulibugula inermis*) [6]. These isoquinolinequinones showed a potent antimicrobial activity against Gram positive bacteria and yeast (*Candida albicans*) and a pronounced cytotoxicity against L1210 and other cell lines with IC_{50} values as low as 30 ngmL^{-1} [7,8]. In 1998, we isolated mansouramycin A (**5**) as a trace component from the marine derived *Streptomyces* sp. B3497 [9]. Mansouramycin A (**5**) and the synthetic analogue 3-methyl-7-(methylamino)-5,8-isoquinolinedione (**4a**) were re-isolated from the marine-derived *Streptomyces* sp. isolate Mei37, together with three new mansouramycins B–D (**1a–3a**) [10]. These compounds showed a pronounced selectivity for non-small cell lung cancer, breast cancer, melanoma, and prostate cancer cells. Recently, mansouramycin A (**5**) was also obtained from the marine-derived *Streptomyces albus* J1074 and found to be a potent inhibitor of the methicillin-resistant *Staphylococcus aureus* ATCC 43300 with an MIC of 8 µgmL^{-1}. *S. albus* J1074 produced additionally the novel isoindoloquinone albumycin [11].

While the marine-derived *Streptomyces* sp. isolate B1848 was previously noted as a producer of 6-hydroxy-isatine and several other known compounds [12–14], further fermentations led now to the isolation and characterization of three unusual mansouramycins

E–G (**1a–3a**) along with mansouramycins A (**5**), D (**6**) (Figure 1) and 13 known metabolites [12–14]. The chemical structures of **1a–3a** were elucidated by NMR (1D, 2D) and HRMS, by comparison with related compounds and by computer-assisted methods. The cytotoxic activity of the isolated isoquinolinequinones was determined.

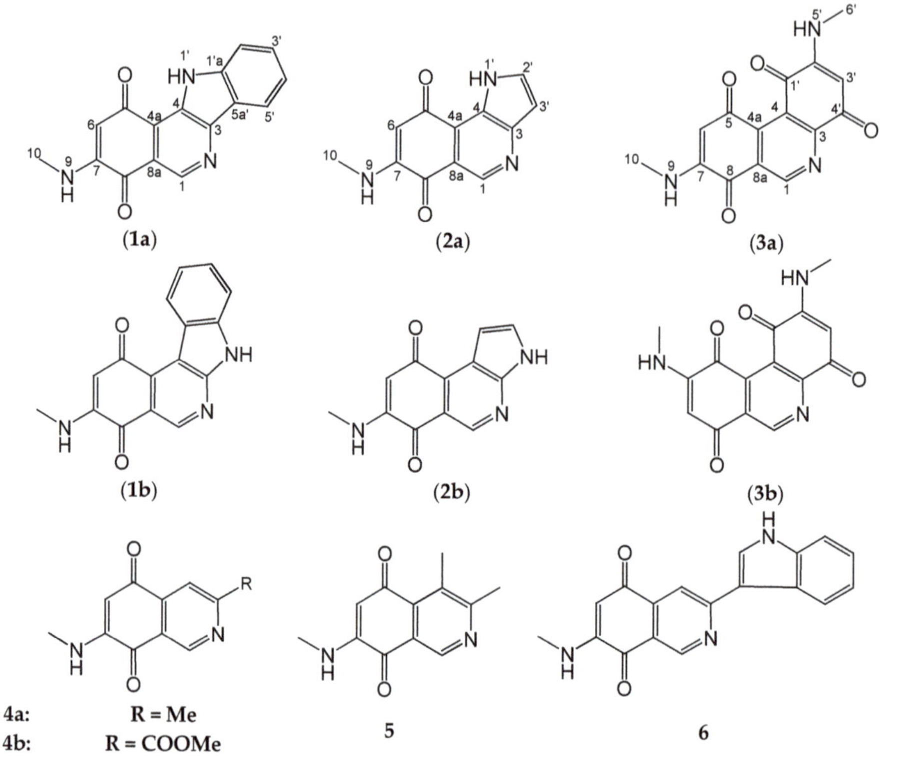

Figure 1. Chemical structures of isoquinolinequinones **1–6** produced by *Streptomyces* sp. B1848, and alternative structures **1b–3b**.

2. Results and Discussion

With a malt extract medium with 50% synthetic seawater (M$_2^+$ medium), the marine-derived *Streptomyces* sp. isolate B1848 produced only traces of mansouramycins A (**5**) and D (**6**), along with the zizaene derivative albaflavenol [9], 6-hydroxy-isatine [13,14], 2′-deoxythymidine, 2′-deoxyuridine, 2′-deoxyadenonsine, anthranilic acid, tyrosol, indolyl-3-acetic acid, phenyl acetamide, indolyl-3-carboxylic acid, N^β-acetyltryptamine, *N*-acetyltyramine, and *p*-hydroxybenzoic acid [12]. Better yields of the mansouramycins and further red pigments were obtained now on a meat extract medium in a fermentation with a 50 L shaker culture. After extraction and chromatographic separation, the strain B1748 afforded under these conditions the mansouramycins A (**5**) and D (**6**) and three new congeners, the mansouramycins E–G (**1a–3a**) as dark red solids. The isoquinolinequinones gave brown-red zones on TLC, with UV absorptions in solution similar as of *peri*-hydroxyquinones. Their reversible color change with sodium dithionite from orange to nearly colorless confirmed quinones; *peri*-hydroxyquinones were excluded, however, by the missing bathochromic shift with sodium hydroxide. Unlike the orange-red phenoxazinone chromophore of actinomycins and related pigments, which are becoming red with concentrated sulfuric acid, the isoquinolinequinones turned yellow. Further physicochemical properties of compounds **1a–3a** are summarized in Table 1.

Table 1. Physico-chemical properties of mansouramycins E–G (**1a–3a**).

Analytical Methods	Mansouramycin E (1a)	Mansouramycin F (2a)	Mansouramycin G (3a)
Appearance	Red powder	Dark red solid	Red solid
R_f [a]	0.76 (CH$_2$Cl$_2$/7% MeOH)	0.50 (CH$_2$Cl$_2$/7% MeOH)	0.23 (CH$_2$Cl$_2$/7% MeOH).
Anisaldehyde/H$_2$SO$_4$ reagent	yellow	yellow	yellow
Staining with NaOH	no color change	no color change	no color change
Molecular Formula	C$_{16}$H$_{11}$N$_3$O$_2$	C$_{12}$H$_9$N$_3$O$_2$	C$_{15}$H$_{11}$N$_3$O$_4$
UV/vis λ_{max} (log ε)	(MeOH): 244 (4.17), 264 (4.20), 287 sh (4.17), 314 sh (3.71), 377 (3.94), 448 sh (3.28), 509 sh (3.17); (MeOH + 1N NaOH): 243 (4.16), 263 (4.21), 287 sh (4.17), 313 sh (3.77) 378 (3.94), 449 sh (3.47), 508 sh (3.17); (MeOH + 1N HCl) 245 sh (4.04), 267 (4.14), 284 sh (4.02) 314 sh (3.31) 387 (3.91), 510 (3.17) nm	(MeOH): 234 (3.66), 288 (3.26), 373 (3.22), 481 sh (2.38); (MeOH+ 1N HCl): 237 (3.53), 313 (3.38), 378 (3.12), 485 sh (2.38) nm; (MeOH+1N NaOH): 233 (3.64), 289 (3.28), 375 (3.19), 485 sh (2.38) nm	(MeOH): 244 (4.13), 299 sh (3.65), 382 (3.42), 435 nm (3.47); (MeOH + 1N HCl): 243 (4.05), 303 (3.65), 377 (3.46), 436 (3.47); (MeOH + 1N NaOH): 245 (4.07), 302 sh (3.57), 384 (3.35), 437 (3.35) nm
IR (KBr) ν_{max} (KBr)	3434, 2925, 2855, 1672, 1625, 1598, 1510, 1491, 1412, 1384, 1354, 1311, 1268, 1208, 1050 cm^{-1}	3419, 2926, 2856, 1669, 1595, 1543, 1515, 1489, 1420, 1384, 1336, 1264, 1097, 1028, 764, cm^{-1}	3426, 2925, 2855, 1616, 1559, 1544, 1458, 1412, 1384, 1325, 1261, 1028 cm^{-1}
CI-MS: *m/z* (%)		245.0 ([M+NH$_4$]$^+$, 5), 228.0 ([M+H]$^+$, 100)	
(+)-ESI-MS: *m/z* (%)	278 ([M+H]$^+$)		320.2 ([M+Na]$^+$, 31), 617.0 ([2M+Na]$^+$, 100)
EI-MS: *m/z* (%)	277 [M]$^+$ (84), 256 (8), 249 (12), 236 (15), 220 (11), 195 (8), 192 (13), 179 (9), 166 (24), 138 (13), 102 (8), 97 (15), 82 (28), 73 (36), 69 (42), 57 (72), 43 (76), 44 (100)	227 ([M]$^{+\cdot}$, 100), 199 ([M-CO]$^{+\cdot}$, 8), 186 (16), 145 (9), 116 (8), 59 (12), 43 (8)	
(+)-ESI-HRMS: *m/z*		228.07663 [M+H]$^+$	298.08203 [M+H]$^+$
Calcd.	277.0846 for C$_{16}$H$_{11}$N$_3$O$_2$	228.07667 for C$_{12}$H$_{10}$N$_3$O$_2$ [M+H]$^+$	298.08223 for C$_{15}$H$_{12}$N$_3$O$_4$ [M+H]$^+$
EI HRMS: *m/z*	277.0848		

[a] Silica gel G/UV$_{254}$; **1b**, **2b**, **3b** (CH$_2$Cl$_2$/7% MeOH); sh = shoulder.

2.1. Structure Elucidation

Compound **1a** was obtained as red powder of moderate polarity. The molecular formula was determined as C$_{16}$H$_{11}$N$_3$O$_2$ by EI-HRMS, indicating 13 double bond equivalents (DBE). The color change to yellow with concentrated sulfuric acid and the characteristic UV curve with a flat absorption at λ_{max} 509 nm as for **5, 6** pointed to an isoquinolinequinone moiety as well [10] (Table 1). The ^{13}C NMR spectrum (Table 2) showed six aromatic/olefinic methines and one methyl signal. Furthermore, signals of nine non-protonated carbon atoms were observed, of which two at δ 182.8 and 181.4 pointed to carbonyl groups of a quinone. In the proton NMR spectrum (Table 2), the CH singlets at δ 9.01 (H-1) and 5.71 (H-6), in addition to a broadened NH signal at δ 7.83 and a methyl doublet at δ 2.85 of the CH$_3$NH fragment were typical for mansouramycins.

The proton H-6 showed HMBC correlations (Figure 2) with C-4 (^4J), 4a, 5, 7, and C-8; correlations of the N-methyl signal with C-7, and of H-1 with C-3, 4a and 8 resulted in a 3,4-disubstitued 7-methylamino-isoquinolinequinone skeleton as in **4a–6**; unfortunately, NH HMBC correlations were not visible for **1a**.

A 1,2-disubstituted benzene ring was deduced from the typical signal pattern of four *o,m*-coupled protons at δ 8.23 (d), 7.79 (d), 7.61 (td), and 7.31 (td) ppm and from the expected HMBC correlations (Figure 2). A further broadened NH signal was seen at δ 11.98, which formed with the remaining atoms an aniline residue. With respect to the two open valencies in both fragments, the isoquinoline and the aniline unit can be merged only in two ways under formation of structures **1a** or **1b**. The more in-depth analysis of the NMR data by means of the structure elucidation program COCON [15] confirmed isomers **1a** and **1b** as allowed structures, but delivered >7600 additional alternatives! Most of them were highly strained (cyclobutenes, non-linear allenes, or bridged aromatic systems) and therefore excluded.

Table 2. ^{13}C (150 MHz) and ^{1}H NMR spectroscopic data of compounds **1a–3a** in DMSO-d_6 (δ in ppm, J in [Hz]).

Position	Mansouramycin E (1a)		Mansouramycin F (2a)		Mansouramycin G (3a)	
	δ_C, Type	δ_H (Mult, J in [Hz]) [a]	δ_C, Type	δ_H (Mult, J in [Hz]) [b]	δ_C, type	δ_H (Mult, J in [Hz]) [a]
1	139.0, CH	9.01 (s)	140.7, CH	8.90 (s)	150.4, CH	9.31 (s)
3	149.9, C		153.5, C		153.4, C	
4	127.6, C		122.4, C		127.0, C	
4a	120.3, C		121.9, C		141.5, C	
5	182.8, C		182.8, C		178.2, C	
6	99.6, CH	5.71 (s)	99.0, CH	5.62 (s)	100.8, CH	5.75 (s)
7	149.9 [c], C		149.9, C		148.9, C	
8	181.4, C		181.2, C		179.8, C	
8a	120.4, C		117.3, C		126.7, C	
9		7.83 (brs)		7.81 (brq, 5.2)		7.84 (brq, 5.1)
10	29.0, CH$_3$	2.85 (d, 4.9)	28.9, CH$_3$	2.83 (d, 5.2)	28.9, CH$_3$	2.82 (d, 5.1)
1'		11.98 (brs)		11.88 (brs)	177.6, C	
1'a	144.7 [c], C					
2'	113.3, CH	7.79 (d, 7.9)	137.6, CH	7.93 (t, 3.05)	99.6, CH	5.77 (s)
3'	129.9, CH	7.61 (td, 7.9, 1.3)	102.7, CH	6.70 (dd, 3.05, 1.83)	152.2, C	
4'	120.6, CH	7.31 (td, 8.1, 1.4)			180.9, C	
5'	120.8, CH	8.23 (d, 8.1)				
5'a	120.1, C					
5'						8.04 (brq, 4.9)
6'					29.0, CH$_3$	2.83 (brd, 4.9)

[a] 300 MHz; [b] 600 MHz. See Supplementary Materials for NMR spectra. [c] Small signals; the assignment was confirmed by their HMBC correlations.

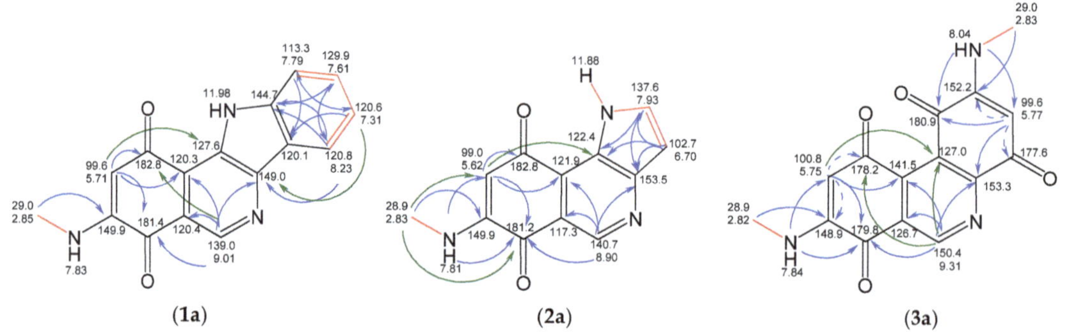

Figure 2. 2D NMR correlations of mansouramycins E–G (**1a–3a**). Blue arrows = ^2J, ^3J HMBC correlations; green arrows = ^4J HMBC correlations, red bonds = COSY correlations.

Amongst 24 plausible indoloquinoline- and indoloisoquinoline-quinones, only 6 (**1a, 1b, S1c, S1e, S1f, S1h**) * were found by COCON and therefore only these are in agreement with the COSY and HMBC correlations. For mansouramycin E, the isomer **1a** showed the best agreement of experimental NMR data with shifts calculated by SPARTAN'20 [16] using ab initio methods on a high level of theory. This structure was therefore assumed for mansouramycin E (see Supplementary Materials). Further applications of this technique have been described previously [17]. * Formula numbers with a leading bold letter "**S**" are refering to structures in Supplementary Materials.

For the dark red mansouramycin F (**2a**) the molecular formula $C_{12}H_9N_3O_2$ (ESI-HRMS) was determined, which entails 10 DBE. The ^{1}H and ^{13}C NMR shifts (Table 2), as well as the HMBC couplings, confirmed again an *N*-methyl-isoquinolinequinone substructure as in all other mansouramycins (Figure 1). According to the chemical shifts and 2D correlations, the unassigned residual atoms C_2H_3N were belonging to an annulated pyrrole ring, which was confirmed by the 1H triplet at δ_H 7.93 and the dd signal at 6.70 with the expected small coupling constants (~5 Hz). The pyrrole ring can be fused with the isoquinolinequinone core in three different ways, yielding structures **2a, 2b, S2m**, and the respective isomers with the N-methyl group at C-6 instead at C-7 (see Supplementary Materials).

With COCON using atom types, 19 isomers were found. Four of them were quinones (**2a**, **2b**, **S2c**, **S2d**). The other structures were azepin-2-ones or highly strained bridged systems. Isomers of type **S2m** were excluded by COCON as well, and also *o*-quinones were not predicted for mansouramycin F.

H-3′ in **2b** should show a ^3J correlation with C-4a, which is missing in the experimental spectrum and therefore better fitting on **2a**. In **S2c**/**S2d**, the quinonoid proton H-7 (δ ~5.6) should show a ^3J correlation with C-8a at δ ~147. However, this was also not observed, so that only structure **2a** was left. For further confirmation, we compared the experimental with calculated shifts of all possible pyrrolo-quinoline- and pyrroloisoquinoline-5,8-quinones. The results (Table S2) pointed again clearly to structure **2a** for mansouramycin F. This conclusion was further confirmed by comparison with similarly fused pyrrolo-pyridine skeletons [18,19].

Compound **3a** was obtained as a red solid as well, which displayed isoquinoline-quinone-like UV/vis and other physicochemical properties. The molecular formula of **3a** was established as $C_{15}H_{11}N_3O_4$ by ESI-HRMS and ^1H and ^{13}C NMR analysis, entailing 12 DBE. The ^1H and ^{13}C NMR spectra confirmed a further isoquinoline-quinone, which showed, however, remarkable differences compared with **1a** and **2a**. Instead of one *N*-methyl residue (7-NHCH$_3$) and one quinonoid proton (6-H), as in the other mansouramycins, the ^1H NMR spectrum showed each two of these signals. In addition, the ^{13}C NMR spectrum displayed four carbonyl groups (δ_C 178.2, 179.8, 177.6, and 180.9) instead of two carbonyls as in **1a** and **2a** (Table 2). Interpretation of the HMBC spectrum of **3a** (Figure 2) revealed an isoquinoline-quinone and an *N*-methylaminobenzoquinone substructure, which can be connected in two different ways only, resulting in **3a** or **3b**, respectively (Figures 1 and 2). The alternative **3b** was excluded, however, based on the significant ^3J HMBC correlations of H-6 (δ_H 5.75) and NH-9 (δ_H 7.84) with CO-8 (δ_C 179.8) and not with CO-5 (δ_C 178.2); the position of the second *N*-methyl group at C-3′ was determined in a similar way. All the remaining HMBC and COSY correlations (Figure 2) were in full agreements with structure **3a**, a novel azaphenanthrene diquinone, which we named mansouramycin G (see also Supplementary Materials).

2.2. Biological Activities

Isolated compounds were evaluated in cytotoxicity assays against the same 36 cancer cell lines as published before [10]. Consistent with results previously reported herein for other members of the group, cytotoxicity profiling of the new mansouramycin **2a** revealed good anti-tumor activity in vitro with a mean IC$_{50}$ value of 7.92 μM (1.797 μgmL^{-1}). Furthermore, **2a** showed good tumor selectivity across the panel of 36 cell lines. Mansouramycin E (**1a**) was less active and selective [mean IC$_{50}$ = 23.10 μM (6.398 μgmL^{-1})]. Previously reported mansouramycins C (**4b**) and A (**5**) exhibited mean IC$_{50}$ values of 0.089 μM (0.022 μgmL^{-1}) and 13.44 μM (2.902 μgmL^{-1}), respectively (Table 3). Mansouramycin G (**3a**) was not tested, due to a lack of material. In the agar diffusion test, crude extracts of *S.* isolate B1848 exhibited high bioactivity against *Mucor miehei* (Tü 284) and *Candida albicans*, and moderate activity against *Escherichia coli* and the alga *Chlorella vulgaris*. The samples of **1–6** were nearly consumed in the cytotoxicity assays and therefore not tested for their antimicrobial activity.

Table 3. In vitro cytotoxic activities of mansouramycins A (**5**), C (**4b**), E (**1a**), and F (**2a**).

Compound	Potency		Tumor Selectivity			
	Mean IC$_{50}$ μM (μgmL^{-1})	Mean IC$_{70}$ μM (μgmL^{-1})	Selectivity */Total	% Selectivity	Rating **	Internal Code
Mansouramycin A (**5**)	13.44 (2.902)	26.26 (5.671)	4/36	11%	++	MNSG078
Mansouramycin C (**4b**)	0.089 (0.022)	0.167 (0.041)	10/36	28%	+++	MNSG091
Mansouramycin E (**1a**)	23.10 (6.398)	33.95 (9.405)	0/18	0%	-	MNSG089
Mansouramycin F (**2a**)	7.92 (1.797)	15.19 ((3.449)	7/36	19%	++	MNSG090

* individual IC$_{70}$ < 1/3 mean IC$_{70}$; e.g., if mean IC$_{70}$ = 2.1 μM the threshold for above average sensitivity was IC < 0.7 μM, ** - (% selective = < 4%); + (4% > %selective >= 10%); ++ (10% > %selective >= 20%); +++ (% selective > 20%).

3. Materials and Methods

3.1. General Procedures

NMR spectra were measured on Varian Unity 300 and Varian Inova 600 spectrometers. The spectra were referenced to the signals of partially deuterated solvents (δ_{Chl} 7.270, 77.000; δ_{DMSO} 2.500, 39.510). Electron spray ionization mass spectrometry (ESI HRMS): Finnigan LCQ ion trap mass spectrometer coupled with a Flux Instruments (Basel, Switzerland) quaternary pump Rheos 4000 and a HP 1100 HPLC (nucleosil column EC 125/2, 100-5, C 18) with autosampler (Jasco 851-AS, Jasco Inc., Easton, MD, USA) and a Diode Array Detector (Finnigan Surveyor LC System). High resolution mass spectra (HRMS) were recorded by ESI MS on an Apex IV 7 Tesla Fourier-Transform Ion Cyclotron Resonance Mass Spectrometer (Bruker Daltonics, Billerica, MA, USA). EI mass spectra (70 eV) were recorded on a Finnigan MAT 95 spectrometer (Thermo Electron Corp., Bremen, Germany) with perfluorokerosene as reference substance for EI HRMS. IR spectra were recorded on a Perkin-Elmer 1600 Series FT-IR spectrometer from KBr pellets. UV/vis spectra were recorded on a Perkin-Elmer Lambda 15 UV/vis spectrometer. Flash chromatography was carried out on silica gel (230–400 mesh). R_f-values were measured on Polygram SIL G/UV$_{254}$ (Macherey-Nagel & Co., Düren, Germany). Size exclusion chromatography was carried out on Sephadex LH-20 (Lipophilic Sephadex; Amersham Biosciences, Ltd., purchased from Sigma-Aldrich Chemie, Steinheim, Germany).

3.2. Isolation and Taxonomy of the Producing Strain

The marine *Streptomyces* sp. strain B1848 was isolated and deposited in the Actinomycetes culture collection of the Alfred-Wegner Institute for Polar- und Marine Research, Am Handelshafen, Bremen, Germany. The taxonomy of the strain has been described previously [12].

3.3. Fermentation and Working Up

The *S*. sp. isolate B1848 was previously cultivated on M$_2$$^+$ medium with 50% seawater in a 25 L jar fermenter (72 h at 28 °C) [12,13]. Optimization of the culture conditions has been performed now using six different media [14] at two pH values (6.5, 7.8), temperatures (28, 35 °C), and shaking rates (110, 95 rpm) for four days. TLC analysis and antimicrobial screenings indicated that medium C (meat extract medium: 10 g glucose, 2 g peptone, 1 yeast, 1 g meat extract, pH 7.8) gave the best yield of mansouramycins.

A 50-L jar fermenter with C-medium was inoculated with strain B1848 and stirred for 4 days at 28 °C with 120 rpm. The resulting pale yellow culture broth was mixed with diatomaceous earth (Celite, ca. 1.8 kg), and filtered-off under pressure. The mycelial cake was extracted with ethyl acetate (3×), and then with acetone (2×). The acetone extract was concentrated under reduced pressure, and the aqueous residue was extracted once more with ethyl acetate. The combined organic phases were concentrated in vacuo, yielding 4.8 g of reddish-orange residue. None of the compounds of interest were detected in the aqueous phases, and therefore they were discarded.

3.4. Isolation and Purification

The mycelial cake extract (4.8 g) was applied to flash silica gel column chromatography (3 × 60 cm) using a CH$_2$Cl$_2$-CH$_3$OH gradient. After monitoring by TLC (CHCl$_3$/5; 10% MeOH), four fractions were obtained. Purification of fractions II-IV, using PTLC and Sephadex LH 20, led to isolation of five dark red compounds: mansouramycin A (**5**; 3.0 mg), D (**6**, 8.0 mg), E (**1a**; 4.1 mg), F (**2a**; 6.0 mg), and mansouramycin G (**3a**; 4.2 mg); for the physico-chemical properties and NMR spectral data of mansouramycins E–G (**1–3**), see Tables 1 and 2, respectively.

Mansouramycin C (3-Carbomethoxy-7-methylaminoisoquinoline-5,8-dione; **4b**): During this investigation, we realized two errors in the previously reported ^{13}C NMR data of mansouramycin D [10]: (CDCl$_3$, 150 MHz): δ 180.6 (C$_q$-8), 179.8 (C$_q$-5), 164.3 (C$_q$-9),

153.3 (C$_q$-3), 148.8 (C$_q$-7), 148.0 (CH-1), 140.5 (C$_q$-4a), 126.0 (C$_q$-8a), 120.7 (CH-4), 101.5 (CH-6), 53.4 (9-OCH$_3$), 29.3 (NCH$_3$).

Mansouramycin F (7-Methylamino-3*H*-pyrrolo [2,3-c]-isoquinoline-6,9-dione, **2a**): ^{1}H NMR (CDCl$_3$, 300 MHz): δ 10.25 (s br, 1H, NH-1′), 9.12 (s, 1H, 1-H), 7.79 (t, ^3J = 2.6 Hz, 1H, 2′-H), 6.81 (t, ^3J = 2.9 Hz, 1H, 3′-H), 6.25 (s br, 1H, 9-NH), 5.69 (s, 1H, 6-H), 2.98 (d, J = 5.1 Hz, 3H, CH$_3$-10).

Data of mansouramycins E–G (**1a–3a**) are listed in Tables 1–3, and spectra are depicted in the Supplementary Information. The working up and isolation of mansouramycins E–G (**1a–3a**) was carried out in August 2004. Spectral measurements, structural interpretation, and biological activity testing of **1a–3a** were achieved in the beginning of 2005.

3.5. Cytotoxicity Assays

A modified propidium iodide assay was used to examine the antiproliferative activity of the compounds against human tumor cell lines. Cell lines tested were derived from patient tumors engrafted as a subcutaneously growing tumor in NMRI nu/nu mice or obtained from American Type Culture Collection, Rockville, MD, USA, National Cancer Institute, Bethesda, MD, USA, or Deutsche Sammlung von Mikroorganismen und Zellkulturen, Braunschweig, Germany, and details of the test procedure have been described previously [20–22]. For the results, see Table 3.

3.6. DFT-Calculations

The calculation of NMR shifts was performed in a sequence of six calculation steps implemented in SPARTAN'20 [16]: (1) for all molecules of interest, the least energy conformers were determined using the "systematic approach" of the Merck Molecular Force Field program (MMFF). Up to 500 MMFF conformers within 40 kJ/mol above the global minimum were kept; in step (2), geometries were further optimized with a Hartry-Fock calculation (HF/3-21G); up to 200 conformers with <40 kJ/mol above the global minimum the energies were kept and (3) optimized (energies) with the DFT functional ωB97X-D and the 6-31G* basis set; (4) for up to 100 conformers within a window of 15 kJ/mol, the geometries were calculated now with the same functional and basis set; up to 50 conformers with <10 kJ/mol were kept for step (5); for the remaining conformers, energies and Boltzmann factors (300 K) were calculated with ωB97X-V/6-311+G(2df,2p) [6-311G*]; (6) for up to 30 resulting conformers with <10 kJ/mol the NMR data were calculated with ωB97X-D/6-31G* using the geometries from step (4). The conformer shifts were averaged with the Boltzmann factors from step five.

4. Conclusions

Isoquinoline-quinones from marine invertebrates and associated streptomycetes attracted scientific attention due to their strong anticancer activities [1–6]. From marine-derived *Streptomyces* spp., we isolated recently five isoquinoline-quinone derivatives, the mansouramycins A-D and 3-methyl-7-(methylamino)-5,8-isoquinolinedione (**4a**), which showed significant cytotoxicity in a panel of up to 36 tumor cell lines, with pronounced selectivity for non-small cell lung cancer, breast cancer, melanoma, and prostate cancer cells [10]. After a culture optimization, we succeeded now to isolate three further mansouramycins E–G (**1a–3a**) from the same marine *Streptomyces* sp. strain B1848, used optimized culture conditions. Their structures were elucidated by computer-assisted interpretation of 1D and 2D NMR spectra, high resolution mass spectrometry, by comparison with *ab initio*-calculated NMR data and by comparison with related compounds. Cytotoxicity profiling of the mansouramycins in a panel of up to 36 tumor cell lines indicated only a moderate cytotoxicity and tumor selectivity for the new quinones E (**1a**) and F (**2a**). The novel azaphenanthrene-diquinone mansouramycin G (**3a**) was not tested due to insufficient material.

Supplementary Materials: The following are available online at https://www.mdpi.com/article/10.3390/md19120715/s1, NMR spectra and other supplementary data.

Author Contributions: Conceptualization, M.S., K.A.S. and H.L.; methodology, M.S., K.A.S. and G.K.; validation, M.S., K.A.S., G.K., H.H.F. and H.L.; formal analysis, M.S., K.A.S., G.K., H.H.F. and H.L.; investigation, M.S. and K.A.S.; resources, H.L.; data curation, M.S., K.A.S. and H.L.; writing—Original draft preparation, M.S. and K.A.S.; writing—Review and editing, M.S., K.A.S., G.K., H.H.F. and H.L.; visualization, M.S., K.A.S., G.K., H.H.F. and H.L.; supervision, H.L.; project administration, H.L.; funding acquisition, H.L. All authors have read and agreed to the published version of the manuscript.

Funding: The financial support of this work was funded by a grant from the Bundesministerium für Bildung und Forschung (BMBF, grant 03F0415A).

Institutional Review Board Statement: Not applicable.

Informed Consent Statement: Not applicable.

Data Availability Statement: Further data are available on request from the corresponding author.

Acknowledgments: We thank R. Machinek for the NMR measurements, H. Frauendorf for the mass spectra, and F. Lissy and A. Kohl for technical assistance.

Conflicts of Interest: The authors have no competing interest to declare.

References

1. Fukumi, H.; Kurihara, H.; Hata, T.; Tamura, C.; Mishima, H.; Kubo, A.; Arai, T. Mimosamycin, a novel antibiotic produced by *Streptomyces lavendulae* No. 314: Structure and synthesis. *Tetrahedron Lett.* **1977**, *18*, 3825–3828. [CrossRef]
2. Kubo, A.; Kitahara, Y.; Nakahara, S.; Iwata, R.; Numata, R. Synthesis of mimocin, an isoquinolinequinone antibiotic from *Streptomyces lavendulae*, and its congeners. *Chem. Pharm. Bull.* **1988**, *36*, 4355–4363. [CrossRef] [PubMed]
3. Pettit, G.R.; Collins, J.C.; Herald, D.L.; Doubek, D.L.; Boyd, M.R.; Schmidt, J.M.; Hooper, J.N.A.; Tackett, L.P. Isolation and structure of cribrostatins 1 and 2 from the blue marine sponge *Cribrochalina* sp. *Can. J. Chem.* **1992**, *70*, 1170–1175. [CrossRef]
4. Pettit, G.R.; Knight, J.C.; Collins, J.C.; Herald, D.L.; Pettit, R.K.; Boyd, M.R.; Young, V.G. Antineoplastic agents 430. Isolation and structure of cribrostatins 3, 4, and 5 from the republic of maldives *Cribrochalina* species. *J. Nat. Prod.* **2000**, *63*, 793–798. [CrossRef] [PubMed]
5. Frincke, J.M.; Faulkner, D.J.J. Antimicrobial metabolites of the sponge *Reniera* sp. *J. Am. Chem. Soc.* **1982**, *104*, 265–269. [CrossRef]
6. Milanowski, D.J.; Gustafson, K.R.; Kelley, J.A.; McMahon, J.B. Caulibugulones A–F, novel cytotoxic isoquinoline quinones and iminoquinones from the marine bryozoan *Caulibugula intermis*. *J. Nat. Prod.* **2004**, *67*, 70–73. [CrossRef] [PubMed]
7. Kubo, A.; Nakahara, S.; Iwata, R.; Takahashi, K.; Arai, T. Mimocin, a new isoquinolinequinone antibiotic. *Tetrahedron Lett.* **1980**, *21*, 3207–3208. [CrossRef]
8. Mckee, T.C.; Ireland, C.M. Cytotoxic and antimicrobial alkaloids from the Fijian sponge *Xestospongia caycedoi*. *J. Nat. Prod.* **1987**, *50*, 754–756. [CrossRef] [PubMed]
9. Speitling, M. Vergleich der Metabolischen Kapazität Mariner und Terrestrischer Mikroorganismen—Isolierung und Strukturaufklärung von Branimycin, Brom-alterochromid A/B und weiteren Stoffwechselprodukten. Ph.D. Thesis, Georg-August University, Göttingen, Germany, 1998.
10. Hawas, U.W.; Shaaban, M.; Shaaban, K.A.; Speitling, M.; Maier, A.; Kelter, G.; Fiebig, H.H.; Meiners, M.; Helmke, E.; Laatsch, H. Mansouramycins A-D, Cytotoxic Isoquinolinequinones from a marine Streptomycete. *J. Nat. Prod.* **2009**, *72*, 2120–2124. [CrossRef] [PubMed]
11. Huang, C.; Yang, C.; Zhang, W.; Zhu, Y.; Ma, L.; Fang, Z.; Zhang, C. Albumycin, a new isoindolequinone from *Streptomyces albus* J1074 harboring the fluostatin biosynthetic gene cluster. *J. Antibiot.* **2019**, *72*, 311–315. [CrossRef] [PubMed]
12. Shaaban, M.; Schröder, D.; Shaaban, K.A.; Helmke, E.; Wagner-Döbler, I.; Laatsch, H. Flazin, Perlolyrin, and other new β-Carbolines from marine-derived Bacteria. *Rev. Latinoam. Quim.* **2007**, *35*, 58–67.
13. Shaaban, K.A.; Shaaban, M.; Nair, V.; Schuhmann, I.; Win, H.Y.; Lei, L.; Dittrich, B.; Helmke, E.; Schüffler, A.; Laatsch, H. Structure Elucidation and Synthesis of Hydroxylated Isatins from Streptomycetes. *Z. Naturforsch. B* **2016**, *71*, 1191–1198. [CrossRef]
14. Shaaban, M. Bioactive Secondary Metabolites from Marine and Terrestrial Bacteria: Isoquinolinequinones, Bacterial Compounds with a Novel Pharmacophor. Ph.D. Thesis, Georg-August University, Göttingen, Germany, 2004.
15. Lindel, T.; Junker, J.; Koeck, M. 2D-NMR-guided constitutional analysis of organic compounds employing the computer program COCON. *Eur. J. Org. Chem.* **1999**, *1999*, 573–577. [CrossRef]
16. *SPARTAN'20*; Wavefunction, Inc.: Irvine, CA, USA, 2020.
17. Shaaban, M.; Abou-El-Wafa, G.S.E.; Golz, C.; Laatsch, H. New haloterpenes from the marine red alga *Laurencia papillosa*: Structure elucidation and biological activity. *Mar. Drugs* **2021**, *19*, 35. [CrossRef] [PubMed]

18. Arzel, E.; Rocca, P.; Marsais, F.; Godard, A.; Quéguiner, G. First total synthesis of cryptomisrine. *Tetrahedron* **1999**, *55*, 12149–12156. [CrossRef]
19. Kim, J.S.; Shin-ya, K.; Furihata, K.; Hayakawa, Y.; Seto, H. Structure of mescengricin, a novel neuronal cell protecting substance produced by *Streptomyces griseoflavus*. *Tetrahedron Lett.* **1997**, *38*, 3431–3434. [CrossRef]
20. Dengler, W.A.; Schulte, J.; Berger, D.P.; Mertelsmann, R.; Fiebig, H.H. Development of a propidium iodide fluorescence assay for proliferation and cytotoxicity assays. *Anticancer Drugs* **1995**, *6*, 522–532. [CrossRef] [PubMed]
21. He, J.; Roemer, E.; Lange, C.; Huang, X.; Maier, A.; Kelter, G.; Jiang, Y.; Xu, L.; Menzel, K.-D.; Grabley, S.; et al. Structure, derivatisation, and antitumor activity of new griseusins from *Nocardiopsis* sp. *J. Med. Chem.* **2007**, *50*, 5168–5175. [CrossRef] [PubMed]
22. Fiebig, H.H.; Maier, A.; Burger, A.M. Clonogenic assay with established human tumour xenografts: Correlation of in vitro to in vivo activity as a basis for anticancer drug discovery. *Eur. J. Cancer* **2004**, *40*, 802–820. [CrossRef] [PubMed]

MDPI

St. Alban-Anlage 66

4052 Basel

Switzerland

Tel. +41 61 683 77 34

Fax +41 61 302 89 18

www.mdpi.com

Marine Drugs Editorial Office

E-mail: marinedrugs@mdpi.com

www.mdpi.com/journal/marinedrugs